US market size $132 bn, 143 mi

Telecom 101

6th Edition 2022

Eric C. Coll, M.Eng.

Teracom Training Institute
www.teracomtraining.com

The publisher offers discounts on this book when ordered in quantity.
For more information, or to place an order, please contact the publisher:
Teracom Training Institute, Ltd.
Publishing Division
1-877-412-2700
www.teracomtraining.com

Softcover print edition T101 ISBN 9781894887113

Sixth Edition 2022 R7

A little learning is a dangerous thing;
Drink deep or taste not the Pierian spring:
There shallow draughts intoxicate the brain,
And drinking largely sobers us again.
– Alexander Pope

Contents

1 Introduction to the Book

In this chapter, we discuss the approach taken in organizing the topics in this text and provide suggestions for how to use it. The chapter is completed with the Answers to all questions about telecommunications.

1.1 Our Approach

Our approach in organizing the topics and the order we present them in can be summed up with a simple philosophy: Start at the beginning of the story. Progress in a logical order, building one concept on another. Finish at the end of the story. Avoid jargon. Speak in plain English.

We've been applying this approach to telecommunications training for more than twenty-five years, and it is the philosophy behind all of Teracom's training products and services, including online courses, certifications, live instructor-led training, self-paced instructor-led training, eBooks and printed textbooks.

Our objective is to fill in the gaps and build a solid base of knowledge, put a structure in place, and show how everything fits together.

This guide is not intended to be an authoritative reference on any particular technology; specifications and low-level details are best found in the official standards documents.

This is a practical guide, and we hope it will serve as a valuable resource in navigating the world of telecommunications, and in building knowledge and understanding that lasts a lifetime.

1.2 How the Text is Organized

This book is based on Teracom's famous three-day instructor-led core telecommunications training Course 101 Broadband, Telecom, Datacom and Networking for Non-Engineers, plus additional material going beyond Course 101.

Following this Chapter 1, the book is organized into four parts, plus appendices:

Part 1: Fundamentals

2. Introduction to Telecommunications
3. Telecom Fundamentals
4. Network Fundamentals
5. The Internet, Cloud Computing, Web Services and Data Centers
6. Telecom Services Overview
7. Digital Media: Voice, Video, Images, Quantities, Text
8. Fundamentals of Voice over IP

The first part of the book is seven chapters that cover the fundamentals of telecom, filling gaps, explaining concepts and establishing a solid knowledge base.

First is a high-level pass with a big-picture view and introducing all of the topics that will be covered in detail in subsequent chapters.

Then we cover telecom fundamentals like multiplexing, modems and pulses; and network fundamentals like IP packets and Ethernet switches.

Next, you'll learn about the Internet as a business: ISPs, web services like AWS, cloud computing and data centers. We'll review today's services in the residential, business and wholesale categories.

The fundamentals are completed with digital media: how voice is digitized, digital video, digital images, digital quantities and digital text, and the fundamentals of VoIP and SIP.

Part 2: Telecom Technologies

9. Wireless
10. Fiber Optics
11. Copper

In the second part of the book, we explore the main technologies for communicating information, grouped into wireless, fiber and copper.

We'll cover wireless spectrum, mobile network components and operation, 4G LTE, 5G, fixed wireless broadband home Internet, Wi-Fi and satellites.

Then you'll learn optical basics, and how networks are built with point-to-point fibers running Optical Ethernet, wave-division multiplexing, fiber in the core, metro and to the premise. We'll finish with copper-wire technologies: DSL and POTS on twisted pair, Hybrid Fiber-Coax cable TV systems, T1 and the categories of LAN cables.

Part 3: Equipment, Carriers and Interconnect

12. Telecom Equipment
13. Carriers and Interconnect

In the third part of the book, we explore the equipment that is connected by the fiber, copper and wireless of Part 2 to form networks, and the place and purpose of each. Then we understand where and how connections physically take place for PSTN phone calls, for Internet and CLEC services.

Part 4: Networking

14. The OSI Layers and Protocol Stacks
15. Ethernet, LANs and VLANs
16. IP Networks, Routers and Addresses
17. MPLS and Carrier Networks
18. Wrapping Up

The fourth and final part of this book is devoted to IP networking and MPLS. We begin with the OSI Reference Model and its layers to provide a structure for the discussion: what a layer is, what the layers are, the functions of each, and the standard protocols at each layer.

Next is a chapter on Layer 2: Ethernet, 802 standards, broadcast domains and VLANs. Then, Layer 3: IP routers, IP addresses, DHCP, public and private addresses, Network Address Translation and IPv6.

After IP, we examine the traffic management system MPLS, and how MPLS is used to manage IP packet flows, implement VPNs, classes of service, service integration and traffic aggregation.

We'll conclude with a top-down review and roundup of technologies and a peek at the future of telecommunications.

Telecommunications technology is in constant change – and technologies that used to be in wide use are no longer. Along with some technical discussions, they've been moved to appendices to make room for the new:

1.3 How to Use This Text

There are two ways to use this text, and we suggest that you try both.

The first way is to read it sequentially, from beginning to end, ensuring that you understand the concepts in each section before moving on to the next. This is what we do in instructor-led seminar and online course versions of this material.

A professional instructor leading you through this material would require four to five days of intense, fairly fast-paced discussion; so it might be reasonable to think that you could read this text in a week full-time or a month part-time.

However, no one expects anyone to absorb all of this information at once. Sometimes it takes time to really understand various concepts. Also, you may run into the technologies discussed in text this some time after you first read it, and will need a reference and refresher.

This leads to the second way to use this text: as a day-to-day reference handbook and glossary. The Table of Contents has been constructed to allow pinpoint navigation to important topics, and is hyperlinked and searchable in the eBook editions.

Thousands of people in the telecom, networking, government, military, educational, financial, insurance, health, entertainment and other sectors, including people working for Cisco, Juniper, Intel, Microsoft, the CIA, FAA and IRS, all branches of the US Armed Forces, AT&T, Verizon, Lumen, T-Mobile, Sprint, Global Crossing, Bell, TELUS, Transamerica Insurance, Oneida tableware, the San Francisco Giants and countless others have benefited from this knowledge and our approach to transferring it.

We hope you will too.

1.4 The Three Answers

Telecom 101 is an easy course. There are only three possible answers to any question anyone asks:

1) Money.

2) History.

3) It's all pretty much the same thing.

1.4.1 Answer Number 1: Money

When someone asks, "Why was it designed that way?" or "What are we really discussing?", the answer usually boils down to money, and a tradeoff between cost and performance.

The reason for this is that it is very easy to measure results in the telecommunications business, unlike the psychology business, where it is very difficult to measure results. In the telecommunications business, we can measure how many bits per second are transferred from one place to another – and not surprisingly, there is a strong correlation between that and how much it costs.

An example is network service cost vs. the probability of being blocked, i.e., failure to communicate.

Dedicated lines, where there is capacity reserved whether it is being used or not, are very expensive; but the probability of being blocked with reserved capacity is zero.

Packet networks are overbooked, and capacity is not reserved for particular users. Knowing that users normally do nothing, and only actually transmit once in a while, more users are connected than there is network capacity for them to all transmit at the same time.

Packet network services are cheaper than dedicated lines; but the probability of being blocked is higher, when here is more demand than the actual capacity of the overbooked network.

Another example is the different versions of Optical Ethernet, which trade off bit rate, maximum fiber length, and cost of the optical transceivers. Why is there a standard specifying 10 wavelengths at 10 Gb/s each to make a 100 Gb/s system, instead of one wavelength at 100 Gb/s?

Answer Number 1: Money. It's currently cheaper to make it that way.

1.4.2 Answer Number 2: History

Those who do not learn history are doomed to repeat it.

You may be interested in learning about Voice over IP (VoIP). You probably should be interested in learning about it, as all telephone calls are VoIP at some point in today's telecommunications network.

If you want to understand Voice over IP, there are a number of technologies that come in to play. One is Voice. Another is IP.

Voice must be digitized to be carried in IP packets.

Voice digitization involves three functions:

1) Taking samples of the value of the analog voltage coming out of the microphone at regular intervals,

2) Quantizing the range of possible values of the sample into fixed increments, and

3) Coding the resulting quantized value into binary.

Questions that arise are: how often do samples have to be taken, what are the quantization increments and what coding method is used to represent the quantized value in binary?

A fellow by the name of Nyquist realized in the late 1950s that it is necessary to take samples more than twice as often as the width of the frequency bandwidth of the analog voltage to be able to reproduce it.

Now we're back to the summer of 1874 when Alexander Graham Bell made some design decisions, and the early 1900s when loading coils were deployed on long-distance trunks, directly affecting the analog signal bandwidth, which directly affects how many bits per second are required to digitize voice, which directly affects Voice over IP today.

We would claim that if you don't understand the progression of technologies, one built on top of another, you will never really understand today's telecommunications, and won't be ready to understand where we are going tomorrow.

In this text, and in our seminars and online courses, we start at the beginning, progress in a logical order, and finish at the end – to build structured knowledge so you can understand how everything fits together.

1.4.3 Answer Number 3: It's All Pretty Much the Same Thing

We could simplify "telecommunications" by claiming that there are two kinds of traffic or information to be communicated: information that happens in continuous streams, and information that happens in bursts.

Video is a good example of information that happens in continuous streams; sending video to someone means transmitting pictures 60 times a second, second after second, minute after minute.

E-mail is a good example of information that happens in bursts: you send e-mail to someone, and then you don't.

Telecommunication service providers have two basic kinds of services: they have services that allow their customers to transmit continuously, and services that allow their customers to transmit in bursts.

If we look just one level of complexity deeper in the network, we find that the way that a service provider offers to its customers the possibility of communicating information in bursts, is to take a circuit that actually communicates all the time, attach boxes called *routers* or *Layer 2 switches* to each end, connect many users to each end, and let them transmit whenever there's a free spot.

At the circuit level, it's all pretty much the same thing.

Moreover, a carrier has only one *network core*: its transmission backbone, made of fiber connecting routers in different cities.

It's not like the carrier has one physical backbone for voice, a different one for business data, a third one for television and a fourth for the Internet.

Everything communicated over the same piece of glass, by the same light flashing on and off 10 billion times per second.

There are a few methods to get *access* to this fiber backbone: wireless, fiber and copper.

There are many, many ways of billing you for using it.

Now you know everything there is to know about telecommunications.

Not.

We're not going to simplify things quite that much, but...

- Once you achieve spiritual nirvana in telecommunications, you will become aware that all of the services you hear about like broadband, Internet, POTS, MPLS VPNs, SIP Trunking, T1 and all the rest are all really just different billing plans for using network capacity in different ways.

Those are the Answers.

The rest of this book is devoted to understanding the questions.

2 The Broadband Converged IP Telecommunications Network

This chapter begins with a history lesson, then a comprehensive big-picture introduction to telecom: the concepts of convergence and broadband, today's telecom network, the parts of the network, and the three key technologies: Ethernet, IP and MPLS, what they are and what each does, and how a service is implemented end-to-end. The chapter ends with an overview of standard residential, business and wholesale services.

2.1 History of Telecommunications

Telecommunications began not with telephones, but with telegraphs. Telegraph systems were the command-and-control systems for railways: used to communicate information about trains from one end of the line to the other. Railways and their telegraph systems were deployed across North America in the first half of the 1800s, and these were the first communication networks.

2.1.1 Invention of the Telephone

The telephone was invented by Alexander Graham Bell between 1874 and 1876, with most of the work done on his father's homestead near Brantford, Ontario in the Niagara region, and some of the work done at his winter job at a school for deaf children in Boston.

It was in Brantford, in the summer of 1874, that Bell told his father how he proposed to build a telephone, and there in the summer of 1875 that he drew up the patent application.

Bell demonstrated the telephone apparatus over short distances of wire with the words "Mr. Watson, come here I want you!" on March 10, 1876 in Boston, and again at the Centennial Exposition in Philadelphia in June 1876… but communications across distance remained elusive.

Returning to Brantford in the summer of 1876, Bell refined his apparatus and made three successful tests of communication across distance. This is generally considered to be the first long-distance phone call.

The article *The Human Voice Transmitted by Telegraph* in the September 1876 issue of Scientific American magazine outlined these experiments.

FIGURE 1 ALEXANDER GRAHAM BELL

In August 1876, Bell successfully demonstrated speech communication across wires of the Dominion Telegraph Company between telegraph offices in Brantford and Mount Pleasant Ontario; then between the Bell homestead and the telegraph office in Brantford, a distance of four miles; and on August 10, 1876 over the eight-mile telegraph line between Brantford and Paris, Ontario with the battery 58 miles away in Toronto.

If the Dominion Telegraph Company had been able to foresee that Bell's company (that to this day bears his name) would eventually put them out of business, they might not have been so cooperative in hosting the trials!

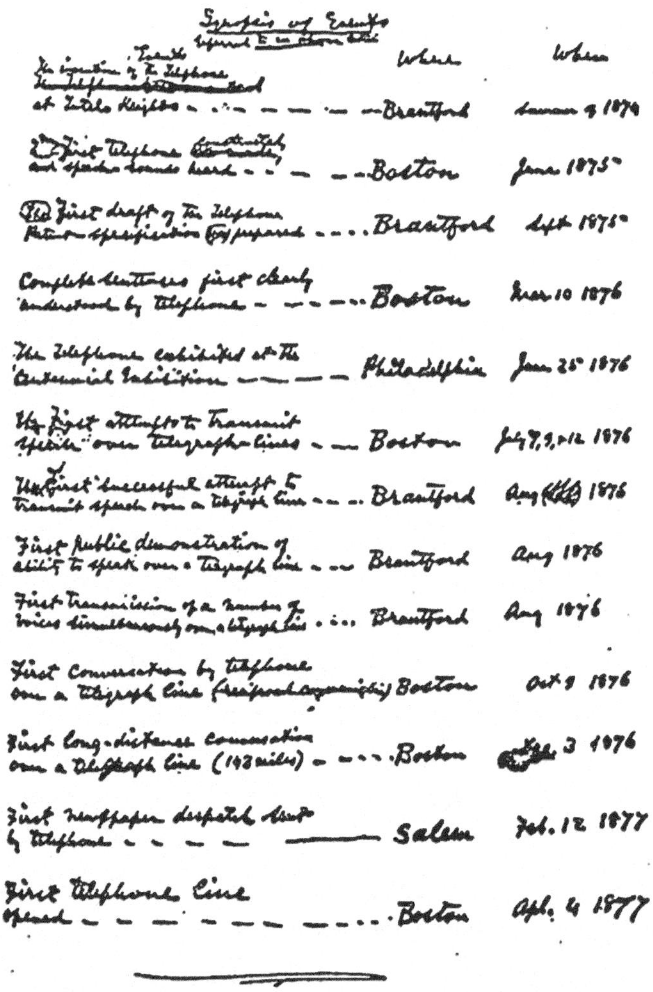

Synopsis of Events

	Where	When
The invention of the Telephone at Tutelo Heights	Brantford	Summer of 1874
First telephone constructed and speech sounds heard	Boston	June 1875
First draft of the telephone patent specification prepared	Brantford	Sept 1875
Complete sentences first clearly understood by telephone	Boston	Mar. 10 1876
The telephone exhibited at the Centennial Exhibition	Philadelphia	June 25 1876
First attempt to transmit speech over telegraph lines	Boston	July 7, 9, & 12 1876
First successful attempt to transmit speech over a telegraph line	Brantford	Aug [illegible] 1876
First public demonstration of ability to speak over a telegraph line	Brantford	Aug 1876
First transmission of a number of voices simultaneously over a telegraph line	Brantford	Aug 1876
First conversation by telephone over a telegraph line ([illegible])	Boston	Oct 9 1876
First long-distance conversation over a telegraph line (143 miles)	Boston	[illegible] 3 1876
First newspaper despatch sent by telephone	Salem	Feb. 12 1877
First telephone line opened	Boston	Apr. 4 1877

FIGURE 2 CHRONOLOGY OF THE INVENTION OF THE TELEPHONE IN ALEXANDER GRAHAM BELL'S HANDWRITING

Bell patented his device in 1876. Subsequently, it became a national sport to challenge his patent in court. There were over 600 court challenges to Bell's patent – every one unsuccessful.

The many notes and diagrams produced in Brantford in the summers of 1874 and 1875, along with his father's diary were used to prove Bell's claim to the invention of the telephone.

Figure 2 is an image of one of the many memorabilia residing in the Bell Homestead Museum at Tutelo Heights, Brantford, Ontario, Canada. It was composed by Bell following the opening of the Bell Memorial at Brantford 24 October 1917. Credit to telecommunications.ca for the image.

Claims are made for both Boston and Philadelphia as being the place where the telephone was invented.

In 2002, the US Congress passed a resolution claiming that Italian-American Antonio Meucci had in fact invented the telephone and Bell had taken his lab notes and patented the idea.

None of these claims are consistent with the serious, repeated, detailed investigations by hundreds of people alive at the time of the events leading to court decisions what was invented where and by whom (the telephone, in Brantford Ontario Canada, by Alexander Graham Bell) during challenges of Bell's patent.

2.1.2 Local Phone Companies

Telephone service began with connections within cities. A company would establish a Central or Central Office (CO) downtown, and connect *subscribers* to their communication service to the CO using pairs of copper wires to carry the electrical signals representing speech.

These subscribers would alert an operator in the CO that they wanted to establish a connection by cranking a handle that caused a bell to ring at the CO, and then telling the operator the name of the person to whom they wished to be connected.

The operator would use a cord to connect the two subscribers via a large patch board. This was the first kind of telephone switch.

Since copper is a good, but not perfect, conductor of electricity – it has some resistance to the flow of electrons through it – the copper wires could only be a certain maximum length before it would not be possible to hear what the other person was saying.

Thus, local phone companies providing service in a radius of a few miles around a Central Office sprung up in major cities across the continent beginning in 1878.

Inter-city long-distance communications was not technically possible yet.

2.1.3 The Bell System

In the USA, these local phone companies were either part of, owned by or licensed by the Bell Telephone Company which became the American Bell Telephone Company in 1880.

Its Chief Operating Officer and later president, Theodore Vail, began creating the Bell System, composed of regional companies offering local service, a long-distance company and an equipment manufacturing arm.

FIGURE 3 MAKING THE FIRST CALL FROM NEW YORK TO SAN FRANCISCO: THEODORE VAIL, THE MAN WHO GOT IT BUILT (LEFT), AND FINANCIER WILLIAM ROCKEFELLER (SEATED). STANDING ARE NETWORK ARCHITECTS AND BANKER J. P. MORGAN JR. (CENTER).

The American Telephone and Telegraph Company (AT&T) was incorporated in March, 1885 as a wholly-owned subsidiary of American Bell, with the initial business plan of providing long-distance service for the Bell System: connecting the local companies.

Building out from New York, its initial goal Chicago was reached in 1892, and San Francisco in 1915. AT&T continued as the "long-distance company" until Dec. 30, 1899, when it changed its business model to be vertically integrated: local and long distance, by acquiring the assets of the American Bell Telephone Company and becoming the parent company of the Bell System.

2.1.4 The Public Switched Telephone Network (PSTN)

Until Bell's patent expired in 1894, only licensees of American Bell could legally operate telephone systems in the United States. Between 1894 and 1904, over six thousand telephone companies, called independents (that is, not part of the Bell System) went into business, and the number of telephones increased from some 250,000 to over 3,000,000… but in many cases, there was no interconnection between the independents.

Several court decisions forced the opening of AT&T's network from a technical point of view. These included the Carterphone decision, which allowed customers to use their own terminal equipment on the Bell System, and a judgment that allowed MCI to connect to AT&T's long-distance network.

The United States government sued AT&T three times under antitrust laws: 1913, 1949 and 1974. The 1974 suit was settled when AT&T agreed to divest itself of local operating companies in January 1984 in exchange for loosening of regulation. The ownership of AT&T's local operations was transferred into seven holding companies, known as the Baby Bells: US West, Pac Bell, Southwestern Bell, Bell South, Bell Atlantic, NYNEX and Ameritech. The remaining operations were the long lines, owned by an Interexchange Carrier (IXC) now called AT&T Corp.

In 1996, the federal government's Telecommunications Act removed many of the remaining obstacles at the federal level to wide-open competition for both local and long-distance telecommunications. Significant obstacles remained at the state Public Utility Commission level, which were slowly overcome.

2.1.5 Consolidation

1996 saw the beginning of consolidation of the Baby Bells with the purchase of NYNEX by Bell Atlantic. SBC Communications Inc., owner of Southwestern Bell, purchased Pac Bell in 1997, SNET in 1998 and Ameritech in 1999. In 2000, Bell Atlantic merged with GTE, owner of many independents, and baptized the result Verizon. US West was purchased by Qwest, later merging with Century Tel to form CenturyLink, which later changed its name to Lumen.

Once the LECs could be IXCs, local and long-distance operations were merged back together by the LECs purchasing the IXCs.

SBC purchased AT&T Corp and Verizon purchased MCI, reconstituting most of the Bell System in two geographic landline pieces.

To get the valuable brand name "AT&T", SBC in 2005 implemented a reverse takeover of AT&T, purchasing AT&T Corp. then changing its corporate name from "SBC" to "AT&T".

Each company provides services in the other's territory via fiber, collocations and especially wireless.

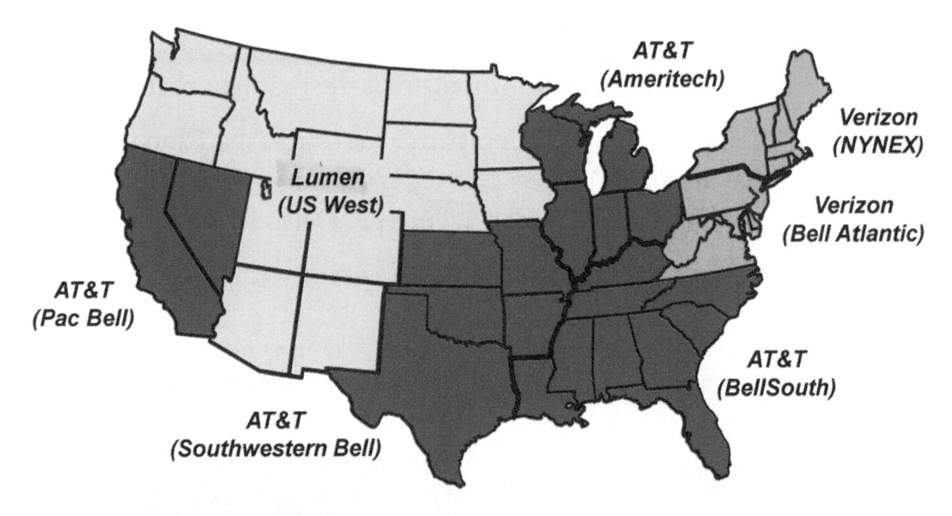

FIGURE 4 CONSOLIDATION: VERIZON AND AT&T

In addition to Lumen, many other independent companies own and/or operate local networks and fiber backbones.

2.1.6 Broadband Carriers

Companies with a *coaxial* entry cable to residence were historically called Community Antenna Television (CATV) or *cable* companies.

For each city, one antenna farm would pick up broadcast television radio signals, which would be distributed to users as electricity vibrating in different frequency channels on coaxial cables right to the television.

Since coaxial cable supports the use of a much broader frequency band than twisted pair, these companies are also called *broadband carriers*.

Cable companies offer Internet access and telephone service for residences and business using modems operating at 1 Gb/s or more on their last mile, in parallel with the cable television programming.

In the USA, cable companies initially captured a majority share of residential Internet access. These companies also offer fiber business services, and have progressed to full telecom service providers.

2.1.7 Canadian Telegraph Companies

Telecommunications everywhere began with telegraph companies, fifty years before the telephone was invented. Telegraph lines were strung up alongside railways, to carry expensive telegrams, as well as a vital part of the operation of the railway.

Railways were built out in North American in the first half of the 1800s.

By 1847, the Montreal Telegraph Company was established and providing telegram service in the Quebec City - Windsor corridor on railway telegraph lines, with a link to Western Union in Detroit.

In 1886, Canadian Pacific Railways Telegraphs came online as a competitor.

FIGURE 5 RAILWAY TELEGRAPHS WERE THE FIRST DOMESTIC TELECOMMUNICATIONS.

After World War I, most of Canada's smaller railways were in serious financial difficulty. A bailout by the federal government saw the merger of these railways into the Canadian National Railway, and their telegraph lines became the CN Telegraph Company.

During the period from 1932 to 1964, these two railway telegraph companies both competed and jointly offered services. In 1932 they provided national network services for the Canadian Radio Broadcast Commission. In 1939, national weather service; and after the Second World War, private wire services.

In 1956 they provided the first telex services in North America, and in 1964 a cross-Canada microwave radio transmission network that was the longest in the world.

These two railway telegraph companies were fused to form CNCP Telecommunications in 1980. In 1988 Canadian Pacific bought out CN, sold 40% of the company to Rogers and renamed the company Unitel.

Decision 92-12 by the Canadian Radio-Television Telecommunications Commission, the federal regulatory agency, allowed Unitel to provide competitive long-distance services. In 1993, 20% of Unitel was sold to AT&T Corporation of the United States.

Even though the regulators gave Unitel a discount on payments to the telephone companies for using their access wires for the first five years, a number of factors made Unitel unprofitable.

These included the necessity to build and maintain a transmission network 7,200 kilometers (4,500 miles) long to connect Victoria to St. John's, plus the costs of the POPs and interconnection in toll centers, customer care and billing systems, the people to run it all and the natural competitive practices by incumbent carriers.

After several years, Rogers Communications Inc. abandoned its interests in Unitel and through a Canadian Creditors' Protection Act bankruptcy-like proceeding, Unitel's ownership was reduced to AT&T Corporation and three Canadian banks. The reorganized company became AT&T Canada Long Distance Services Company.

Subsequently, AT&T Corporation of the USA bought out the rest of AT&T Canada through a holding company.

However, the geographic and market factors that made Unitel unprofitable had not changed, and AT&T Canada continued to lose money. AT&T Canada sold its residential long-distance operations to Primus Telecommunications, and in 2003 went through a second bankruptcy-like reorganization.

The resulting company was re-baptized Allstream, providing corporate and data services. Allstream was subsequently purchased by MTS of Manitoba, then sold to Zayo Group, which has extensive fiber routes across North America.

Zayo Group's 5,000-mile-long Canadian fiber backbone, depicted in Figure 63, is the legacy of the railway telegraph companies that went into the telephone business in competition with the telephone companies.

2.1.8 Canadian Telephone Companies

In 1880, the Bell Telephone Company of Canada was established in Montreal, and other companies providing local service in other cities sprung up across the country.

At the time, long-distance inter-city communications was not technically possible, so these companies each provided telephone service in a local area.

In 1921, the Telephone Association of Canada was formed to promote the construction of a national network.

In 1931, the Trans-Canada Telephone System, again an association of "local" telephone companies, began the development of a national network. The facilities that made up the national network were owned and operated by member companies including Bell Canada and BC Tel.

In 1958, TCTS inaugurated its 158-station cross-Canada microwave radio network to carry long-distance phone calls - the world's longest at the time.

In 1983, TCTS changed its name to Telecom Canada and in 1992 to Stentor.

The Stentor Alliance was terminated effective December 31, 1999.

The Canadian business model mostly evolved to large holding companies owning local operations in many areas and being CLECs in many others, plus long-distance transmission facilities, and in some cases, operating agreements between the companies for interconnect.

BCTel, AGT, Ed Tel and Quebec Telephone merged to form TELUS.

Bell Canada and the four telephone companies in the maritime provinces were reorganized as Bell Canada in metropolitan areas of Ontario and Quebec and Bell Aliant elsewhere, both majority-owned by BCE.

These companies both provide IPTV on VDSL or fiber, and are both expanding to provide national service via fiber, collocations and wireless.

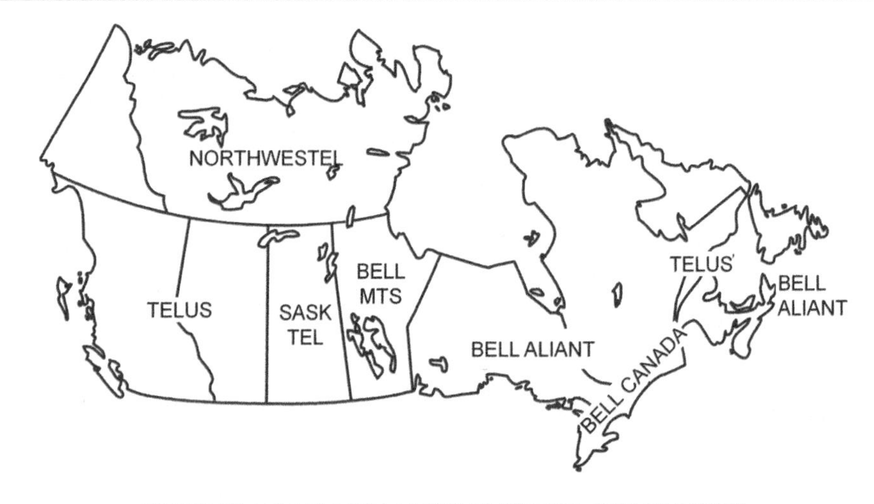

FIGURE 6 CANADIAN TELEPHONE COMPANIES

Rogers Communications is another major player in the telecom business. Rogers started in the cable TV business, with the creation of Rogers Cable by Ted Rogers in 1967 in Ontario.

In 1979, Rogers acquired Canadian Cablesystems and became the largest cable company in Canada.

In the 1980s, Rogers entered the cellular market under the Cantel brand name and later acquired Microcell and its Fido brand.

After its first venture in telecom with Unitel ended, in 2004 Rogers re-entered the business, acquiring Sprint Canada and Callnet, operating as a facilities-based Inter-Exchange Carrier, an ISP, a reseller and as a Competitive Local Exchange Carrier (CLEC) with landlines in Richmond BC. Rogers expanded with the acquisition of Shaw in 2022.

2.1.9 The Rest of the World

In many European countries, the government operated a Post, Telephone and Telegraph (PTT) company that was a government-owned monopoly.

Competition for both local and long-distance voice and data communications has been introduced at different rates in different countries. In all cases, we see the progression of telecommunications service characterized by:

- Monolithic organizations holding a monopoly and the mandate to provide universal service under government ownership or regulation,
- Then the breakup of the monopoly to introduce competition in inter-city and long-distance communications,

- Followed by competition in providing phone service on the last mile,
- Then Internet overwhelming circuit-switched telephone calls, and the focus of competition changes to providing Internet access
- Mobile Internet and PSTN access exploding.

In many parts of the world, particularly in developing areas without usable existing infrastructure, mobile wireless is a more popular method of accessing the Internet - since it is far simpler and less expensive to set up radio base stations than it is to wire or fiber neighborhoods.

The disadvantage of wireless access to the Internet is that radio is an unguided transmission system, sending energy over public airspace, so the available capacity has to be shared with other members of the public. This reduces the capacity available to each user.

If the access is in a guided system, like a fiber or copper cable, it does not have to be shared, each user gets full line-speed Internet access.

In suburban areas of North America, as of the beginning of the 2020s, the phone companies were typically about halfway through installing fiber to the residence in serving areas where there was a threshold of at least 500 homes per square kilometer (2 homes per acre). Some companies have already completed the transition from copper to fiber, others have much more expensive work to do.

In addition to services for individuals, providing high-capacity and high-availability voice and data services for large customers like banks, data centers and government is a viable business everywhere in the world.

2.1.10 New Technologies and New Players

The last decade of the 20th century saw an explosion of new technologies and players, including mobile, broadband to the home with DSL and Cable modems, telephone service over cable modem and Internet phone service.

The first decade of the 21st century saw the beginning of the integration of the Internet and the telephone network, the beginning of mass deployment of fiber to the home, broadband wireless, and a never-ending stream of new players with new capabilities and new market niches.

It also saw the realization of a long-held goal in the telecommunications business: convergence, where telephone calls, television and data are all carried on the same network.

2.2 Convergence

Historically, telephone calls, data communications and television were carried on separate networks using different technologies.

It has always been a goal in the telecommunications business to carry these types of traffic on the same network using the same technology, for the obvious cost savings, and new services and business opportunities for various players.

FIGURE 7 CONVERGENCE: EVERYTHING ON THE SAME NETWORK

Convergence or *service integration* means carrying all kinds of traffic: telephone calls, business data, Internet traffic, music, television, video on demand and everything else on the same network... getting away from separate networks for telephone, television and Internet, with separate entry cables, separate equipment, separate services and bills for each.

Internet, telephone and TV are carried on a single access circuit to the customer, and carried together on the same network circuits.

2.2.1 Network Sharing Strategies

On a network, there are two main methods of dividing the available capacity into usable portions and assigning capacity to particular users:

Dividing the capacity into fixed fractions called *channels* and assigning channels to users for the duration of their communication session whether they actually have anything to transmit or not; or

Allowing users to employ transmission capacity on demand, that is, only when they actually have something to transmit.

These two strategies are called *channelization* and *bandwidth on demand* respectively. The terms *channelized multiplexing* and *statistical multiplexing* or are also used to describe the two strategies respectively.

Telephone calls were historically carried on channelized systems as a natural evolution from physical trunk circuits, and data carried on packetized systems for efficiency. Television programs were delivered in frequency channels on cable TV.

To implement convergence, integrating everything on a single network, the two choices are to either treat everything like voice and video and carry it in channels, or treat everything like data and carry it in packets.

2.2.2 ISDN: Fail

Beginning in the 1960s, Integrated Services Digital Network (ISDN) was viewed as a solution for convergence using channelized multiplexing for all applications: carry everything like voice.

In theory, users would communicate data by establishing circuit-switched connections end-to-end for the duration of a data communication session, then disconnect.

For many reasons, ISDN did not gain momentum and the technologies developed for access to the ISDN, such as Basic Rate Interface, are now obsolete. This was a first attempt at "convergence".

2.2.3 ATM: Fail

In the 1980s, significant effort was expended in product and service development of a technology called Asynchronous Transfer Mode (ATM), which would carry everything: voice, video and data over a single statistically-multiplexed network in packets called cells, essentially treating everything like data.

ATM was deployed to manage data traffic on carriers' backbones, and large organizations deployed ATM, but it was so complicated and expensive that it remained a technology mostly limited to the core of the network. ATM was not deployed on the telephone network to carry POTS trunking. ATM is now an obsolete legacy technology.

2.2.4 IP: Third Time is the Charm

The third attempt, carrying everything in IP packets plus a traffic management technology called MPLS in the network core has proven to be the charm, and the Holy Grail of convergence: integration of all services on a single network, appears to have been achieved.

With IP and MPLS, we are treating voice and video like data, carrying everything in data packets and providing bandwidth on demand to users.

Since data communications used to be a completely separate topic from voice and video communications, with its own paradigm, jargon, buzzwords and technologies, understanding the nuts and bolts of the converged network begins with lessons on data communications.

2.3 Broadband

Bandwidth is a term commonly used to mean capacity, originally referring to the width of frequency band available for use, and now used to also refer to the available number of bits per second.

FIGURE 8 HOW MUCH IS "BROADBAND"?

Broadband is the term used to mean “high capacity”. The term broadband is usually used in the context of end-user, last-mile connections and access speeds, since the capacity internal to the network is already exponentially more (four or five orders of magnitude) than what is called “broadband”.

2.3.1 How Many Bits per Second is "Broadband"?

The question is then, how much is "high" capacity?

Aside from consulting Tommy Chong, one way of defining "high capacity" would be by comparing it to "normal" capacity.

Dial-up modems on a voiceband POTS line, the standard for Internet access in the early days, can achieve about 50 kb/s.

Broadband at that time was defined as "substantially more than 50 kb/s". Services for business customers at 1.5 Mb/s were branded "high-cap" services.

Then things started getting faster:

- In 1996, the US FCC defined broadband as 200 kb/s uploading and downloading.
- In 2010, the FCC updated their definition of broadband to 4 Mb/s downloading and 1 Mb/s uploading (4/1).
- In 2015, this was updated to 25/3.

In 2016, the Canadian regulator defined broadband as 50/10.

2.3.2 4K Guilty Dog Videos

Another way to determine how much broadband is would be to consider utility: what are the most bandwidth-consuming things users want to do? Answer: watch YouTube and Netflix.

Conventional wisdom holds that a "two TV" solution is the minimum requirement: two people streaming HD video plus high-speed web surfing at the same time, without any delays or packet loss.

HD video requires 6 to 8 Mb/s; 4K requires 40 Mb/s.

A reasonable calculation would be 25 Mb/s if based on Full Quad HD (1920x1080), and 100 Mb/s if based on 4K definition video streams.

It is important to note that end-users download content, and very rarely upload anything more than keystrokes and mouse clicks.

Therefore, the user can get the best experience for the cost by allocating most of the available capacity to downloading and only a small amount to uploading.

2.3.3 Universal Service

Regulators will inevitably move toward requiring broadband as a minimum service level for their citizens, just as electricity and telephone.

After it was available in big cities, getting electricity became a right, and telephone service became a right.

In many areas, cooperatives were formed to provide these new, expensive services. In other cases, private companies and government expanded to provide the service.

As part of a social bargain, service providers are obliged to subsidize service provisioning to rural areas with profits made in very densely populated areas.

This trend will continue with broadband.

2.4 Today's Converged Telecom Network

This section is an introduction to the topics that will be covered in detail in subsequent sections and chapters.

Figure 9 provides a framework and reference for organizing discussion of all the different aspects of telecom.

Telecom networks are built with network equipment in buildings in different locations connected by long-distance transmission systems.

The equipment in each building is connected on one side to the long-distance transmission system, and on the other side to access circuits leading to customers.

2.4.1 Common Carriers

Many users' traffic is carried together over the network, leading the owners of the networks to be called *common carriers*. We call them *common* not to disrespect them, but to mean traffic carried on common facilities like transmission lines and switching centers.

The physical network is a major capital investment. The reason for building all of this is, of course, to make money, with many subscribers who pay for services with recurring billing month after month.

A carrier network is traditionally described as being made of three parts: network access, network edge and network core.

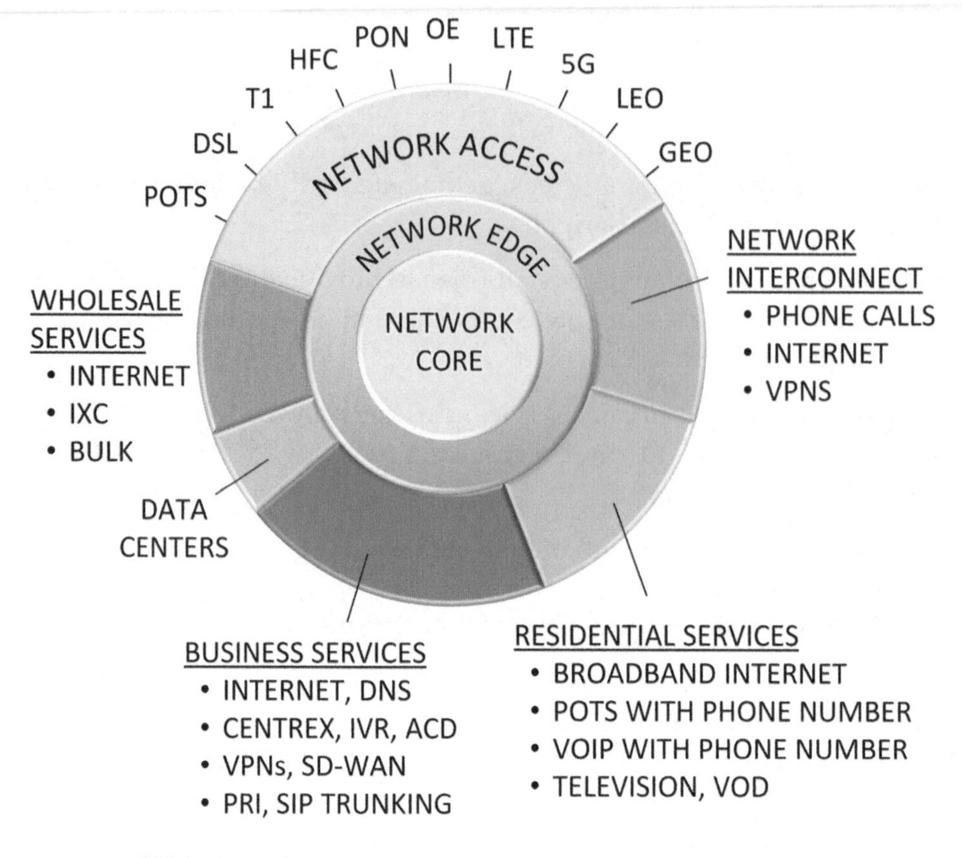

FIGURE 9 TODAY'S CONVERGED TELECOM NETWORK

2.4.2 Core

The network core is also colloquially referred to as the *backbone*. It provides high-capacity, high-availability communications, typically constructed of fiber optic cables between cities.

2.4.3 Access

The network access is the equipment and cabling used to connect the customer to the network, historically to a Central Office. This is also referred to as the "last mile", though of course the people who work in this part of the business prefer to call it the "first mile". These circuits may be much more or much less than a mile long.

Today, there are many kinds of network access, including Plain Ordinary Telephone Service (POTS), Digital Subscriber Line (DSL), Trunk Carriers (T1), Hybrid Fiber-Coax (HFC) Community Antenna Television (CATV) systems, Passive Optical Network (PON), Optical Ethernet (OE), 4th-generation Long Term Evolution (LTE) for both mobile and fixed wireless, 5G and New Radio (NR), Low Earth Orbit (LEO) and Geosynchronous Earth Orbit (GEO).

2.4.4 Edge

Between the core and the access is the network edge. The network edge is buildings and equipment that connect access circuits to the network core. This equipment performs a concentration function, funneling users onto the core, and a physical standards conversion, for example, copper to fiber.

All of these networks must interconnect to allow communications between users who are on different networks. Interconnect between carriers for telephone calls, for Internet traffic and for business communications are each implemented in a different way.

2.4.5 Residential, Business and Wholesale Services

Telecom services have historically been grouped into residential, business and wholesale.

"Residential" of course includes the mobile and the homeless. "Business" includes government, education, military and corporate customers. Carriers sell telecommunication services to each other wholesale, and to aggregators, integrators and resellers.

The main residential services are:

- Broadband Internet
- POTS with phone number
- Voice over IP cellular mobile service with phone number
- Internet VoIP telephone service with phone number
- Television channels and Video on Demand (VoD)

Business services include:

- Internet access and services like Domain Name System (DNS) services
- Centrex, Interactive Voice Response (IVR) and Automated Call Distributor (ACD) services
- Virtual Private Networks (VPNs)
- Software-Defined Wide-Area Networks (SD-WANs)
- Private Branch Exchange (PBX) trunks and Primary Rate Interface (PRI) service

- SIP Trunking service

Wholesale services include:

- Internet transit services
- IXC services carrying telephone calls, and
- Bulk services: dark fiber, wavelengths, VPNs and Carrier Ethernet.

2.4.6 Data Centers

In Figure 9, between wholesale services and business services lie Data Centers. Data centers are buildings with vast numbers of computers and hard drives that run applications like Instagram and are the source of content like YouTube video streams.

2.5 The Network Core

The network core provides high-capacity, high-availability connections between cities, and between districts of a big city.

The core is implemented with point-to-point connections between different switching and access aggregation centers, which can include Central Office (CO), Toll Center, Point of Presence (POP), CATV Head End, Mobile Telephone Switching Office (MTSO), Internet Exchange (IX), Data Centers and others.

Fiber is used for the point-to-point connections since it can support very high numbers of bits per second. Lower-speed (and lower-cost) circuits are used to provide access to this core to users.

A method of organizing the bits for transmission, physically transmitting them on fiber, plus monitoring, alarming, testing and automatic protection switching are required for reliable service.

In the vast majority of cases, Optical Ethernet (Section 10.5), that is, 802 MAC frames (Section 15.1) signaled on fiber point-to-point between physical Optical Ethernet ports is used.

The simple example of Figure 10 shows two or three fibers terminating on each location; in reality, tens if not hundreds of thousands of fibers terminating at each location would be more accurate.

Layer 2 switches at each location terminate these fibers; they front-end and concentrate data into the *routing* function inside each location. The routing function relays the data onto a particular outbound fiber.

Layer 2 switch is another term for a LAN switch or Ethernet switch, the functions corresponding to Layer 2 of the OSI 7-Layer Reference Model.

Routers in different cities connected point-to-point with Optical Ethernet on fiber is the plumbing of the network core.

In the past, the most popular technology for these functions in North America was a standard called Synchronous Optical Network (SONET). In the rest of the world, the similar Synchronous Digital Hierarchy (SDH) was employed. There is an installed base of SONET and SDH systems.

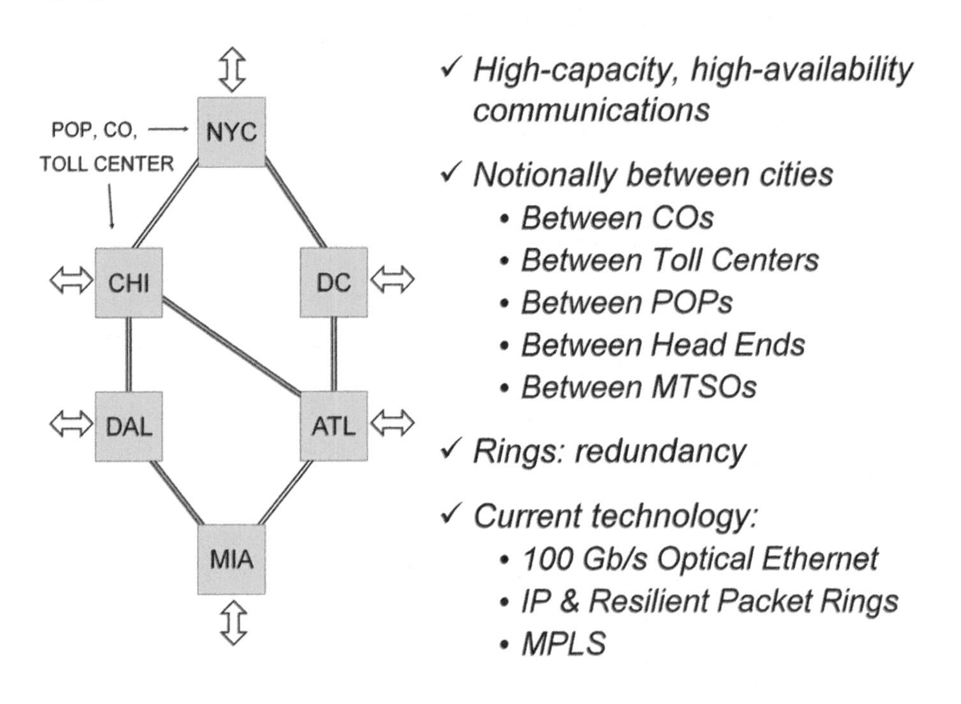

FIGURE 10 THE NETWORK CORE

2.5.1 Rings

To ensure high availability, that is, the possibility of communicating even if a line is cut, it is necessary to provide multiple redundant paths between each point.

The cheapest way to implement redundancy is to connect locations neighbor-to-neighbor in *ring* patterns. This way, there are two connections to every location, but only one extra circuit.

Rings are used to connect COs, POPs and other buildings in a city. Rings are also installed to connect cities in regions, and these regional rings are interconnected at multiple places for long-distance communications.

Short-cuts, i.e., connections between non-adjacent points on the ring, are implemented as traffic dictates.

The end result is a *semi-meshed network,* where some locations are directly connected and others are reached via intermediate stations.

Resilient Packet Ring (RPR) and/or MPLS technologies are used to implement protection against data loss in the event of broken connections by rerouting traffic on another of the redundant paths.

2.6 Network Protocols: Ethernet, IP and MPLS

A *protocol* is a method of performing a procedure. In telecommunications, network protocols are those that enable the communication of bits from one user to another across a network.

It is desirable to use *standard* protocols so that equipment and software from different vendors can be used to build and access the network.

The definition of a *network* is circuits connected so that there are multiple possible routes to the final destination, and a decision is made to choose one route over another. The information being transmitted is not broadcast to every station on every circuit in the network; it is switched or routed to get to the single final destination circuit.

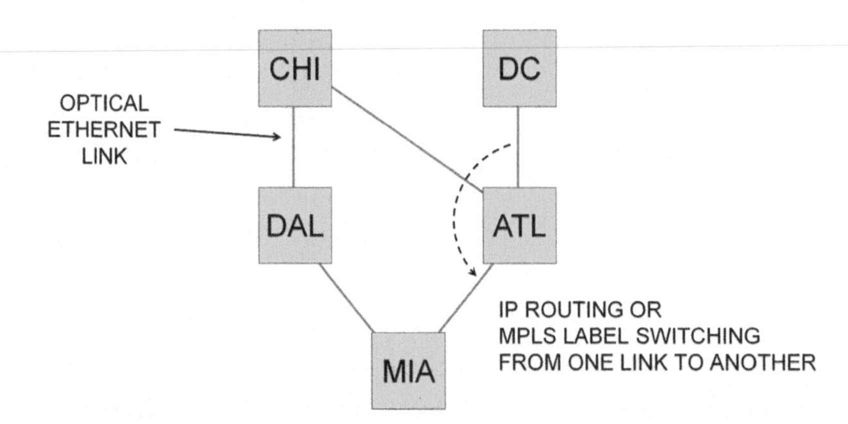

FIGURE 11 NETWORK PROTOCOLS: ETHERNET, IP AND MPLS

Three main things are required to implement a network:

- Links between different geographic locations,
- Devices to terminate links at these geographic locations, and to relay information from one link to another, and
- A method of managing the flow of information.

2.6.1 Optical Ethernet Point-to-Point Links

Today's IP telecom network uses Optical Ethernet (OE) to implement links between locations. Optical Ethernet is bits organized into groupings called Media Access Control frames or MAC frames transmitted on point-to-point fibers between locations.

The frames are transmitted by flashing a laser on and off to represent the value of each bit in the frame to the next location, one after another in time. In the simplest systems, light on means 1 and light off means 0.

2.6.2 IP Packet Routing Between Links

The devices that are connected by these point-to-point links are routers. Routers move information from one link to another. Knowing **which** link to move the information to is the routing part of the story.

To enable routing, every device on the network is given a network address. The network address is like a telephone number, used to indicate the final destination. The Internet Protocol (IP) is the world's most popular standard for network addresses.

To communicate, a message is broken into segments, and the IP address of the desired final destination is placed at the beginning. The segment of data with the IP address at the beginning is called an IP packet.

IP routers examine the destination address on an incoming packet to decide where to go next, then implement the decision by transmitting the packet on the selected outgoing link.

2.6.3 MPLS Traffic Management

For end-user networks, Ethernet and IP are usually sufficient.

Common carriers, who sign contracts with banks to guarantee service quality, go a step further and use an overlaid traffic management system called Multiprotocol Label Switching (MPLS) internal to their networks.

MPLS allows the network operator to define traffic classes. A traffic class is a stream of packets coming from the same place, going to the same place, and requiring the same treatment in terms of delay and packet loss.

On entry to the carrier network, an MPLS label is placed before the IP address on the packet, indicating the traffic class the packet belongs to. The routers in the core then look at the label and not the IP address to make forwarding decisions.

This allows the carrier to treat a stream of a single kind of traffic as a single item and assign bandwidth and relative priority to the stream to guarantee the service.

The MPLS label is removed at the far end before the IP packet is relayed to the user's access circuit. MPLS is the basis of an internal network traffic management system that is invisible to end-users.

2.7 Network Access: The Last Mile

The access is the physical connection between the user and the network. This circuit is the face of the telecom network to users. Service is usually billed on a per-access basis, and usually coupled with a network address.

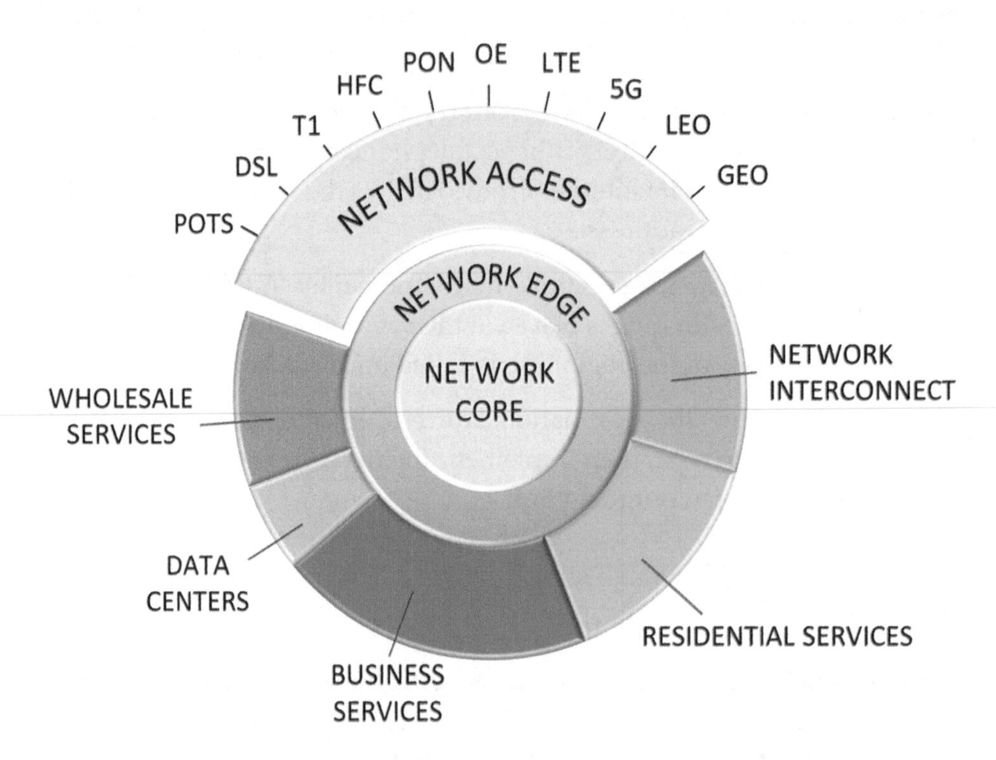

FIGURE 12 NETWORK ACCESS: THE LAST MILE

There are many types of access circuits. To get started, we list the access circuit types by technology: copper, fiber and wireless, with a brief description. Every one will be explained in detail in subsequent chapters.

2.7.1 Twisted Pair Loops

It all started with telephone companies and two copper wires called a *loop*, insulated and twisted together to minimize noise pickup. Plain Ordinary Telephone Service (POTS) is implemented using voltage analogs of speech on the loop. Digital Subscriber Line (DSL) implements broadband communications on the loop using modems. T1 business services use voltage pulses on twisted pair.

2.7.2 Coax

Coaxial cable or *coax* is also two copper wires; one inside the other, instead of side by side. The outer wire is like a pipe, through which the wire runs. This geometry allows the transmission of very high frequency electrical signals, so compared to twisted pair, much more information including television signals and modem signals can be communicated.

Of course, nothing is free: the maximum practical length of coax is about 1/10 that of twisted pair. The solution is to transport the signal on fiber to the neighborhood then on coax copper to the premise, called a Hybrid Fiber-Coax (HFC) system, to implement a Community Antenna Television (CATV) service, commonly called cable TV, and run modems on it.

2.7.3 Fiber to the Residence & PONs

With the coming advances in bandwidth and coverage of HFC and wireless access, telephone companies with an installed base of twisted-pair copper loops, called Incumbent Local Exchange Carriers (ILECs), had to decide if they wanted to continue in that business, as the twisted pair loops cannot support enough bandwidth to compete. ILECs continuing in the wired-access business means pulling a fiber past every residence.

Pulling a fiber past every residence is an expensive and time-consuming project, but leapfrogs all other technologies in bandwidth, erecting a business barrier against any competition being able to deliver more bandwidth for the foreseeable future.

Telephone companies are provisioning Optical Ethernet for business customers and Passive Optical Network (PON) links to residences.

When a neighborhood is "fibered", many dark fibers for future use as backhaul for future short-range ultra-high-capacity wireless access are installed. The incremental cost is low, and having the wiring for such a system pre-installed may be a cost differentiator between the ILEC and a wireless competitor in the future.

2.7.4 Wireless: Fixed and Mobile

Wireless systems can be fixed or mobile. Fixed wireless is access using an antenna on a building communicating at frequencies like 2.5 and 3.5 GHz. Wi-Fi is fixed wireless, but mostly fits into the customer premise equipment category rather than telecom services.

Mobile wireless means many radio base stations, each with multiple antennas pointing in different directions so that there is radio coverage in a large geographic area, and a system of handing off the user from one base station to the next as either the user moves around (cellphones) and/or the base stations move around (Low Earth Orbit LEO satellite phones).

The Universal Terrestrial Radio Access Network Long-Term Evolution (LTE) standard is also referred to as the fourth generation (4G) of mobile cellular. LTE radio is also used in fixed wireless.

The fifth generation (5G) brings 40% more spectral efficiency (bits per Hertz) plus the beginnings of massive bandwidth for end-users using Multiple-Input, Multiple-Output (MIMO) a type of spatial multiplexing, heading toward 1 Gb/s download speed, as well as new frequency band allocations.

The longest access circuit, Geosynchronous Earth Orbit (GEO) is 22,300 miles above the surface of the earth above the equator. At this radius, the orbital velocity is the same as rotational velocity of the earth, so the satellite appears to be stationary to an earthbound antenna.

Medium Earth Orbit, with GPS and other satellites is approximately 12,000 miles above the earth, with an orbital period of about 12 hours.

Low Earth Orbit, hosting communication and sensing satellites, ranges from 250 miles (400 km) to at most 1200 miles (2000 km) above the earth.

2.8 Anatomy of a Service

As illustrated in Figure 13, any service provided by a telecommunication service provider is made up of three components: access circuit, network connection type and billing agreement.

2.8.1 Access Circuits

The access circuits are physical lines with circuit terminating equipment at each end. These lines run from a user's site to the nearest physical access point to the network. The location containing this physical attachment point is usually called a Central Office (CO), but could equally be a CATV Head End, or a wireless base station.

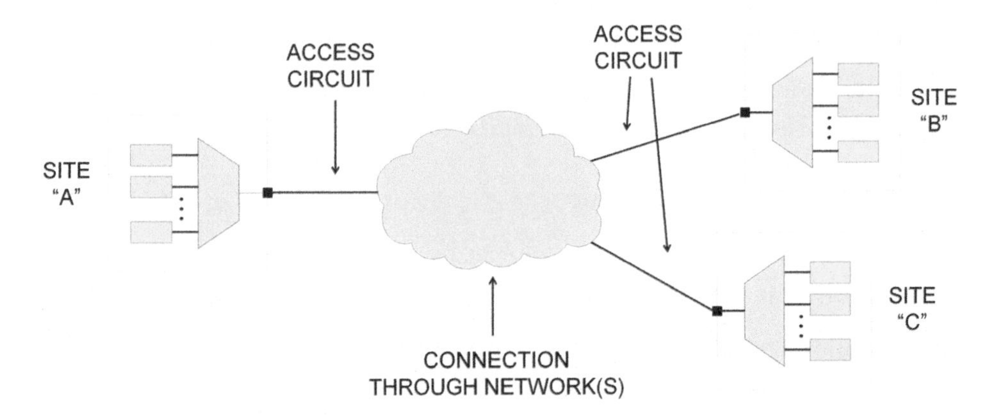

FIGURE 13 ANATOMY OF A SERVICE

There are many different technologies for access circuits, including

- Plain Ordinary Telephone Service (POTS) lines
- Older-style digital data circuits at up to 56 kb/s,
- ISDN BRI digital telephone lines at 128 kb/s,
- T1 digital access circuits at 1.5 Mb/s,
- xDSL technology at up to 200 Mb/s,
- Cable modem technology at up to 1 Gb/s.
- Passive Optical Networks at 1 Gb/s or more,
- Mobile Cellular radio at 5 Mb/s and up,
- Fixed wireless at 25 Mb/s and up,
- Optical Ethernet from 1 to 100 Gb/s
- Satellite modem download at 12 to 100 Mb/s

With the exception of the satellite link, these are short circuits, hopefully less than a few miles long.

2.8.2 Circuit-Terminating Equipment

Each type of access circuit must have a specific type of Data Circuit-terminating Equipment (DCE) attached at each end to be able to transmit data from one end to the other.

This hardware, also called Data Communications Equipment and Customer Premise Equipment, physically generates the representation of the information on the access circuit.

Some examples are:

- Modems, used on wireless, coax cable modems and twisted pair DSL,
- LAN Network Interfaces: copper, fiber and wireless implementations,
- Optical Network Units (ONUs), Optical Network Terminals (ONTs), and Optical Line Terminals (OLTs) used on fiber circuits,
- Data Service Units (DSUs), Channel Service Units (CSUs) and CSU/DSUs used on T1 and related circuits.

2.8.3 Network Connection Type

Network connections between the access circuits are made over high-capacity circuits that are owned and managed by the network provider.

Many options for connection through their networks are offered.

These can be summarized into three fundamental choices:

- Full period: connected all the time, with a fixed amount of capacity, billed as a monthly fixed charge.
- Circuit-switched: connected on demand, with a fixed amount of capacity, billed as a monthly fixed charge for the access circuit plus a per minute usage-sensitive charge.
- Bandwidth on Demand or "packet-switched": available all the time but used only when needed, with capacity up to a maximum of the access line speed. Billed as a monthly fixed charge for the access plus a usage-sensitive charge based on the amount of data transmitted, often with a threshold before usage-based pricing is applied.

There is a monthly charge for each access. Each location needs an access.

A combination of access circuits with their circuit-terminating equipment, method of connection, and of course, billing plan make up a *service*.

2.9 Inside the Network Cloud

When ordering and troubleshooting services from service providers, it is useful to know what is going on inside their network "cloud".

The telecommunications network has historically had two parts from a network planning engineer's point of view: *core* and *edge*.

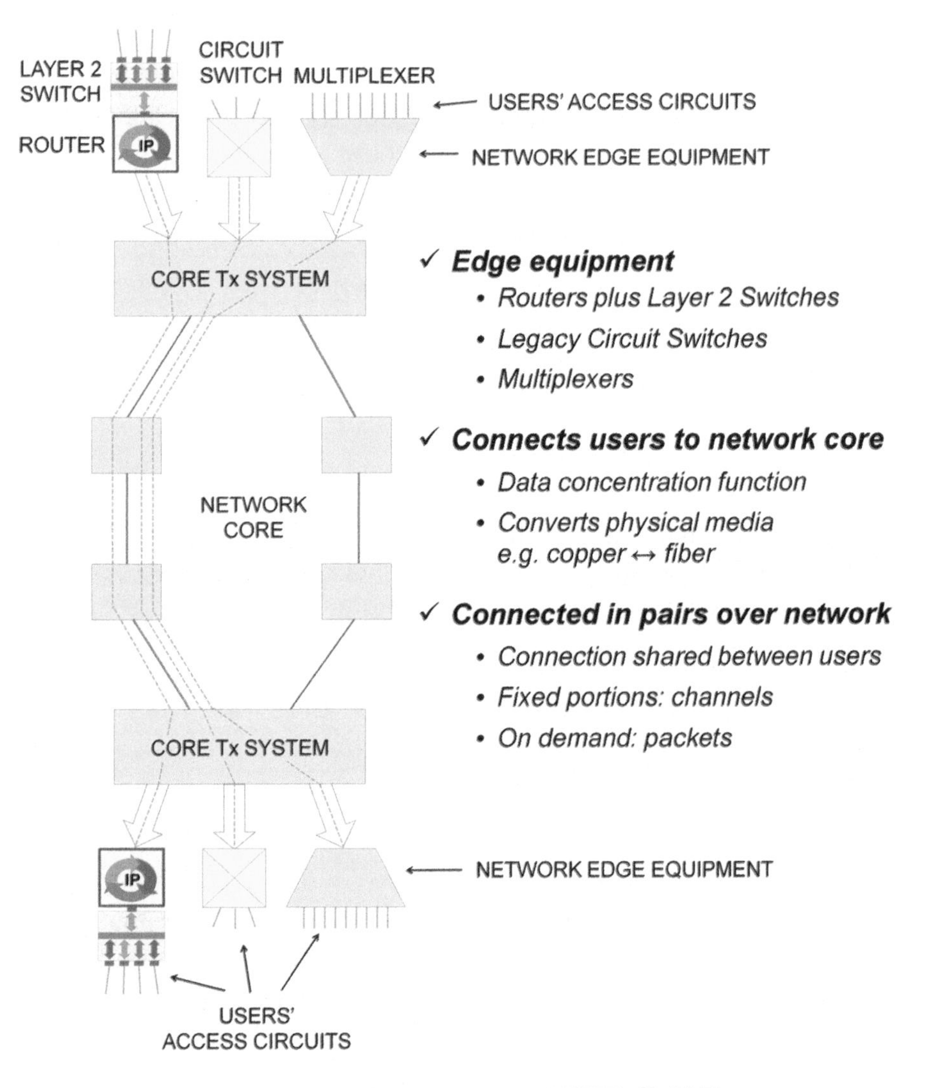

FIGURE 14 INSIDE THE NETWORK CLOUD

2.9.1 Core

As discussed in Section 2.5, the *core* provides high-capacity, high-availability communications between different geographic locations. Facilities-based carriers each build their own core networks.

Physically, the core is routers in buildings in different locations connected with point-to-point fibers. The buildings are connected with fiber neighbor-to-neighbor, to form what looks like a *ring* from above to implement redundancy. Shortcuts across the ring are implemented where traffic warrants, as was illustrated in Figure 10.

The router and/or the building it is in can be called a network *node*, after the French word for knot. It is where many fibers are tied together.

These buildings are Central Offices, Mobile Telephone Switching Offices, CATV Head Ends, Toll Centers, Internet Exchanges, Data Centers, Collocation Facilities, and anywhere else there is a convergence of a lot of circuits and traffic. Regional rings connecting a group of cities are interconnected at multiple places to cover the entire country.

Dense Wave-Division Multiplexing (DWDM) of fiber is the "Core Tx (Transmission) System" in Figure 14. It is used to implement links of 100 Gb/s or more to build the ring.

2.9.2 Edge

From the network point of view, access circuits are connected at the *edge* of the core network, using appropriately-named *edge equipment*. As illustrated in Figure 14, the edge equipment acts as a data concentrator, with multiple lower-speed circuits on the access side, and one high-speed *aggregate* circuit on the network side.

The edge equipment is connected in pairs across the ring, and each connection is assigned a portion of the capacity of the underlying circuits. The three different kinds of edge equipment, described in the next section, each share the capacity assigned to the pairwise connection between the lower-speed circuits in a different way.

The edge equipment also provides physical ports to terminate access circuits, typically Optical Ethernet on fibers going to buildings, to neighborhoods, to cell sites and everything else geographically nearby.

The edge equipment at one location aggregates the traffic of the lower-speed circuits into a high-speed stream of bits.

The stream of bits is transmitted to its opposite number, which distributes the traffic to the correct far-end lower-speed circuits.

2.10 Network Edge Equipment

There are three basic kinds of edge equipment: multiplexers, circuit switches, and routers front-ended with Layer 2 switches. They are connected in pairs across the core.

Each of these partitions the capacity of the connection between the paired edge equipment in a different way, and so each is used to implement a different kind of network service.

2.10.1 Multiplexers: TDM, FDM and WDM

In a bottom-up worldview, a multiplexer combines the data from many lower-speed inputs into a high-speed aggregate stream to be carried on a high-capacity connection to its pair at the far end.

In a top-down worldview, a multiplexer is a device that implements multiple *channels* on a physical transmission medium. Channels are fixed amounts of capacity, a subset of the total capacity.

Each channel can be used to carry a different bit stream to the far end; or multiple channels can be used in parallel to transmit one bit stream.

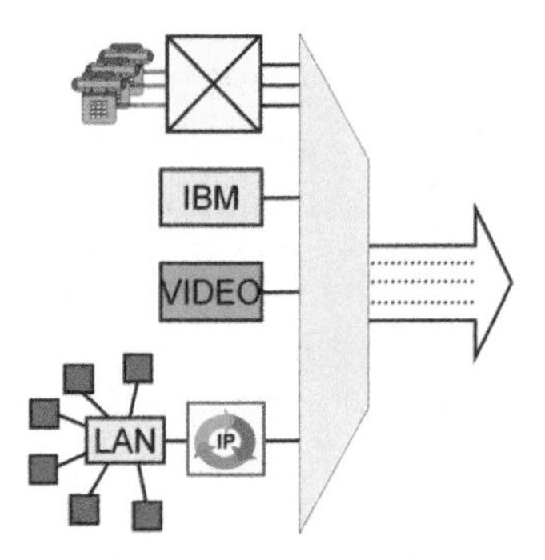

Multiplexer

- *Static, non-intelligent device*
- *Carries multiple circuits on one*
- *Channel = fixed amount of capacity*
- *TDM: PON, legacy 64 kb/s channels*
- *FDM: Radio, CATV channels*
- *WDM: Optical wavelengths*
- *Cross-connect: change channels*

FIGURE 15 MULTIPLEXER

To implement an end-to-end connection, the user's access circuit is connected to a hardware port on the multiplexer at each end, and a channel is assigned to connect the two hardware ports across the core.

Their data is then moved, along with that of other users, between the hardware ports at each end, connecting the access circuits and thus implementing a service.

Multiplexers are fixed, static devices, that implement a full-time connection with dedicated capacity, called a *full-period service*, also called a dedicated line, private line or leased line.

Different ways of implementing channels include Time-Division Multiplexing (TDM), Frequency-Division Multiplexing (FDM), and Wave Division Multiplexing (WDM).

Time-Division Multiplexing means sharing a single circuit in time amongst users, one after another in a specific order that repeats over and over.

Examples of TDM include:

- The upstream connection on Passive Optical Networks, where typically 32 users on the "lower-speed" side share one high-capacity fiber for uploading one after another in time;
- Legacy 2G GSM cellular, where users time-share radio channels; and
- The large installed base of legacy systems that began in the 1970s to use TDM multiplexers to create 64 kb/s DS0 channels to carry phone calls and data.

Today, *statistical* Time-Division Multiplexing, where users share capacity in time, but only when they need it, is most common. Users do not get fixed amounts of capacity like a channel, but rather capacity or bandwidth on demand. This is implemented with routers and Layer 2 switches, discussed in Section 2.10.3 below, and in more detail in Section 3.8.

On other transmission systems, the multiplexer implements Frequency-Division Multiplexing (FDM), dividing the available frequency band into smaller frequency channels.

This technique is used to create radio channels and channels on CATV systems, which are assigned to users. Modems are typically deployed on a channel at each end to communicate bits that represent voice, video, Internet traffic or anything else.

In fiber optics, the center of a frequency channel is referred to as a *wavelength* (L in English, λ or lambda in Greek), and the device that implements optical channels is called a *Wave-Division Multiplexer* (WDM).

This multiplexer represents bits on the fiber at specific wavelengths using simple schemes like light on means "1" and light off means "0", or more sophisticated modulation schemes to increase the bit rate.

A *cross-connect* is like a multiplexer, except that it has the same speed circuit on both sides, and can move information from one channel to another.

2.10.2 Legacy Circuit Switches

A *legacy* is something left to us by a previous generation; for example, the large number of telephone circuit switches and related transmission systems that stopped being deployed around year 2000 but are still in use.

In time, these will all disappear, replaced with Voice over IP and routers. In the meantime, is necessary to understand the principles to deal with the installed legacy base.

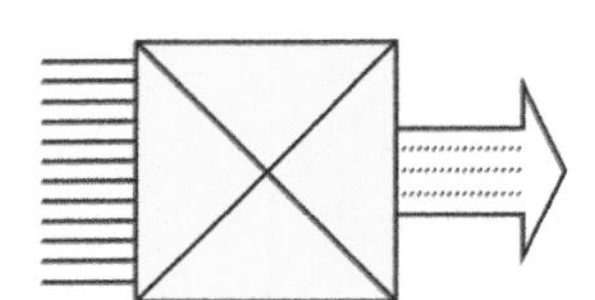

Legacy Circuit Switch

- *Legacy telephone circuit switch: seize a trunk (64 kb/s channel) for duration of phone call, then release*

FIGURE 16 CIRCUIT SWITCH

A *circuit switch* is a device that reserves a circuit for a user for a period of time, then releases it, so a different user can use the circuit.

In the telephone business, these circuits were called *trunks,* carried on 64 kb/s channels on fiber or copper between circuit switches called *CO Switches* at the neighborhood level, and *Toll Switches* at the city level.

Private circuit switches that manage phone calls in and out of office buildings are called Private Branch Exchanges (*PBXs*).

To make a phone call, the user dials a telephone number, which identifies the far-end telephone line, then the telephone switch they are connected to *seizes* a trunk going in the right direction, reserving it for use for the duration of the phone call.

The telephone circuit switch at the far end connects the trunk to the user identified by the telephone number. When the call is completed, the trunk is *released* for someone else to use.

This is called *circuit-switching* and covered in more detail in Section 11.1.

A circuit switch is edge equipment that reserves a trunk for a user's phone call on a per-call basis. They are used for POTS and ISDN services.

2.10.3 Router + Layer 2 Switch

On today's telecom network, built using Ethernet, IP and MPLS, the most common network service is communication of IP packets. Packets of many users are interspersed on network circuits.

Unlike circuit-switched phone calls and dedicated lines that reserve channels, in an IP packet network, there is no capacity reserved or guaranteed for any particular user.

IP packets are created by a user's device, like a phone, or a computer, then transmitted to a near-end network router over the user's access circuit.

The network router relays the packet via the network core to a network router at the far-end location. The far-end network router relays the packet to the far-end access circuit, where it's delivered to the far-end device.

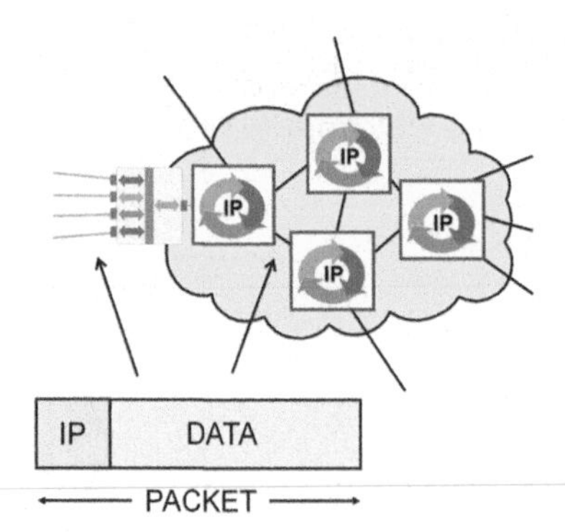

Router + Layer 2 Switch

- *Layer 2 switch aggregates traffic*
- *Router relays packets between circuits*
- *Knowing which circuit = routing*
- *First-come, first-served**
- *"Bandwidth on Demand"*

* Maybe not, if prioritization for CoS is implemented.

FIGURE 17 ROUTER AND LAYER 2 SWITCH

IP routers are stateless devices that treat each packet they receive individually. An IP router doesn't establish connections, and doesn't reserve capacity for a particular user.

In the simplest terms, an IP router receives a packet on a fiber plugged into one of its hardware ports, then sends it back out on a fiber plugged into a different hardware port. Deciding on *which* fiber the packet should be sent back out is the "routing" decision.

Fiber is usually used in pairs: one for transmit, the other for receive. The fibers are plugged into a physical Optical Ethernet port, which has sockets to receive the fibers. Fiber is covered in detail in Chapter 10.

In most locations, there are not only the intercity fibers, there are also thousands of access fibers connecting to buildings, neighborhoods, cell sites, data centers and everything else in the geographic area.

For scalability and cost reasons, in big networks, the physical fiber ports and the routing function are implemented in different devices.

Fiber ports for the access circuits are implemented in rack-mount hardware called *Layer 2 switches,* with up to hundreds of OE ports each.

The routing function, along with a relatively small number of OE ports, is implemented in a separate device called a *router.*

Multiple Layer 2 switches physically terminate the access circuit fibers, and perform a data concentration function, efficiently aggregating the traffic from them into a single high-speed stream, which is fed to the router.

So, the third type of edge equipment is a router front-ended by Layer 2 switches. Routers are used for bandwidth on demand services like Internet service and commercial IP packet communication services.

In an IP router, packets are processed and sent out on a first-come, first-served basis, so service characteristics like delivery delay and packet loss are the same for all packets, regardless of what they contain.

If prioritization as a Quality of Service (QoS) mechanism to deliver a guaranteed Class of Service (CoS) has been implemented, then the processing order might change based on the traffic being carried.

2.11 Interconnect to Other Carriers

For certain services, carriers are required to connect other carriers to their networks by regulators. The method, service level and cost of the connection is specified in a legal document called a *tariff.*

For other services, carriers choose to connect for business reasons, and the method or cost is not regulated but instead driven by market forces.

2.11.1 The ILEC

Incumbent Local Exchange Carrier (ILEC) is the term given to the carrier that owns the fiber optic cables and twisted pair copper wire cables that pass everyone's residence and business in a city.

This company owns the buildings where the cables for a neighborhood come together, called Central Offices. They also own a building called the Toll Center, where connections to long-distance networks are made.

2.11.2 Toll Switches, Toll Centers and POPs

The ILEC connects their CO switches to a *toll switch* in the toll center.

Any company wanting to connect to the ILEC's customer, either another LEC like a wireless carrier, or an Inter-Exchange Carrier (IXC), must establish a Point of Presence (POP) in or near the toll center building.

A company's POP is where their network equipment and fiber connections to their core is located.

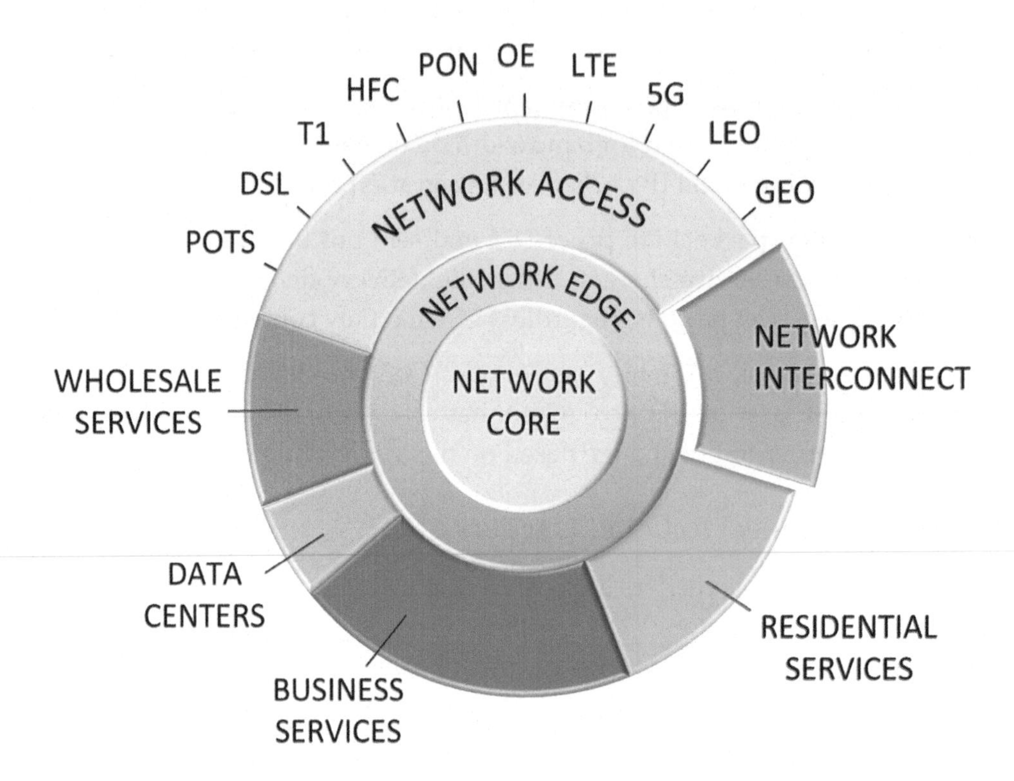

FIGURE 18 INTERCONNECT TO OTHER CARRIERS

2.11.3 Switched Access Tariff

The other LEC or IXC pays the ILEC monthly for (many) Tandem Access Trunks, which connect the equipment at their POP to the ILEC's toll switch. Connections are established to other carriers through the toll switch on a call-by-call basis.

There are usage-based charges to use the ILEC's last mile to connect to the user. For each phone call that the IXC asks the ILEC to connect to an ILEC customer, the IXC must pay the ILEC a *switched access charge,* typically in the one cent per minute range.

Mobile and wireless companies, other landline companies and companies whose backbone is the Internet all connect at the toll center for access to the ILEC's customers and to each other using this method.

In the early 2020s, this is the only tariff available to connect telephone calls between carriers in the USA. Eventually there will be a tariff for SIP and VoIP phone calls between competing carriers. But it is so complex, and there is so much money involved, it will take time to come up with a plan that can be agreed to.

2.11.4 Internet Traffic Interconnections

Internet traffic is completely different.

The Internet is a giant amorphous blob of Autonomous Systems (ASs), connected together for business reasons. An AS is a collection of routers managed by the same organization, like a university or an ISP.

Connections between ASs are made at buildings called Internet Exchanges (IXs), run by neutral third parties. Each party is responsible for establishing a physical connection from their network to a port on a router in the IX. They pay the IX owner to establish a route between the two router ports inside the router, and hence connect the two networks.

2.11.5 Business Service Interconnections

Business services generally means Virtual Private Network (VPN) service, which is point-to-point connections between an organization's buildings with specified bandwidth and performance guarantees.

In cases where there are different carriers involved to make an end-to-end connection, it is necessary to coordinate the traffic management system settings end-to-end. This would be done in the context of a partnering agreement between carriers, and could also be managed by a consolidator or aggregator.

2.12 Services

Telecommunications service providers sell services, with an installation charge plus monthly recurring service and equipment charge model. It is a robust business model that has been very successful.

Services are organized in groups by type of customer: residential, business and wholesale.

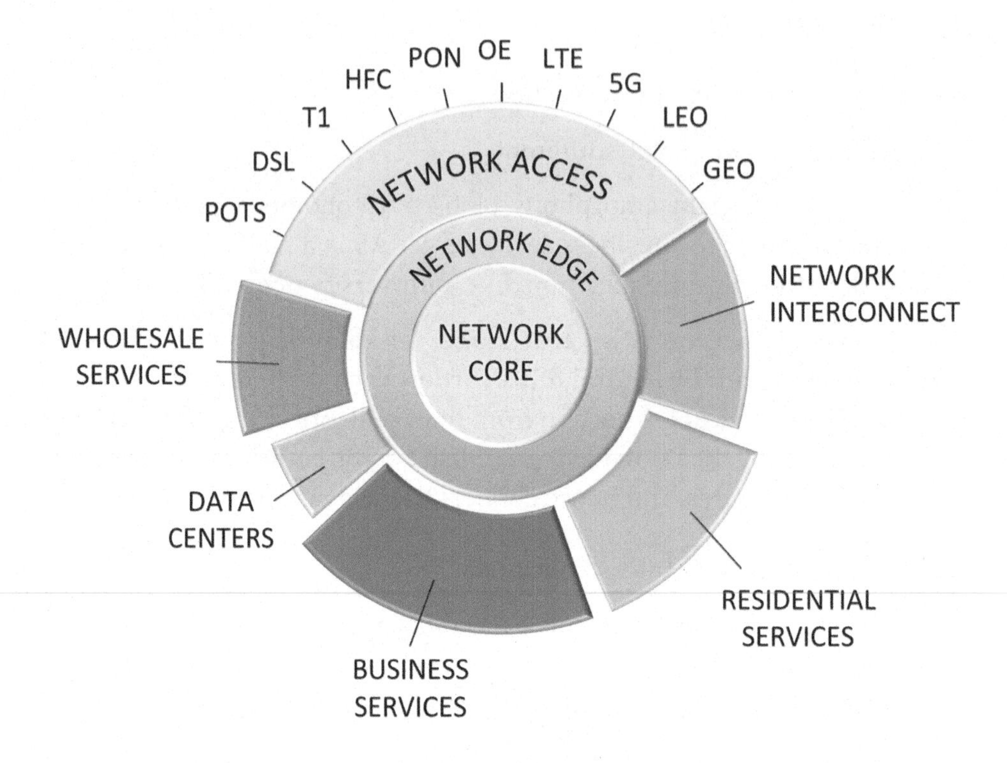

FIGURE 19 SERVICES

2.12.1 Residential Services

Residential services include a triad of broadband Internet, telephone service and television content.

Telephone service can be via landline or mobile; both require a POTS telephone number to receive incoming calls from anyone's telephone.

Basic cable is a package of streams from a selection of broadcast and non-broadcast channels.

Video on demand can mean a separate communication service from the same company that is providing Internet service.

It can also mean Netflix and YouTube, along with AppleTV and AmazonTV, which are all kinds of Internet traffic to be carried on the residential broadband Internet connection.

2.12.2 Business Services

Business "data" services include Internet access and MPLS VPN service: point-to-point IP packet communications backed up with a service level guarantee. These are also called *MPLS services* since MPLS is used to implement the guarantee in the network.

Internet access is another business service. Secure point-to-point connections over the Internet called *tunnels* and *IP VPNs* are implemented using encryption.

Multiple tunnels implemented pairwise between an organization's locations, that can be remotely configured, is called a Software-Defined Wide Area Network (SD-WAN).

With Internet browsing, IP VPN and SD-WAN, there is usually no performance guarantee, as communication is over the Internet, where it is unpredictable what circuits the data will traverse. It will probably work fine... but there is no guarantee.

A guarantee requires the telephone company do it on their non-Internet circuits, called an MPLS VPN.

Business voice services include Centrex, where the phone company supplies lines and phones and manages everything, which is reliable but expensive.

Functions like Interactive Voice Response (IVR) (press 1 to make a new reservation), and Automated Call Distribution (ACD) (please wait for the next available operator) can be supplied as Centrex services by the ILEC.

Another business voice service is PBX trunks. A Private Branch Exchange (PBX) is a telephone switch at a business customer premise. The connections from the CO to the business are not individual lines, but trunks connected to business phones by the PBX on a call-by-call basis.

A service called ISDN Primary Rate Interface (PRI) is required in addition to the trunks for the exchange of call control messages like "disconnect" and caller ID.

PBX trunks are replaced by SIP trunking, where telephone calls are carried not in ISDN channels, but rather as Voice over IP (VoIP) between a business's locations and a PSTN gateway for dial-out to the public.

SIP trunking is much less expensive than PBX trunks.

2.12.3 Wholesale Services

Wholesale means business-to-business sales between carriers.

This includes one facilities-based carrier (one that owns transmission facilities) leasing capacity from another on a particular fiber route; and resellers, who lease all of their capacity.

Bulk carriers are those whose business model is to put fiber infrastructure in the ground between cities and sell connectivity over this infrastructure to carriers. A wireless carrier might use this type of service for backhaul from base stations and connections between mobile switches in different cities.

Connections between different carriers for Internet traffic means transmitting an IP packet to a different company and hoping they will cause it to be eventually delivered, and vice-versa.

There are two flavors of wholesale Internet service: peering and transit. Peering is a cooperative connection where no money is exchanged, called settlement free. Peering generally does not come with quality guarantees.

Transit is a service sold by big players to other players, giving the buyer really high bitrate Internet access, backed up with a service level guarantee.

3 Telecom Fundamentals

3.1 Communication Circuit Model

Since the Holy Grail of convergence and service integration has been achieved by treating voice and video like data, we begin understanding the nuts and bolts of the network with lessons on what used to be called *data communications,* historically a different topic than voice or video communications.

3.1.1 Information Theory

Data communications as we know it today started with the publication of a set of papers by Claude Shannon in the Bell System Technical Journal in July and October 1948, entitled "A Mathematical Theory of Communication".

This was the beginning of what is known today as *information theory.* The core premise of information theory is the desire to communicate *information* across a *physical medium.* To do this, the information is coded in binary digits or binits.

The physical medium, being real, has a maximum capacity, a maximum rate at which information can be transmitted across it.

The code rate, i.e., the number of binary digits required to represent the information, can never be less than the information content of the signal measured in bits.

That was the original definition of the word "bit", a measure of information content in a signal. This original meaning has been lost in the mists of time. "Bit" is now used to mean logic level, and Information Theorists have to make do with "binits". The information transfer rate in binits/second can never be faster than the circuit capacity. The capacity is directly related to the signal-to-noise ratio.

In this course, we explore the practical aspects of communications, and leave the study of information theory as a homework assignment for the interested reader.

3.1.2 ITU Model: DTEs and DCEs

To start our journey through the world of data communications and networking, we'll establish a model for circuits at their lowest level.

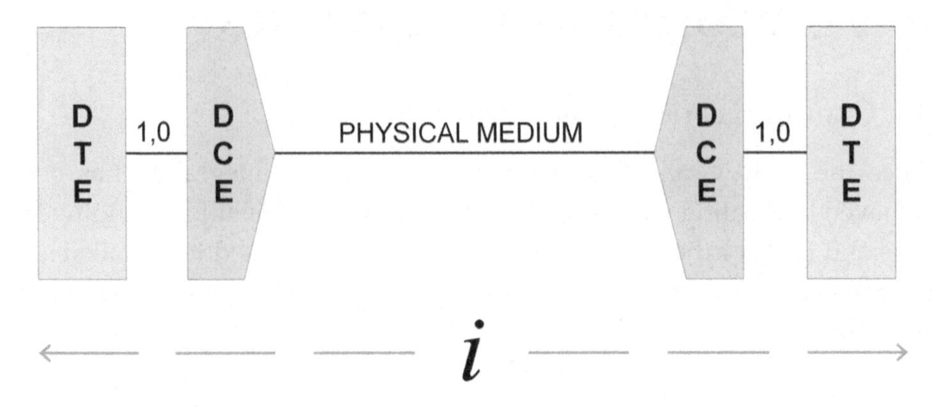

FIGURE 20 COMMUNICATION CIRCUIT MODEL

A model or *paradigm* widely used to discuss some of the more fundamental ideas of data communications comes from an organization that used to be known as the CCITT (le Comité Consultatif International de Téléphone et de Télégraphe). The name was changed to the Telecommunications Standards Sector of the International Telecommunications Union, and is abbreviated ITU-T, or simply ITU.

This is an international treaty organization, which has general meetings, committees, study groups and working groups and that historically operated on a four-year study cycle. Each member country sends delegates, who study and sometimes decide on international standards for telecommunications and data communications.

The ITU model has three components: *Data Terminal Equipment* (DTE), the devices between which one wishes to communicate information; the *physical medium* over which the information will be communicated; and the *Data Circuit-terminating Equipment* (DCE), the devices placed at each end of the physical medium.

To communicate the information, it must be coded into binary digits (1s and 0s), which are then represented on the physical medium. The data circuit could be a guided system such as wires or optical fiber, or a non-guided system such as radio.

The DTE is the source of the 1s and 0s. The DCE represents them on the physical medium. There are specific types of DCE for each type of physical medium.

3.2 Terminals, Clients, Servers and Peers

There are many kinds of Data Terminal Equipment (DTE). The word *terminal* comes from the Latin terminus, and is used to indicate that these are the devices at the ends of the circuit.

3.2.1 Dumb Terminal and Remote Host

Historically, there were two types of terminal equipment: dumb terminals and host computers.

The dumb terminal, also called a display terminal, ASCII terminal, Video Display Terminals (VDT) or even Cathode Ray Tube (CRT) display, performs only input/output functions: keyboard and screen. All of the computing functions are performed on the remote host computer.

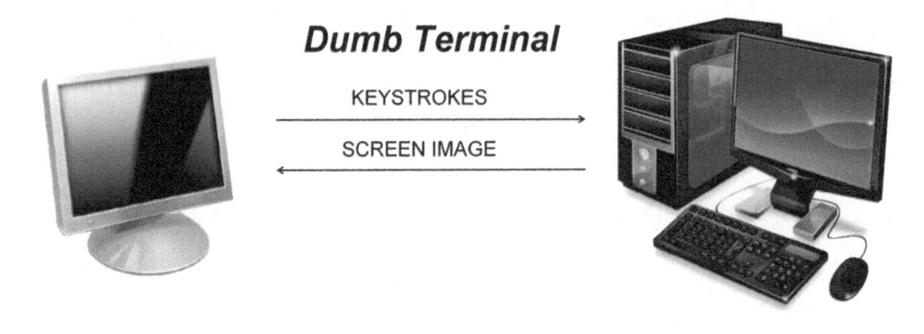

FIGURE 21 DUMB TERMINAL AND REMOTE HOST

A Personal Computer (PC) is often used as a dumb terminal by running *terminal emulation* software. Emulation means taking on the attributes of something else.

Popular software for dumb terminal emulation includes telnet, allowing remote login to a computer running UNIX such as a web server; SSH, a secure (encrypted) version of telnet, and Remote Desktop Connection from Microsoft for remote login to a desktop computer running Windows.

3.2.2 Client-Server

Of course, today, most communications are computer to computer.

The most important type is client-server computing: where a client software program running on a user's computer accesses centralized resources (servers), and much of the processing - like creation of a graphics screen - is performed in the client, not on the server.

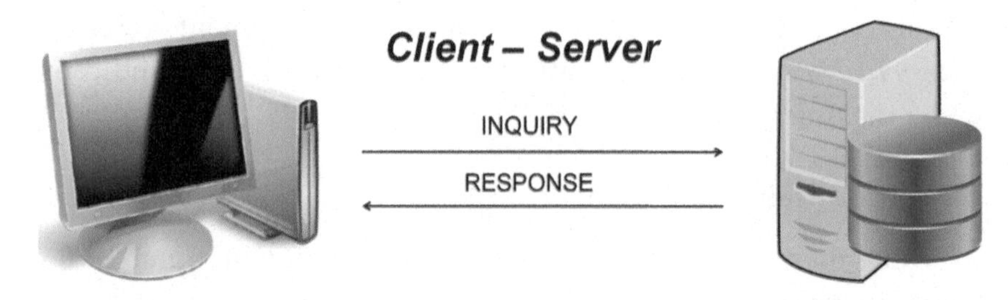

FIGURE 22 CLIENT-SERVER COMMUNICATIONS

Point of sale terminals usually have other functions incorporated, such as chip card readers and magnetic stripe readers. Examples of point of sale terminals are the "cash registers" in department stores, and the credit authorization terminals at gasoline stations.

Many other devices use data circuits, and can be considered as DTE. Examples include printers, bar-code readers, and FedEx data terminals.

3.2.3 Peer-to-Peer

Once a Voice over IP (VoIP) phone call is established, the two telephones exchange packets with digitized speech directly one to the other.

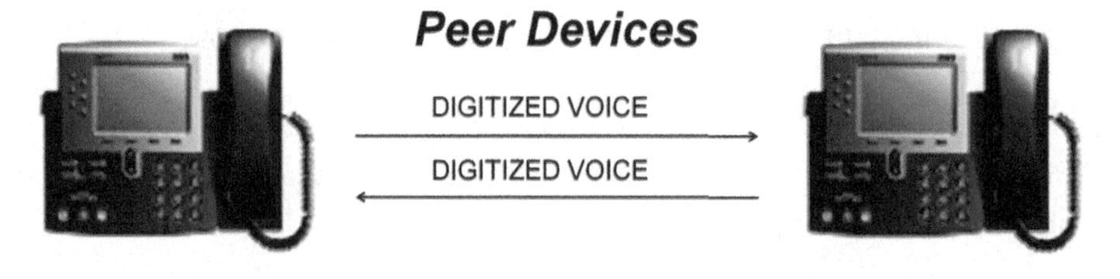

FIGURE 23 COMMUNICATIONS BETWEEN PEERS

In this case, the communication is not the asymmetric client-server relationship, but communication between equals, the VoIP software on the phones. This could be called *peer-to-peer* communications to differentiate it from client-server.

3.3 Representing Bits on Digital Circuits: Pulses

"Digital" transmission means applying energy to the communication circuit, or not, for a predetermined length of time to represent a 1 or a 0.

The burst of energy is called a *pulse*, and the strategy for representing 1s and 0s using pulses of a particular technology is called its *digital line code*.

3.3.1 Two-State Transmission Systems

Most wired digital transmission systems are two-state systems – the laser is either on or off; either there is a pulse happening on the line, or there is not – and so are actually *binary* systems.

There is a huge advantage in terms of noise performance using a binary signaling scheme compared to using an analog signal: as long as the pulse can be reliably detected, however diminished or corrupted it may be, communication is achieved.

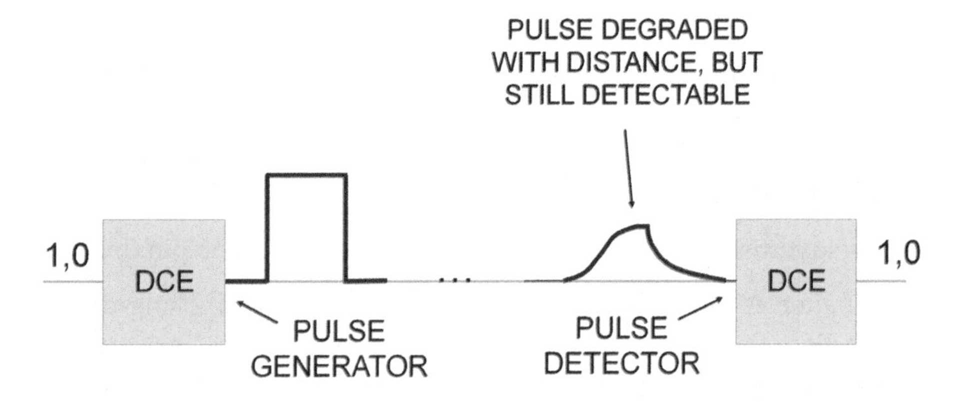

FIGURE 24 PULSES REPRESENTING BITS ON DIGITAL CIRCUITS

3.3.2 Range Limiting Factors

The factor limiting range on copper wires is *attenuation*: the signal diminishes in value with distance away from the transmitter due to resistance of the copper to the flow of electricity through it.

This limits the distance that pulses can be reliably detected. Pulses of electricity on copper wires will be attenuated with distance, and will also have noise added. The square corners will also be rounded off due to capacitance of the wires.

The limiting factor on fiber is *dispersion*: the pulse becomes less powerful and longer in duration.

3.3.3 Repeaters

For long-haul circuits, it is necessary to insert repeaters to regenerate the pulses before they become so distorted that they cannot be reliably detected. Repeaters are typically 40 to 80 km (50 miles) apart on fiber.

Repeaters do not boost the incoming signal: repeaters are binary devices that make a decision. If a repeater decides it detects an incoming pulse, however degraded, it regenerates a new clean copy of the signal: a new noiseless square pulse on its output to be carried on the next segment of cable.

Repeaters are required every mile or two on copper wire systems to be able to regenerate the pulses while they are still detectable without errors, for example, every 6,000 feet on T1 and every 12,000 feet on High-Speed Digital Subscriber Line (HDSL). [Note that HDSL is not related to residential ADSL or VDSL].

3.3.4 Comfort Noise Generation

Using this digital (actually binary) technique, information is coded into 1s and 0s, which are represented as pulses, which can be reliably transmitted across long distances, allowing the eventual reconstruction of the signal at the far end with no added noise.

Sprint's advertising tag line was "so quiet, you could hear a pin drop".

In fact, after AT&T introduced this technology to carry long-distance phone calls in the 1970s, it was so quiet on the long-distance lines that users complained it was too quiet: they couldn't tell if it was silent or broken; if they were still connected or not.

A very subtle hissing called "comfort noise" had to be added as a feedback signal to the users that they were connected.

It is the codec, the device that creates the 1s and 0s that are transmitted using the methods described in this chapter, that adds the noise... not the binary transmission system.

3.4 Representing Bits in Frequency Channels: Modems

3.4.1 Passband Channels

A circuit that is restricted so that only information represented in a particular range of frequencies can be successfully communicated is often called a *pass-band* or *passband* channel.

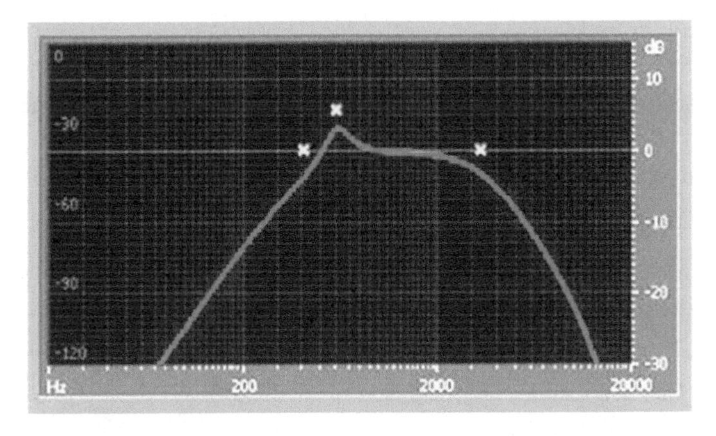

FIGURE 25 A PASSBAND FREQUENCY CHANNEL

A passband channel has a range of frequencies at which information can be passed to the other end, and everything else is suppressed. A passband does not start at zero cycles per second or 0 Hertz (Hz)

Radio channels, cable TV channels and the POTS voiceband (illustrated in Figure 133) are all examples of passband channels.

The challenge is to find a method of successfully representing binary digits on this kind of circuit.

3.4.2 Carrier Frequencies

One question often asked is, "Why not transmit the binary digits 'digitally'?" What is meant by this is, "Why not use pulses to represent bits?".

For example, a pulse could be +3 volts applied to the line for a short time to represent a "1", and nothing to represent a "0".

The answer is that pulses need a frequency band that includes zero Hertz, that is, zero changes per second, to implement the "+3 volts applied to the line for a short time" of the previous sentence.

Holding the voltage at 3 V is zero changes per second, and that cannot be represented on the circuit. The voltage must vibrate at least 54 million times per second to pass through cable TV channel 2, for example.

A design that *will* work on a passband channel is one that employs frequencies within the passband of the channel, called *carrier frequencies.*

3.4.3 Modulation

By varying some characteristic of the carrier frequency, such as the *amplitude* (level or volume), its *frequency,* its *phase* (relative position in time), or combinations of these, we can communicate information as signals that will be successfully passed to the other end, thus successfully representing binary digits on a bandwidth-restricted channel.

The technique of representing binary digits by varying one or more characteristics of one or more carrier frequencies is called *modulation,* and the circuit-terminating equipment that performs this function is called a *modem,* a contraction of *modulator* and *de-modulator.*

Modulation is used in all wireless, including 4G LTE, 5G, Wi-Fi; on coax by cable modems; and on twisted pair by DSL modems, fax machines and dial-up modems.

3.4.4 "Press 1 to Understand How Modems Work"

In the vein of the sort of humor that an Engineer would enjoy, the operation of modems can be explained by an Interactive Voice Response system answering a phone call by saying

"Press 1 to understand how a modem works".

"Press 1 to understand how modems work"

FIGURE 26 MODEMS REPRESENT 1S AND 0S USING CARRIER FREQUENCIES

While the user presses 1, the user and the IVR hear some tones.

The user stops pressing.

The IVR says "one", then hangs up.

Pressing the 1 button caused the telephone to generate a specific pattern of frequencies to represent the 1.

The IVR was listening at these frequencies, and decided that what it heard meant a 1. It confirmed this by playing a recording of someone saying "one" to the user.

3.4.5 Radio-Frequency Modems

Modems whose waveforms are used to modulate electrical voltage and/or radio signal power in passbands in the megahertz (MHz) and gigahertz (GHz) bands are called Radio-Frequency or *RF modems.*

3.4.6 CDMA and OFDM

In wireless systems, the challenge is to design a modulation scheme that uses low power, with the pattern so complex that it can be reliably detected by an antenna a mile away picking up a tiny fraction of the transmitted power, with hundreds and hundreds of people all communicating at the same time, in the same place.

Discussion of strategies for complex patterns like CDMA and OFDM continues in Chapter 9.

3.4.7 ASK, FSK, PSK, QAM and QPSK

Regardless of the complexity of a pattern, at the very lowest level, the modem has to represent 1s and 0s using carrier frequencies.

The touch-tones of Section 11.6.4 employed a form of a modulation technique called Frequency Shift Keying (FSK).

Other strategies include

- Amplitude Shift Keying (ASK)
- Phase Shift Keying (PSK)
- Quadrature Phase Shift Keying (QPSK), and
- Quadrature Amplitude Modulation (QAM).

These modulation techniques, along with ideas like baud rate vs. bit rate are covered in detail in Appendix A Modulation Techniques.

3.5 Serial and Parallel

Now, we begin to look at some circuit configurations.

The simplest configuration is when there are only two devices to be connected point-to-point. The term point-to-point is reasonably self-explanatory: the circuit goes from "A" to "B" and nowhere else.

The requirement is usually to communicate groups of bits, for example, groups of eight bits, point-to-point.

There are two different strategies for communicating bits on point-to-point circuits: serial and parallel.

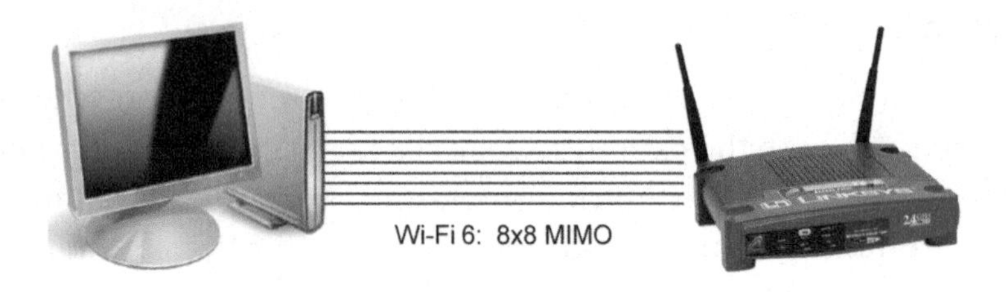

FIGURE 27 PARALLEL COMMUNICATIONS

One choice would be to connect eight circuits, and represent one of the bits on each circuit, then tell the other box to look at the circuits because the data is valid.

This is called *parallel* because the wires implementing multiple circuits would be literally in parallel in a cable.

In the illustration of Wi-Fi 6 implementing Multiple Input - Multiple Output (MIMO) in Figure 27, eight separate radio signals are transmitted in parallel from antennas separated in space.

3.5.1 Serial Ports: USB, LAN, SATA

The other choice would be to use one single circuit, and represent the bits one after another in a sequence in time on the single circuit. This is called *serial*, though a mathematician would prefer to call it sequential.

Inside a computer, data is grouped into bytes, which are grouped into messages or files.

To communicate these groups of bits over a serial line, a function traditionally called a *serial port* is required to transmit the bits by representing them one after another in a sequence in time on the circuit.

Another serial port function is required at the far end to look at the line at the appropriate times to determine the value of or receive the bits.

In days past, this was the serial port on a PC, driven by an inexpensive chip called an Intel 8251 Universal Asynchronous Receiver / Transmitter (UART), which supported bit rates up to about 100 kb/s.

Today, the Universal Serial Bus (USB) or the Ethernet LAN interface on a terminal is used to perform this function for data circuits. The Serial Advanced Technology Attachment (SATA) standard is used to connect drives to motherboards inside terminals.

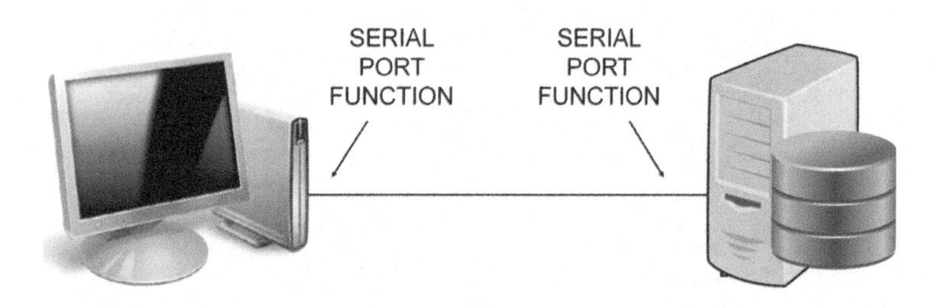

FIGURE 28 SEQUENTIAL OR SERIAL COMMUNICATION

Virtually all communication circuits are serial, as manufacturing costs are lower with only one line driver and one line receiver required.

3.5.2 Serial in Parallel for High Bit Rates

Then, to get more bits per second communicated, serial communications is deployed in parallel.

The bitrate of each is added together to yield a system with a much higher total number of bits per second.

For example, Gigabit Ethernet on copper LAN cables is four 250 Mb/s parallel streams on four pairs of wires.

40 and 100 Gb/s Optical Ethernet on fiber splits the bit stream into *paths* carried on multiple optical wavelengths in parallel.

4G LTE cellular can split the bit stream into 100 or more lower-rate streams communicated on individual frequency channels called *subcarriers*. 5G promises thousands.

3.6 Sharing: Frequency-Division Multiplexing

The term *plex* has the same root as the English word *plait*, as in plaiting or braiding hair. *Multi* is much or many in Latin... so multiplex means to weave many things together.

In a top-down worldview, multiplexing is dividing the capacity of a circuit into smaller portions to be used independently.

In a bottom-up worldview, multiplexing is combining lower-capacity individual circuits into a high-capacity *aggregate* circuit.

FIGURE 29 SPACE-DIVISION MULTIPLEXING LANES REPRESENTING FDM CHANNELS

Frequency-Division Multiplexing means dividing the available frequency band into smaller frequency bands called *channels*. The channels have a fixed bandwidth on a given system and are spaced a bit apart to avoid interference. Frequency-Division Multiplexing is represented in Figure 29 using spatial multiplexing: dividing the wide highway into separate lanes.

3.6.1 Baseband vs. Frequency-Shifted

Baseband signal is the term given to the source signal, i.e., the signal coming out of the television camera or broadband modem; the natural, non-frequency-shifted channel.

An FDM multiplexer shifts the baseband signal on an input in frequency up to the frequency of a particular channel.

For example, shifting a NTSC television signal that naturally occurs in the range 0 to 4 MHz, its baseband, to the range 58 to 62 MHz, is frequency channel 2 on a cable TV system.

The FDM multiplexer can represent as many baseband signals on the physical medium as there are frequency channels available. It shifts each to a different channel and adds them all together on the output.

At the other end, the signals are shifted back down to baseband on separate outputs.

3.6.2 Coax, Radio and Fiber

This technique is used for cable TV on coaxial cable, on radio systems and in fiber optic transmission systems.

In fiber optics, the frequency channels are referred to by center wavelength: (L in English) or lambda (λ in Greek). Using multiple carriers (light) at different wavelengths is called Wave Division Multiplexing (WDM).

Wireless systems – radios – also of course use frequency channels, usually one for each direction.

3.6.3 Parallel

High bit rates on systems that are channelized can be achieved by assigning a user multiple channels, splitting their bit stream into substreams and representing each on a different channel, then combining them back together at the far end.

This technique is used in Optical Ethernet and in LTE and 5G wireless, implementing parallel communications as in the previous section.

In 5G systems, the channel used by a system is split into thousands of smaller frequency channels, each with a subcarrier running a modem.

3.7 Sharing: Time-Division Multiplexing

Channelized Time Division Multiplexing (TDM), also known as Synchronous Time Division Multiplexing, was first developed in 1874 – 75 by Emile Baudot, an engineer at the French Telecommunications Service, for transmitting multiple telegrams simultaneously on a telegraph circuit.

3.7.1 Synchronous TDM Channels

Multiplexing means combining or sharing. Time division means that the sharing of the circuit is in time. Channelized means that each user is assigned a fixed number of bits per second, called a *channel*, out of the total available on the high-speed or *aggregate* circuit.

Synchronous means that data from each user is transmitted on a regular timed basis. On voice trunk carrier systems, one byte is transmitted for each channel 8,000 times per second, whether there is any information to communicate or not.

A familiar example of channelized Time Division Multiplexing is time-share condos. There is one condominium and 52 users. Each user enjoys full use of the condo for one week. At the end of their week, they have to pack up and leave and then the next user moves in and enjoys full use of the condo for the next week, and so on in a strict rotation. The users are time-sharing the condo, and each gets a fixed portion: one week per year.

3.7.2 Trunk Carrier Systems

In conjunction with digitizing voice at the constant rate of 64 kb/s, channelized TDM became the standard method for carrying multiple voice trunks over high-speed long-distance digital transmission circuits in the second half of the 1900s. These were called *trunk carrier systems*.

As illustrated in Figure 30, to implement channels, a multiplexer is attached to each end of a circuit. On one side are the users, each on a separate hardware port.

On the other is the high-speed aggregate port, where the users' channels are interspersed in a strict order in a high-speed bit stream to be carried on the high-speed or aggregate transmission circuit.

Each low-speed port is allocated a fixed time slot to transmit a single byte on the circuit, one after another in a strict rotational order in time.

For voice trunks, this happens 8,000 times per second. One byte 8,000 times per second is 64 kb/s, a bit rate called Digital Service Level 0 (DS0).

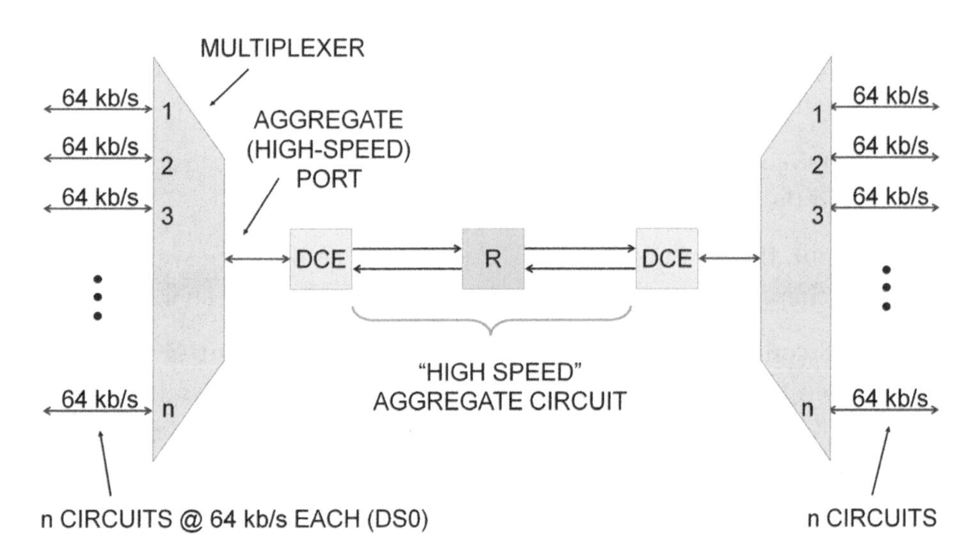

FIGURE 30 SYNCHRONOUS OR CHANNELIZED TIME-DIVISION MULTIPLEXING (TDM)

There is no switching in this system; it is a *carrier system*. What goes in on a particular channel's hardware port at 64 kb/s at one end comes out on the corresponding hardware port at the other end at 64 kb/s, and vice-versa.

3.7.3 T1, SONET and SDH

As an example, the Trunk Carrier System 1 (T1), a technology that was popular from about 1960 – 2000, designed to carry 24 trunks over 4 copper wires using channelized TDM is illustrated in Figure 30.

The Channel Service Unit (CSU) and the Repeaters move the high-speed aggregate long distance on the physical circuit using pulses of voltage.

SONET and SDH fiber systems, widely deployed before year 2000, also employ channelized TDM. These are covered in Appendix B Legacy Channelized Transmission Systems.

T1 was the state of art, achieving 1.5 Mb/s and costing $1000 per month in 1990. Detailed lessons on T1, which used to have their own chapter, are now relegated to Appendix C All About T1.

Though fiber is now routinely used, there remain many T1 circuits installed and in operation.

3.7.4 Other TDM Implementations: PONs, GSM, CAN-BUS

The idea of sharing a resource by taking turns in time is used in many technologies. With fiber to the home, Passive Optical Networks (PONs), typically 32 users are sharing the fiber, and in the upstream direction, the sharing method is fixed time slots, which is channelized TDM.

Earlier versions of cable modems conforming to the DOCSIS 2.0 standard also shared the uploading capacity using TDM.

At one point, the world's most popular cellular technology was 2G GSM, featuring channelized time division multiplexing on radio channels.

Industrial control systems – including those that control rollers in warehouses to move the stuff you ordered to a boxing station then a shipping station, as well as the CAN-BUS connecting parts inside your car, use simple TDM to communicate to multiple devices on a single circuit.

3.8 Efficient Sharing: Statistical Time Division Multiplexing

Statistical TDM is an improvement on the channels in terms of efficiency.

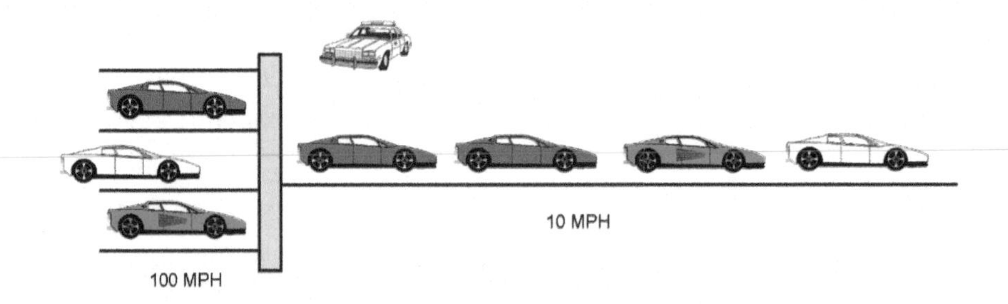

FIGURE 31 STATISTICAL TIME DIVISION MULTIPLEXING

With synchronous or channelized TDM, each user gets to use the high-speed circuit to send a byte, in a strict rotational order like a time-share condo. This has the effect of giving each user a fixed amount of capacity, called a channel.

This was designed for voice communications, since it is easiest to do quality voice communications if there is a constant amount of transmission capacity available for it.

For data communication applications, we don't need fixed amounts of capacity available for transmission, since the traffic isn't fairly constant, but rather bursty, like e-mail messages and web page downloads.

In this case, a more efficient scheme of multiplexing called *statistical TDM* can be employed. This is first-come, first-served: capacity is allocated to users when they demand it; if they do not use any capacity, someone else can use it.

An example is a toll plaza. There are a number of toll booths, with lines of traffic moving slowly through each onto a toll highway, where traffic moves at high speed. Each line sends a car onto the highway as needed. When the highway is busy, lines send a car onto the highway when the next chance arises.

If there are no cars in the line at a particular booth, that lane will not use any capacity on the highway, and other lanes can send cars onto the highway in those spots.

3.8.1 Bandwidth on Demand

Since we use the term bandwidth to mean capacity, this is called a *bandwidth on demand* strategy.

To make the service less expensive for users, we can take advantage of knowing that users are transmitting in bursts and not continuously, to oversubscribe or overbook the circuit, but tell them that they all have full-speed access.

Overbooking the transmission circuit is connecting up far more users than we would were they all sending data constantly.

This technique is also called *statistical multiplexing* because it is necessary to know the statistics of how often users will demand capacity, to know how much to overbook the transmission circuit.

Ideally, the goal is to end up with 100% occupied transmission circuits, even though the users are sending bursts of data. In practice, the goal is 80% occupancy.

The example of the toll plaza shows incoming lanes at 100 miles per hour (MPH), and the outgoing lane at 10 MPH.

Of course, this is exaggerated – but in the direction opposite to what you might think!

In a small office / home office, the "incoming lanes", the LAN, could be 1000BASE-T Gigabit Ethernet and the "outgoing" lane could be "broadband" at 25 Mb/s uploading… 1000 MPH and 25 MPH on the diagram would be closer to reality.

3.8.2 Packet Switching

The upside of overbooking is that the transmission circuit is used efficiently when the traffic is bursty. The downside is that since the users' data is now coming across the transmission system in a random order, control information is needed to determine what goes where at the far end.

The solution is to attach an *address* to the front of each segment of data being sent over the system. The far end uses this address to route the segment of data to the correct output.

A segment of data with an address on it is called a *packet*.

Bandwidth on demand is implemented with network equipment including Ethernet switches, IP routers and MPLS Label Switching Routers.

These are the traffic cops, deciding who gets capacity on the overbooked circuit. IP routers and MPLS LSRs are covered in detail in upcoming chapters.

3.9 Overbooking: Reducing User Cost

How to efficiently use a high-speed circuit that moves bits constantly when the traffic to be communicated arrives in bursts? The short answer: overbook the high-speed circuit.

With channelized multiplexing: each connection is allocated a constant 64 kb/s. This is called a *non-blocking* system.

If a particular device is idle, its channel, its assigned fraction of the high-speed circuit, is nonetheless reserved and cannot be used by any other devices. This makes implementation simple, but is not an efficient use of the high-speed circuit.

The extent of the inefficiency becomes particularly obvious when the application generating the traffic is email or web surfing. The vast majority of the time, when communicating email or web pages, nothing is being transmitted. Occasionally, small file transfers happen.

The solution: knowing that users are normally not going to be transmitting - normally they will be doing nothing - connect up many more devices to the high-speed circuit than with channels, tell the users they've all got a high-speed connection... and pray they don't all try to use it at the same time!

Airlines do this. Knowing that some passengers will not show up to claim their seat, airlines overbook flights. The objective is to end up with 100% full planes: no-one left behind, no empty seats.

Knowing how much to overbook requires knowledge of the historical demand statistics – how often people actually do show up for the flight – hence the name statistical multiplexing.

The same concept is employed to allow the efficient use of communication circuits, which move bits continuously, to carry bursty traffic from multiple users. The total of the line speeds on the access is far greater than the high-speed circuit.

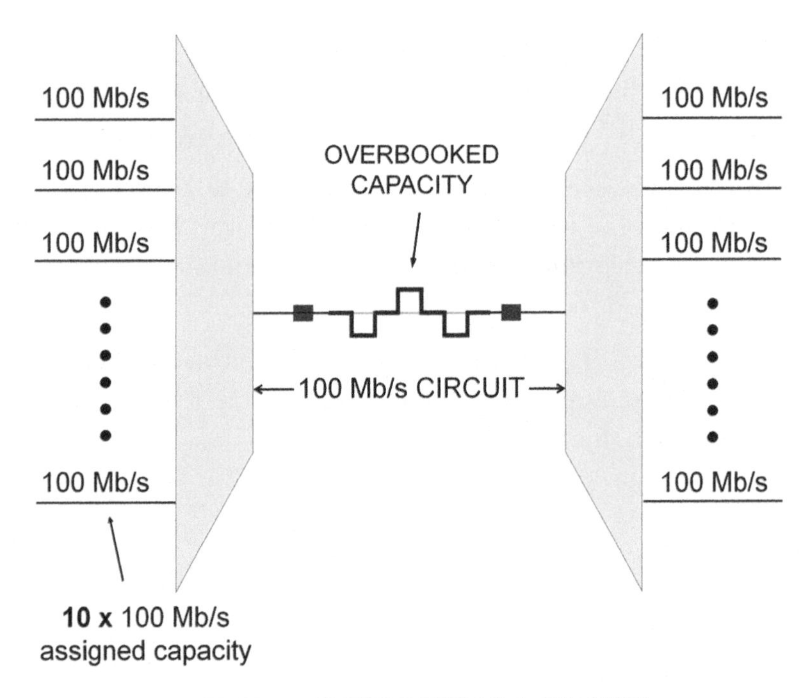

FIGURE 32 OVERBOOKING A CIRCUIT

This is also called bandwidth-on-demand, since devices now have *opportunistic* access to the full bandwidth or capacity of the high-speed circuit. They have the possibility of transmitting, and can use it on demand.

This introduces a complication, however. Channelized TDM systems' overhead is ultra-efficient, requiring only minimal control information, because the channels transmit bytes in a strict rotational order.

However, when the circuit is overbooked, the channels no longer transmit in a strict order. They transmit data only when needed, and only when capacity is available.

Because of this, it is necessary to add extra control information called an address to each segment of data so that the far end knows which output to give it to.

This uses up some of the capacity of the high-speed circuit, but not a significant fraction.

Figure 32 illustrates the concept of overbooking using the simplest example, where the overbooking is done by the multiplexers at each end of the circuit.

If we then extend the concept to a whole network of high-speed circuits, what do we have? IP/MPLS networks including the Internet.

This is the essential idea behind a packet network: take a segment of data, put an address on the front that the far end uses to determine where to route it, then send this data whenever there is a free spot on the overbooked circuits that make up the network.

Overbooking allows the network to give users more for their money: it gives users higher apparent bandwidth compared to a non-blocking channelized system, for the same price.

4 Network Fundamentals

A *network* is routers in different locations connected with *data links*. Data links take care of moving packets from one router to another.

Usually redundancy is implemented, where there are multiple possible routes taking some or all different data links across the network between any two routers.

Users are connected to the network on access circuits, which are data links that can have two stations on the circuit, like a blue LAN cable; or many stations on the access circuit as in the case of Wi-Fi.

In this chapter, we take a first pass through the nuts and bolts of how the data links are implemented, how the routers relay packets carried in frames from link to link to eventually get to the desired destination, how reliability is implemented and how flows of packets are managed.

Many of the topics in this chapter, e.g., Ethernet, IP and MPLS, are covered in detail in subsequent chapters.

4.1 Essential Functions for Communication

We begin by identifying *what* has to be done to communicate:

- Coding the message into bits,
- Framing,
- Error control,
- Link addressing, and
- Network addressing.

4.1.1 Bits and Bytes

The smallest unit in communications is a logic level, a "1" or a "0". In a computer, all information is coded into 1s and 0s, and these are called bits.

As noted at the beginning of Chapter 3, 1s and 0s in a computer is not the original meaning of "bit". The original use of this word comes from information theory, the body of theoretical knowledge that is the base for much of what we are discussing. The term bit was first used in information theory as the unit of measurement for how much information there is.

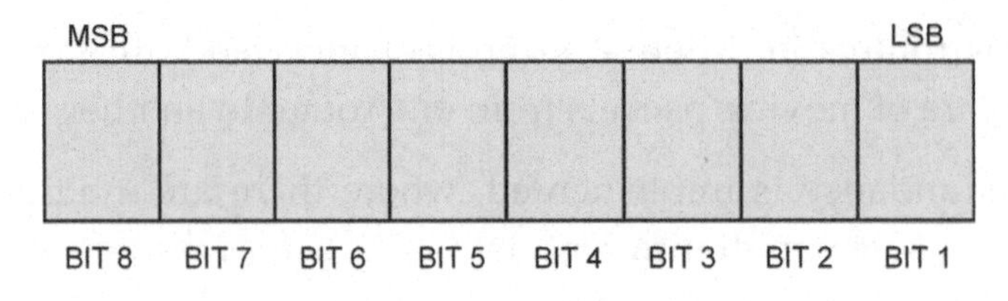

FIGURE 33 BITS AND BYTES

Just as fluid ounces or milliliters are used as the unit of measurement for how much water is in a glass, bits were the unit of measurement of how much information is in a message. This information is coded into binary digits (1s and 0s) using a coding technique. The binary digits are then communicated over a data circuit made of a physical medium.

Because of this, when discussing data communications, we sometimes use the term binary digit instead of bit to discuss logic levels

However, unless you are an information theorist, you can safely ignore it. These days, the term "bit" is used to mean everything that is a 1 or a 0.

Bits are organized into groups of eight, called bytes. Historically, bytes have had various sizes: 6, 7, 8, 12 bits. Today, eight bits per byte is more or less standard. Half a byte is called a nibble. (Really).

To refer to individual bits within a byte, they are given numbers: bit 1, bit 2, ..., bit 8. During serial transmission, there is ambiguity as to which bit is bit 1 and which is bit 8… which end is which.

In the computer, bits might be numbered 1 through 8, but the transmission system transmits bit 8 first, leading the data link to call that bit 1.

To avoid this problem, the terms Least Significant Bit (LSB) and Most Significant Bit (MSB) are used. The LSB is the one which has the least numeric value, the bit that changes the most often when counting; the MSB changes least often when counting.

4.1.2 Coding

Coding means representing data in bits, i.e., 1s and 0s. Different techniques for coding are employed, depending on the nature of the data.

When the data to be conveyed is a quantity, it is coded using a numbering system called binary. For ease of use, a related system called hexadecimal is often employed.

When the data to be conveyed is a character in the English language (and selected others), usually indicated by pressing a button on a keyboard, US ASCII and variations are used. To be able to represent other characters, including Asian scripts, unicode is becoming dominant.

When the data is a continuous signal, such as an analog of speech, music or video, then the coding is performed via a sampling, quantization and coding process in a codec as covered in Chapter 7.

4.1.3 Error Control

All communication is subject to errors during transmission. Normally, methods for error control are implemented to deal with this. Error control consists of *error detection* and *error correction*.

Error detection methods include parity checking and the more reliable Cyclic Redundancy Check, both of which involve adding *redundancy* (extra bits) to the transmitted data so the receiver can determine if an error happened. Error correction is implemented by retransmitting errored data.

Forward Error Correction (FEC) means that a great deal of redundancy is added to the transmitted data so that the receiver (the forward end) can determine if an error occurred, and correct the error without a retransmission.

4.1.4 Framing

Whether pressing buttons on a keyboard or downloading a web page, data happens in bursts; there are times when there is data to be transmitted and times when there is not data to be transmitted.

To indicate to the receiver the start and end of a group of data, markers or delimiters are placed before and after the data. This is called *framing*.

4.1.5 Link Addressing

A single circuit may have more than one terminal connected. In a wireless LAN, there are many devices, and all devices receive all transmissions.

An *address* is required to indicate for whom the data is intended on the link, which station should react to the data, since all of them will hear it.

This is called the *link address,* and in the IEEE LAN standards, called the Media Access Control or *MAC address*. Every LAN interface, whether using copper, fiber or wireless, has a unique 48-bit-long MAC address.

4.1.6 Network Addressing

A *network* is composed of many point-to-point circuits, generically called *links* in data communications, between specific locations. At each location, links are physically terminated on hardware ports on *routers*. In the center or *core* of a network, there is a router at each end of each link.

Routers perform a relay function between links. They receive data on one link and transmit it back out on a different link. Knowing *which* link to transmit it out on is the "routing" part of the story.

An *edge router* also terminates links to user devices, and assigns a unique IP address to each device. IP addresses are much like telephone numbers.

An *IP packet* is a piece of a message being transmitted with the IP address of the destination device pasted on the front.

Routing tables in routers relate blocks of IP addresses, called *subnets*, to the various links connected to the router; identifying what ranges of addresses are reachable via which link.

To determine to which link a packet should be relayed, the router looks up the destination's IP network address in its routing table. This returns the IP address of the router to go to next, its MAC address and hardware port.

To transmit the packet, the router creates a MAC frame with the packet as payload, populates the MAC address and sends it to that hardware port.

Using pulses of light or voltage, the hardware port signals the frame to the hardware port of the router at the other end of the link, one bit at a time. The process repeats until the packet reaches the destination IP address.

> That's the short version. The full discussion follows in subsequent sections, and in some cases, subsequent entire chapters.

4.2 Shared Multidrop Links: Wi-Fi, PONs, CATV, CAN-BUS

A *multidrop* circuit is a data link consisting of two circuits with multiple derivations connected in parallel as illustrated in Figure 34.

4.2.1 Primary Station and Secondary Stations

The link can operate in *unbalanced mode*, where there is a *primary station*, or controller, which controls the link, and *secondary stations* or controllees at the end of each cable.

There is usually one circuit for each direction; on the downstream circuit, the primary station transmits and all of the secondaries listen. On the upstream circuit, the secondary stations can transmit and the primary station is listening.

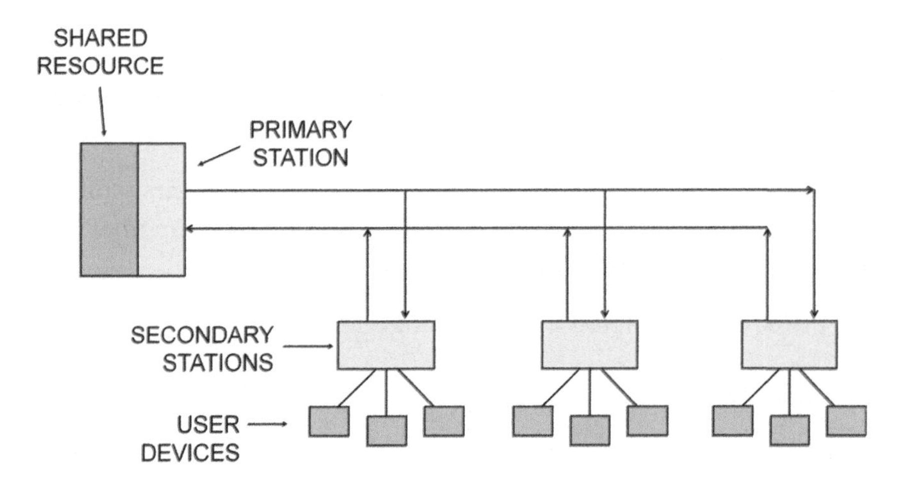

FIGURE 34 UNBALANCED ACCESS LINK

This kind of data link is used to give multiple users with their secondary stations shared access to a primary station with a resource, like an upstream Internet connection, or a server, under the control of the primary station.

The secondary stations are computers, and in turn have Human-Machine Interface (HMI) devices, controllers or sensors attached. Examples of HMI devices are video displays, telephones, and smartphones. An example of a sensor is a water level sensor. An example of a controller is a water valve actuator, or an electrical motor controller.

4.2.2 Wi-Fi

Wi-Fi has an unbalanced architecture, where many secondary stations are connected to the Access Point via space.

The Access Point is the primary station and the "shared resource" of Figure 34 is the upstream Internet connection.

The secondary stations are smartphones, tablets, laptops, and increasingly, sensors and controllers. Wi-Fi is covered in detail in Section 9.14.

4.2.3 PON

This architecture is also evident in a Passive Optical Network (PON) where there one transceiver in an Optical Line Terminal on the network side, the primary station, connected via a passive optical splitter / combiner to 32 last-mile fibers with transceivers on the user side, the secondary stations.

The capacity of the connection upstream from the network terminal is shared by the 32 subscribers via fixed time slots, in other words, the channelized Time-Division Multiplexing of Section 3.7.

The capacity downstream to the subscribers is shared first-come, first-served, which packet will be transmitted on the downstream fiber next. All 32 subscribers will receive it, so encryption is required.

PONs are covered in detail in Section 10.8.

4.2.4 Cable TV

Another common implementation of this architecture is Cable TV distribution in neighborhoods.

On DOCSIS 2 Cable TV systems, the upstream is channelized between up to 500 subscribers. DOCSIS 3 systems use spread spectrum to allow simultaneous transmission. DOCSIS 3.1 uses OFDM and a controller to assign subchannels to users on a dynamic basis.

Cable systems are covered in detail in Section 11.11.

4.2.5 Industrial Controls: CAN-BUS

Industrial controls including the CAN-BUS in automobiles and warehouse conveyors have one primary station connected to dozens to hundreds of secondary stations.

In some of the warehouse and factory "field bus" implementations for sensors and controllers, there are only two wires; the multidrop transmit and receive capacity is implemented via time slots in a synchronous multiplexing system.

Polling is implemented on CAN-BUS. The primary polls the secondaries and gives them permission to transmit by selecting them.

4.2.6 Legacy IBM Mainframes

This multidrop architecture and unbalanced mode of operation was used in large IBM mainframe computer installations. The Front End Processor or Communications Controller is the controller, and Remote Terminal Controllers are the controlees.

4.3 Point-to-Point Links: Ethernet

4.3.1 Ethernet LANs and Balanced Mode

Local Area Networks began as the technology used to connect computers in-building, to exchange data, share printers, servers and Internet access.

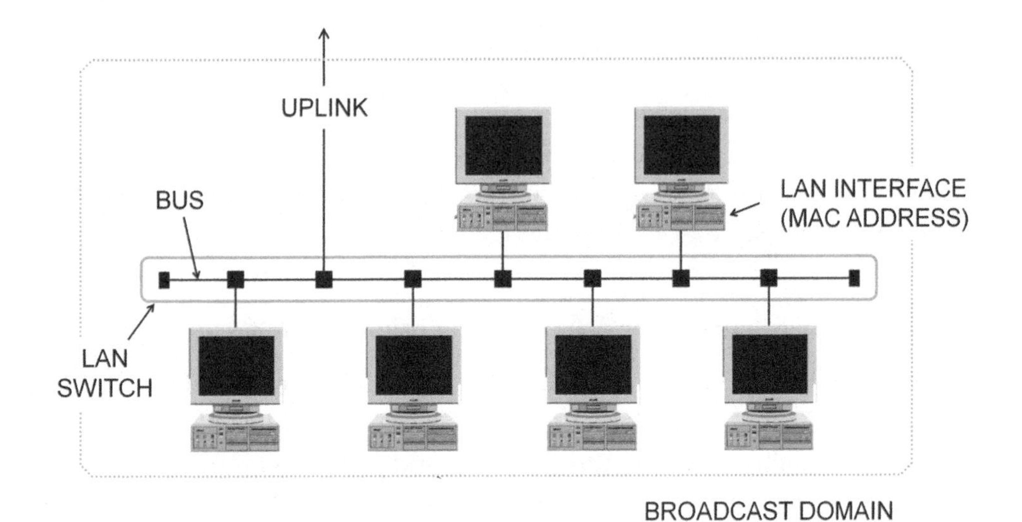

FIGURE 35 ETHERNET

In the very beginning, Ethernet, the first technology for LANs, was a two-wire system where many devices were connected, and communicated in *balanced mode*.

Balanced mode means all stations hear what any station transmits, and stations only transmit when they have something to communicate - and they have to fight it out amongst themselves which station is allowed to transmit next, since only one can transmit at a time.

4.3.2 Transition to Point-to-Point and Switches

With the introduction of the Ethernet Switch, also called a LAN switch or Layer 2 switch, Ethernet changed from being a bus cable with multiple stations connected and taking turns, to point-to-point four-pair cables with only two devices, communicating in both directions at the same time.

As can be seen in Figure 35, all devices have their own LAN cable plugged into a hardware port on a LAN switch, which stores and forwards devices' traffic to other devices plugged into the switch, most often the one that has a high-capacity upstream connection to share, and vice-versa.

Optical Ethernet is two fibers point-to-point between switches, one for each direction to also implement *full duplex* (two-way simultaneous) communications.

Optical and copper Ethernet is now deployed exclusively in point-to-point configurations, with a LAN switch at at least one end.

4.3.3 802 Standards

The original type of LAN technology was branded *Ethernet* by its inventors at Xerox Palo Alto Research Center (PARC). "Ethernet" is now used to refer to technology that follows the IEEE 802 series of standards, (almost but not quite exactly the same as the original Ethernet), which took over.

People say that Xerox never does anything original, but it's not true.

Ethernet subsequently migrated from in-building connections to the telecommunications network core and access circuits. Point-to-point Optical Ethernet, conforming to the 802.2 and 802.3 standards, is now the technology implementing links in the network core, and increasingly used on access circuits.

4.3.4 Buses, NICs, Interfaces and MAC Addresses

As illustrated in Figure 35, LANs were originally multidrop circuits, where all information was broadcast to all stations.

This was accomplished with a *bus cable* running in the ceiling, and each computer connected to the bus via a tail circuit from the bus to the *network adapter* on the PC.

The network adapter or *Network Interface Card (NIC)* was plugged into the motherboard of a computer to implement the physical connector.

Today, rather than plugging in a Card, the LAN function is integrated in devices like a laptop, both wired and wireless – and so the term *LAN interface* or *network interface* is used instead of NIC.

Every interface has a hard-coded serial number called its Media Access Control (MAC) Address. The interface responds to transmissions specifying its MAC address as the destination.

4.3.5 Ethernet LAN Switches

LANs changed topography to have the bus be eight inches long inside a small box called a *switch,* and (blue) LAN patch cables from the PC to a port on the switch. This is illustrated in Figure 195 in the LANs chapter.

A LAN switch, also called an Ethernet switch or Layer 2 switch, is a wiring hub with a processor in it. The processor examines the destination address and directs the transmission to the correct station.

LAN switches come in all different sizes... four ports, eight ports, sixteen ports, 192 ports. Eight is a popular size. If one took eight PCs with LAN interfaces and wired each to a switch, this would form a LAN.

To connect that LAN to the Internet, or any other network, one of the ports is used as an "uplink" connecting to a computer that has additional upstream links. The computers that connect LANs are called routers.

4.3.6 Broadcast Domains and MAC Addresses

Any device in the group of Figure 35 has the possibility of communicating information to any or all other devices in the group. For this reason, the group of PCs and the uplink shown is said to form a *broadcast domain.*

It is necessary to transmit the destination device's MAC address along with data to indicate for whom the transmission is intended, as all stations in the broadcast domain might receive the data.

Dealing with risk related to broadcasting to all stations is an important element of network security.

4.4 Data Link Frames & MAC Addresses

4.4.1 MAC Frames

To communicate bursts of data reliably on a circuit where there may be multiple stations attached, it is necessary to add framing, addressing and error control information to the data to be transmitted.

Adding framing, addressing and error control one byte at time is inefficient. It is more efficient to segment the data to be transmitted into blocks or segments of up to 1500 bytes, and add the framing, addressing and error control to the entire block.

The resulting output data unit is called a *frame*. If the 802 standards are followed, it has a MAC Address and is called a *MAC frame*, illustrated in Figure 36.

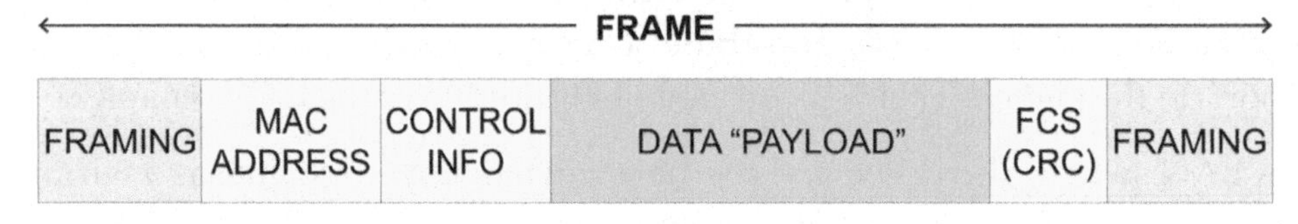

FIGURE 36 MAC FRAME

4.4.2 Transmission Between Devices on the Same Circuit

Successful transmission of a frame results in communication of the payload from one computer to another that are *on the same physical circuit*, or more precisely, in the same broadcast domain. This function corresponds to Layer 2 of the OSI Model.

The method for formatting frames, sending and acknowledging them, or not, is called a *data link protocol.*

The data link protocol in universal use is called Ethernet, and is specified in IEEE standards 802.2 and 802.3.

The *framing* is a special bit pattern one or more bytes long identifying the beginning and end of the frame.

An *address* is usually included at the beginning of the frame. The address on the frame indicates which device on the current LAN cable the frame is intended for, and which device it is coming from. These addresses are described called Media Access Control (MAC) addresses.

A *control field* follows the address. In the 802.2 LAN standard, this contains the length of the payload.

The *data field* or *information field* or *payload* follows the control field. The data field can be any length. In practice, it is about 1500 bytes long.

An error detection scheme called Cyclic Redundancy Checking (CRC) is implemented using a Frame Check Sequence (FCS) appended to the MAC address, control field and data, calculated so that the ensemble satisfies a complicated mathematical formula.

At the receiver, if it still satisfies the formula, there is a very, very high probability that no errors occurred.

If it fails this error check, the frame is discarded by the receiver and may have to be retransmitted (email), or not (VoIP).

4.4.3 Legacy Systems and Terminology

Early designs for transmission of blocks of data by IBM for mainframes called the idea of frames *synchronous communications.*

"Synchronous" should not be used in that context because it also means many other things.

The mother of all data link protocols is the High Level Data Link Control protocol (HDLC).

In HDLC, the control field can identify the type of frame: Information frame, Supervisory frame or Unnumbered frame.

An information frame is used for data transfer. Sequence numbers, acknowledgments and a poll/final indicator are placed in the control field for the information frame.

The other two types of frames are used for data link control, and have codes indicating the action to take.

4.5 Packet Networks

The definition of a Wide Area Network (WAN) is connecting LANs in different locations with data circuits.

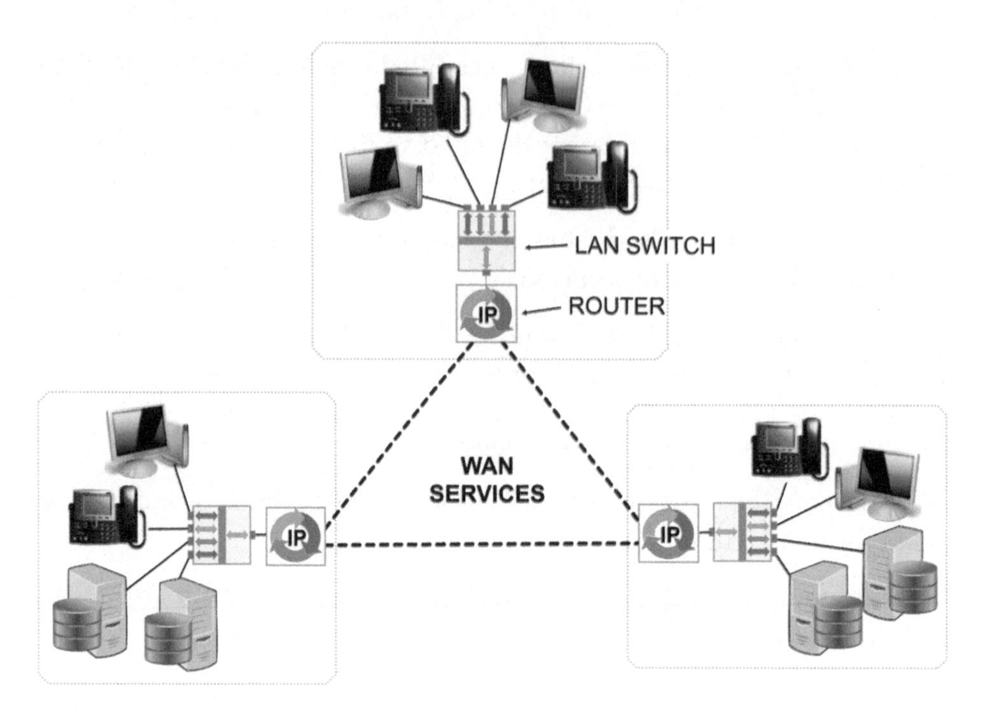

FIGURE 37 SIMPLEST EXAMPLE OF A NETWORK

For reasons of availability, the sites are often connected with redundant paths, so that there is more than one way to get from "A" to "C".

4.5.1 Routers and Network Addresses

For this reason, equipment that is capable of making a route decision is needed at each site: which route to take to get to the desired destination.

Router would be a good name for a box that can make route decisions.

The router needs information to be able to make a route decision. The most popular strategy is to assign *network addresses* to all of the computers and use those as the basis of making route decisions.

The standard for network addresses is IP, the Internet Protocol, regardless of whether used on the Internet or on a private system.

4.5.2 Packets

To send a block of data to another computer, the IP address of the source and destination is placed before the data, forming an IP packet, which is sent to the IP router.

The IP router uses the destination IP address on the packet as the basis of making a route decision.

4.5.3 Network Connections

There are many choices for the telecommunications services used to connect the sites to physically implement the WAN of Figure 37.

Choices include dedicated lines like point-to-point fiber, and the almost retired circuit-switched services like ISDN and dial-up modems.

The most popular, flexible and cost-effective choice for services to connect locations are IP packet bandwidth on demand services, which includes Internet services and private business network services.

4.5.4 Traffic Management

In the core of the network, MPLS is used to implement an overlaid control system that allows management of packets, allowing control of both routing and prioritization across the network. MPLS is usually invisible to users.

4.6 Carrier IP Networks

The word *network* comes from fishing nets, where many strings are tied together to form a mesh, and one could trace many different possible routes between any two knots or *nodes* on the net-work.

In the communications business, the word network is used in the same way. Telecommunication networks are implemented by connecting buildings with high-capacity point-to-point circuits, with multiple routes possible between any two buildings.

As illustrated in Figure 38, a common carrier will build the network, then install access circuits to their network for customers. Customers can transmit IP packets to the carrier for delivery, and receive IP packets over their access circuit. Packets are relayed from building to building to reach the destination.

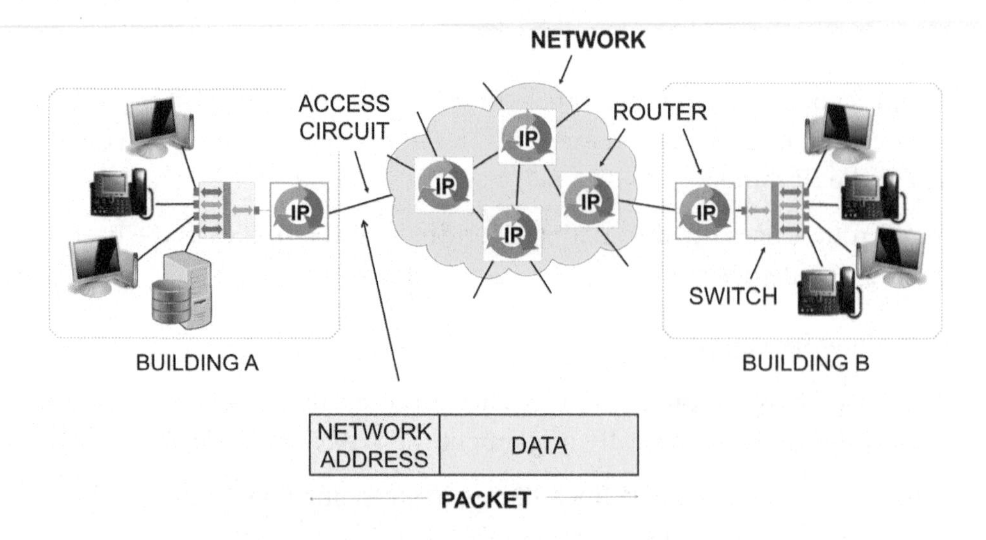

FIGURE 38 COMMUNICATING OVER AN IP NETWORK

4.6.1 Routers and Routing

The academic definition of a network is having to make a route decision: which route to take to get to the destination. The network equipment that deals with packets is the *router*.

Routers perform a *relay* or *forwarding* function: moving packets, one at a time, internally from an input port to an output port.

Knowing *which* output to move the packet to is the "routing".

4.6.2 IP Packets

Data to be transferred over a network is formatted into *packets*, which are sometimes also called datagrams or Network *Protocol Data Units* (PDUs).

Packets are blocks of data with network control information at the beginning. The most interesting type of network control information is the *destination network address*, which indicates the final destination of the packet.

The standard for packet format and network address is the Internet Protocol (IP), both on the Internet and on private networks not connected to the Internet.

4.6.3 Network Routers and Customer Edge Router

In the network core, each router will be connected to multiple other routers on separate point-to-point Optical Ethernet fibers, enabling multiple possible routes across the network.

At the user location there is typically one router called the *premise router* or *edge router* or *Customer Edge (CE).* All of the devices at the user location are connected to the premise router.

Physical access circuits, wired or wireless, are connected from the customer router to the network router.

4.6.4 End-to-End Packet Relay and Routing

Users transmit packets end-to-end across the network by transmitting a packet from their computer to their edge router, which relays it over their access circuit to the network router, whence it is relayed or forwarded from one network router to another until the far end network router is reached, then delivered over a far-end access circuit to the customer edge router then the packet is delivered to the computer at the far end.

IP routers make the route decision by looking up the destination IP address in its routing table, which lists ranges of IP addresses and which router to go to next to get towards the final destination.

The router implements the decision by forwarding the packet to the next hop, in other words, transmitting the packet to the chosen router on the point-to-point Optical Ethernet connection to it.

Then it's a whole new day. IP routers do not have any memory of past events. Packets are routed one at a time.

4.7 IP Packets vs. MAC Frames

It is important to understand how packets and frames are related.

4.7.1 Purpose of Frames

A frame is a low-level idea. Frames are used to communicate between stations on the **same circuit.** The circuit may have multiple stations physically connected onto it, like a wireless LAN, or may have only two stations like a point-to-point Optical Ethernet cable.

A frame has framing to mark the beginning and end, sender and receiver addresses to indicate the stations on the circuit, control information, a payload and an error detection mechanism.

The frame is transmitted on the circuit, and all stations on the circuit receive it. If an error is detected at a receiving station, the frame is discarded and might have to be retransmitted somehow.

If no errors are detected, the end result is that the payload is communicated to the correct station on the same circuit with no errors.

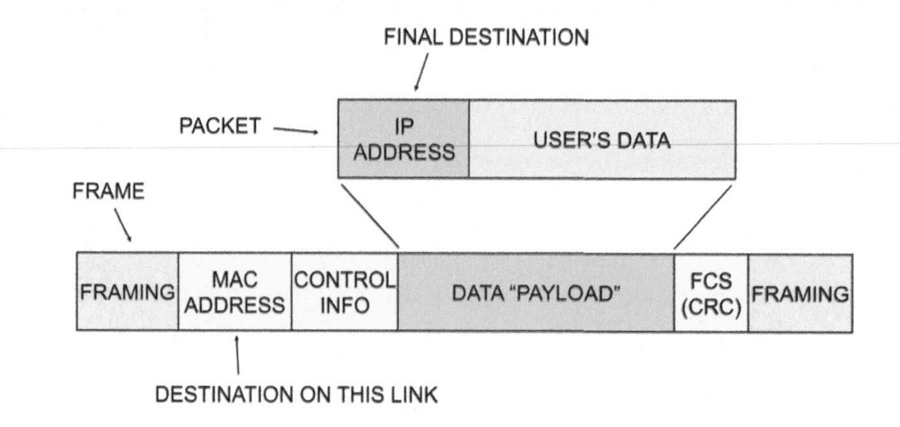

FIGURE 39 IP PACKETS IN MAC FRAMES

4.7.2 Purpose of Packets

Packets are for networks. A packet is a block of user data, such as a piece of an e-mail message, with a network address on the front.

The network address is the final destination.

Network equipment like routers receive a packet on an incoming circuit, examine the network address, use it to make a route decision, then implements the decision by forwarding the packet on a different circuit to the next router.

The main purpose of packets is to append a network address to your data. The network address is used by network equipment to make route decisions: to relay the packet from one circuit to a **different circuit**.

4.7.3 Packets Carried in Frames

To actually transmit a packet on a circuit within the network, the IP packet is inserted as the payload in a MAC frame.

As illustrated in Figure 39, the destination IP address is that of the final destination computer; the destination MAC address is that of the LAN interface at the other end of the circuit.

The bits in the frame are signaled on the circuit by pulses of voltage, pulses of light or radio modems.

4.7.4 MAC Address vs. IP Address

There are two addresses: the network address (IP address) and the link address (MAC address). The network address on the packet is the final destination, and so does not change. The link address on the frame indicates the destination on the current circuit, and so is changed as the packet is forwarded from one circuit to another.

4.8 IP Packet Format

The Internet Protocol (IP) is part of the TCP/IP suite of protocols developed by the US military, now used on the Internet and carrier networks. Eventually, most all communications will take place in IP packets.

IP is a network protocol, defining the network packet format and network addressing scheme.

Packets are called *datagrams* in strict IP terms. In this course, we will use the term packet to avoid getting bogged down in jargon.

4.8.1 Packet Header

Included in the IP packet specification is the IP header, which is the "network control information", encapsulated or placed in front of the block of data to be sent over the network. Figure 40 illustrates the elements in the IP version 4 (IPv4) header.

The Identification, Don't Fragment (DF), More Fragments (MF) and Fragment Offset fields are used when the packet does not fit inside a frame and has to be broken into parts or segments or fragments and carried in separate frames.

When that occurs, a copy of the IP header is repeated at the beginning of each fragment – dramatically increasing overhead – and typically if one of the fragments is lost, all of the fragments are discarded.

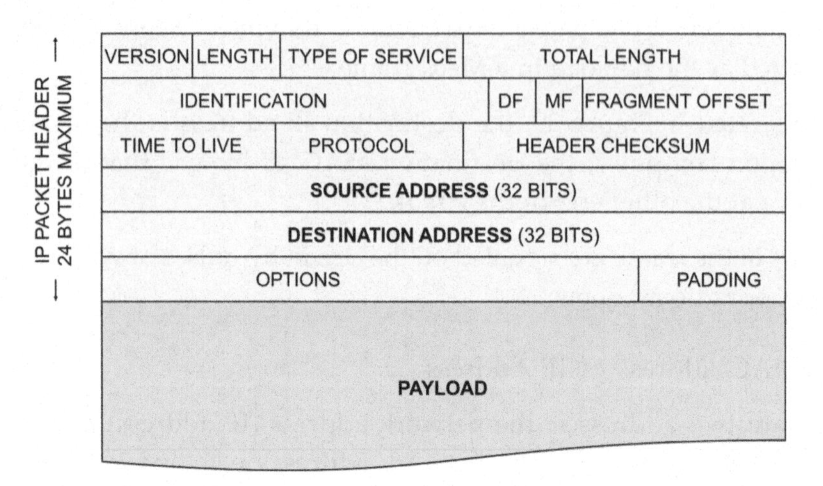

FIGURE 40 IPV4 PACKET FORMAT

The most interesting part of the IP header is the sender IP address and receiver IP address. This is the main purpose of IP.

IP was originally published as military standard MIL-STD-1777, then updated in Request for Comments RFC 0791 Internet Protocol.

IP version 4 is currently in use. IP version 6 (IPv6) is coming.

The myriad aspects of IP are explored in detail in Chapter 16, IP Networks, Routers and Addresses.

4.9 TCP, UDP, Ports and Sockets

4.9.1 Unreliable, Connectionless IP Network

IP implements a *connectionless* network service, which means that at the network level, there is no communication between receiver and sender.

Communicating over the Internet means your ISP giving your IP packet to another ISP, and hoping that ISP actually transmits it onward to another ISP, and so on, to eventually get your packet to the receiver's ISP.

There are no guarantees that a packet will be transmitted, when that might happen, nor how often that might happen. Packets may be corrupted, overwritten or discarded at intermediate nodes, and never delivered.

This is called an *unreliable network service.*

4.9.2 Reliable Communications over an Unreliable Network

The *Transmission Control Protocol* (TCP) is used to guarantee the integrity of file transfers over the unreliable IP network, using error checks and retransmission.

Before data is passed to IP, TCP adds a header that includes an error check and sequence number to the block of data and starts a timer at the sender.

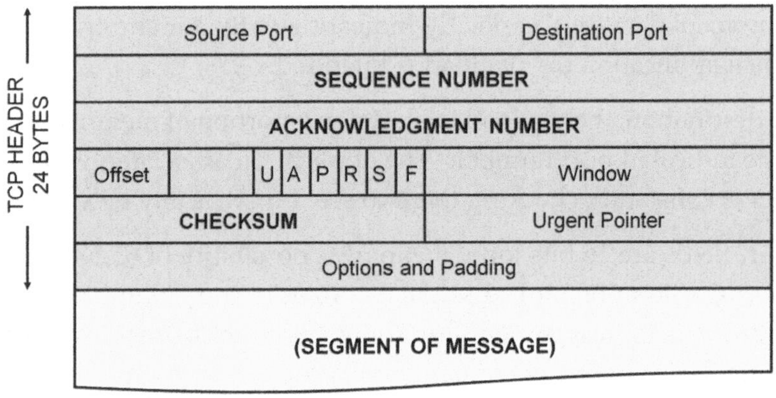

TCP HEADER
24 BYTES

Source Port		Destination Port
SEQUENCE NUMBER		
ACKNOWLEDGMENT NUMBER		
Offset	U A P R S F	Window
CHECKSUM		Urgent Pointer
Options and Padding		
(SEGMENT OF MESSAGE)		

FIGURE 41 TCP PROTOCOL DATA UNIT

The receiver's TCP checks the error check and sequence number. If the data is corrupted, the receiver discards the data. The sender *times out* and retransmits. If the data received OK, the receiver TCP sends an *acknowledgment* to the sender TCP and the sender stops retransmitting.

TCP turns the underlying unreliable connectionless IP network into a reliable transport service for use by upper layers. TCP is for file transfers – when it absolutely, definitely has to get there, and if it doesn't, the missing piece(s) will be retransmitted until they do.

This is not very useful for voice and video. During a live, streaming communication session, there is no time to retransmit missing pieces. For these applications, a different transport-layer protocol called UDP: the User Datagram Protocol is employed. UDP is similar to TCP, but only implements port numbers and error checking. UDP does not implement sequence numbers or retransmission of missing or errored data. It provides best-efforts transport.

4.9.3 Port Number Identifies Application at the IP Address

It is normal to have multiple applications running at the same time, using the same network connection; for example, email and browser.

When a packet arrives at a device, how does the device know which application the data is for?

A "port" is how communications get to a particular application on a device.

Each application is given a number, called its port number. In the TCP or UDP header, the first two fields are the source port field and destination port field. The sender populates these fields when creating the transport layer protocol data unit, explicitly indicating to the far end transport layer for which application the payload is intended.

At the destination, the payload is placed in a portion of memory associated with the indicated port number. The application associated with that port number is constantly checking there to see if there is any new data.

The port fields are 16 bits long, meaning a possibility of 65,536 ports. The Internet Corporation for Assigned Names and Numbers (ICANN) maintains official assignments of some port numbers for specific uses. Many of the port numbers are not reserved, and are assigned and meaningful only when a communication session is established.

Port numbers are grouped into ranges:

The Well Known Ports are 0 – 1023… for well-known applications.

Registered Ports are 1024 – 49151… for less well-known applications.

Dynamic or Private Ports are 49152 – 65535. Port numbers in this range are usually assigned on a per-session basis and sometimes called *ephemeral ports.*

To communicate to an application on a machine, it is necessary to know the IP address of the machine and the port number of the application. When these are concatenated, the result is called a *socket,* uniquely identifying a particular application on a particular machine.

4.10 MPLS Labels

Multiprotocol Label Switching (MPLS) is a traffic management system used on the core of big networks. It is generally invisible to end-users.

TCP is used to make sure data is successfully transferred end-to-end. MPLS is used to manage transmission characteristics like how long it takes to deliver one packet, manage relative priority of different streams of traffic on the same fiber, and allow a hierarchical traffic monitoring system, amongst other uses.

4.10.1 Managing Flows of Packets

MPLS is the basis of traffic management systems managing *flows* of packets in the network core. This is accomplished by identifying the flows with numbers, then on entry to the core, tagging or *labeling* the packet with the number of the flow the packet belongs to.

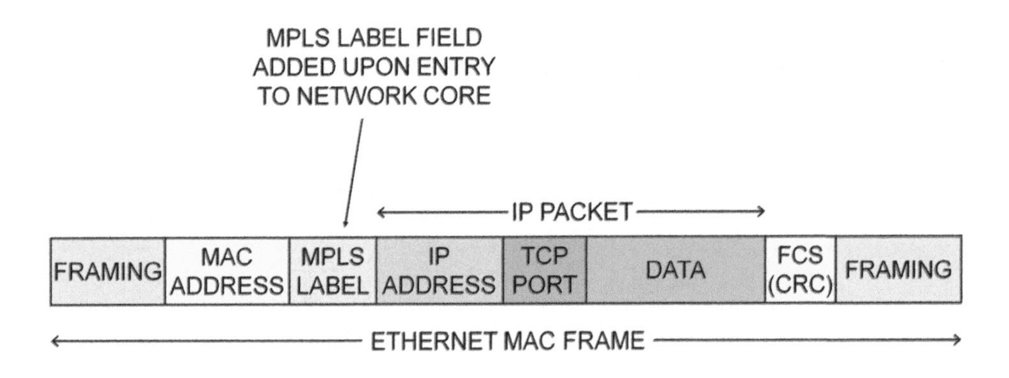

FIGURE 42 MPLS ADDS A FLOW LABEL

This allows all of the packets in the same flow to be managed as a single item in an operator's Network Operations Center (NOC).

4.10.2 Traffic Classes

To set up an MPLS network, *traffic classes,* called *Forwarding Equivalence Classes* (FEC), are defined. A traffic class is the "flow": all of the packets that are all going from the same place to the same place, and that are to be treated the same way in terms of service level: bandwidth, loss and delay.

Each traffic class is assigned a number.

Next, the actual hop-by-hop route for each traffic class is determined. The routing is implemented by populating the routing tables in MPLS routers with two important entries: the traffic class number and the next hop for packets in that traffic class.

On ingress to the core, an IP packet is analyzed to determine the traffic class that the packet belongs to. Once decided, the packet is labelled with the class number.

When a labelled packet arrives at a router, the router looks up the label number in its routing table to find out what the next hop is; where to send the packet. The main difference between this strategy and IP routing is that the contents of the MPLS routing table is populated by a control system at the Network Operations Center. IP routers populate their own tables.

The label field can also be used to indicate relative priority to implement prioritization of different kinds of traffic.

This implements centralized control of traffic on the network core, with status and configuration for each kind of traffic displayed as a single item on a screen.

On egress or exit from the core MPLS network, the label is removed by the MPLS Edge Router, and a normal IP packet is delivered to the end user.

Chapter 17 explains MPLS in detail.

5 The Internet and Cloud Computing

5.1 A Network to Survive Nuclear War

The Internet has its roots in anarchy. It's like some sort of fungus, spreading across the planet. There are theories that it will become self-aware one day. The humans will try to unplug it, and the network will retaliate by nuking the humans and hunting the survivors.

FIGURE 43 A NETWORK TO SURVIVE A NUCLEAR WAR

One of our favorite urban myths is that The Internet started off as a research project funded by the US Department of Defense (DOD) Defense Advanced Research Projects Agency (DARPA) to develop network protocols that would be capable of surviving a nuclear war. Whether that's true or not, it's a useful way of understanding IP.

This research project began before the ISO Reference Model was established and before X.25 packet networks. The philosophy for this work was markedly different than commercial packet services. In contrast to the telephone companies' X.25, which provided reliable packet communication service over links that were mostly stable, the DARPA net design assumed that the links might be totally unreliable, that communications might take place over a number of intervening networks, and that loads upon remaining links might become very high.

5.1.1 Connectionless Network Service

To meet the requirement to survive a nuclear war, and allow the exchange of files and electronic mail, the core idea of the DARPA net was that it provides a *connectionless*, or datagram service, allowing the transfer of packets of data one at a time, store and forward from one computer to another to another across multiple links and intervening sub-networks.

Protocols for exchanging router table update messages ridiculously often (once a minute) were developed. With parallel and diverse links, data could be re-routed around trouble spots. Parts of the network could be terminated, and the rest would continue functioning.

In the beginning, links between existing networks at selected universities and research institutes were established. To access the Internet, it was necessary to physically go to one of those locations and have an account on their network.

5.1.2 Al Gore Invents the Internet

The initial inter-net links were paid for by the government, and the universities and research institutes. One of the government institutes most often mentioned is the National Science Foundation (NSF), which paid for some of the most expensive links, for which Al Gore was apparently instrumental in getting funding approved. The DARPA net was renamed the ARPA net, and then the Internet.

Today, the Internet is no longer just a national security project, but has gone global, and will soon replace the telephone and television networks. What started off as a small group of technically advantaged researchers and computer buffs exchanging 7-bit ASCII text messages has turned into a network reaching literally into your pocket and accessed by billions around the planet.

5.1.3 Who Pays for the Internet?

The government no longer pays for the links; telephone companies and broadband carriers provide the network connectivity and bill for access in a giant pyramid scheme.

5.1.4 Primitive Beginnings

The DARPA net had a Human-Machine Interface (HMI) developed by technically advantaged personnel, and was more or less useless to the general public. This has been fixed with the Web, browsers and apps, giving a Graphical User Interface (GUI).

The DARPA net was based on UNIX computer to UNIX computer communications developed by UNIX computer programmers. At the time, UNIX only handled seven-bit bytes... and to this day, email programs transform files of eight-bit bytes like images, video and spreadsheets into files of 7-bit bytes for backwards compatibility with the UNIX mail system, adding 33% overhead.

The Internet is a system very closely linked to computers, data and computer operating systems, where communications were short text messages. It was not designed for isochronous (continuous bit rate) traffic like voice or video, and there are no mechanisms for implementing or assuring Quality of Service (QoS) in IP.

Those functions are implemented with an MPLS management system.

5.2 The Inter-Net Protocol

The Internet started out as federally-funded data links connecting universities and research institutes like UC Berkeley, UCLA and MIT, illustrated in Figure 44. However, packet networks using protocols like X.25 were already implemented at these locations, so the Internet was originally really the Inter-net, connecting existing packet data networks with data circuits.

5.2.1 Gateways

The devices that connected the local networks to the data circuits were called External Gateways.

Gateway generally means *protocol converter;* presumably the interface devices were called gateways because they had to convert between different packet formats and network addresses used at different universities and research institutes. Today, these devices are called routers.

Once it is possible to exchange packets between networks, the next problem is: to send a message to a user on another network, how to determine their network address? Each institution was already using its own unique addressing scheme, with no way of communicating it outside the local network.

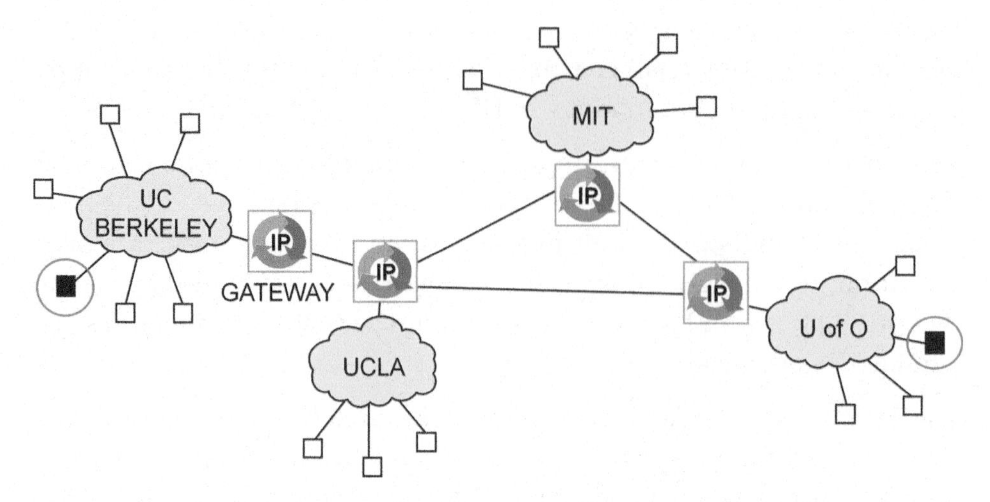

FIGURE 44 NEED FOR AN INTER-NET PROTOCOL

The Internet Protocol (IP) was the solution to this problem: a common network addressing and packet format that would be used by all parties.

5.2.2 IP: Common Packet Format and Address Scheme

To send a message to another user, the message is segmented into chunks and the IP address of the destination is added to form packets. These IP packets are then transmitted to an IP router, originally carried inside whatever packets were being used at each institution.

The IP router looks up the destination IP address in its routing table to determine what neighbor router to send it to next, the *next hop*, then transmits the packet to that router, moving it to the next network in a chain.

5.2.3 Connectionless, Unreliable Network Service

Each IP packet is treated by the router as being completely independent from any other; they might follow different routes across various different networks depending on changing congestion conditions.

There are no guarantees with IP. Data can be corrupted, copied, or thrown away. Bombs might fall. Routers and links might turn into fused glass.

5.2.4 TCP and UDP

Strong end-to-end error control implemented by the users is required to check to see if each packet arrived, and if not, retransmit it. This end-to-end error control is implemented by the users running the Transmission Control Protocol (TCP), which uses error checks, sequence numbers, positive acknowledgments and source timers to decide when to retransmit.

In the case of streaming voice and video, there is no time for retransmitting errored or lost packets, so the User Datagram Protocol (UDP), which does not do retransmissions, is used instead of TCP for those content types.

5.2.5 Routing Protocols

A protocol called the Routing Information Protocol (RIP) was first used to populate routing tables by sharing known routes between routers, then replaced with Open Shortest Path First (OSPF) to exchange routing table update messages between routers within an organization.

The Border Gateway Protocol (BGP) is used to define routing between organizations' networks.

MPLS is used in the core of big networks instead of IP routing.

5.3 Internet Service Providers (ISPs)

5.3.1 Internet Access Providers

Originally, the only way to get on to the Internet was a terminal connected to a computer at a university or research institute. The Internet was mostly circuits paid for by the taxpayers via the National Science Foundation.

Today, commercial Internet access providers, called Internet Service Providers (ISPs) provide the capability for anyone to access and communicate over on the Internet.

These ISPs are for the most part business units of facilities-based carriers, i.e., telephone companies and cable companies.

Internet service providers connect users' physical access circuits to a router with access control and security functions. If permitted, user packets are relayed to the ISP's core network, consisting of high-capacity connections to other routers and to other ISPs.

The routers are housed in data centers in cities or regions, which are interconnected with fiber or leased services. This ensemble of interconnected routers controlled by an ISP is called an Autonomous System (AS).

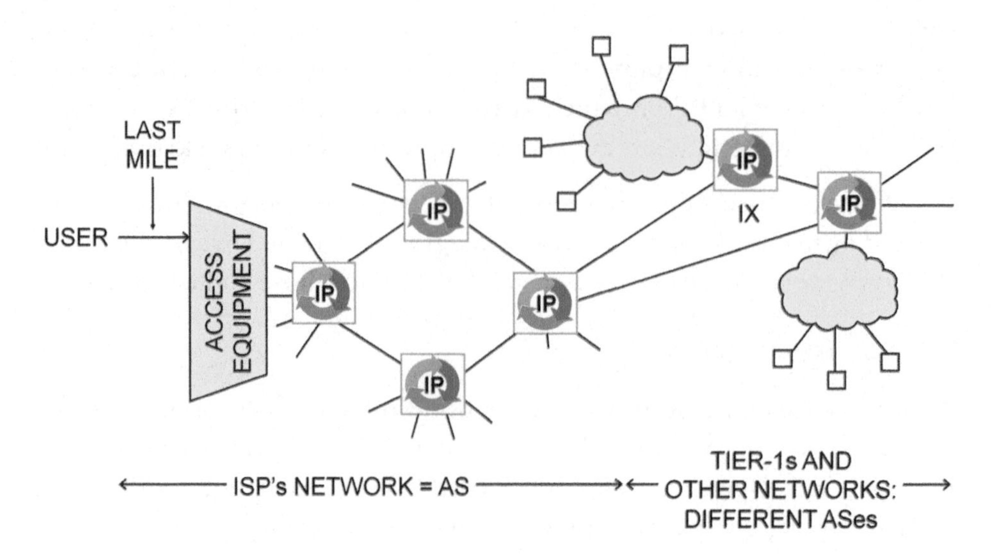

FIGURE 45 INTERNET SERVICE PROVIDERS (ISPS)

5.3.2 The Internet is a Business

The Internet is a vast, unregulated collection of interconnected Autonomous Systems. The connections between ASs are not specified by a central authority or world government, but are implemented on a case-by-case basis by the operators of an AS for business reasons.

The Internet is not free. It is not a public utility. It is a business.

As of 2020, there were 70,000 ASs connected to form "the Internet".

5.3.3 Interconnect, Peering and Transit

ISPs operating ASs will connect to competitors and content providers like Google to exchange traffic terminating on each other's network (called peering), and will connect to larger organizations who will assure delivery of packets to other destinations (transit).

The networks are physically connected at Internet Exchange (IX) centers such as Equinix Chicago at 350 E Cermak Street. These are buildings with equipment implementing network interconnection operated by a neutral third party.

The ASs are responsible for paying for connectivity to the IX building.

Peering is settlement-free, i.e., no money is exchanged. Transit is a commercial service that costs money. Larger ISPs charge smaller ISPs for transit services.

The largest networks are sometimes called Tier-1 service providers. "Tier-1" is not an officially defined term. Some definitions are a network "close to the center of the Internet", or a network that does not pay for transit.

However, there is no "center" to the Internet, and virtually all networks employ a mix of peering and transit agreements to connect to other networks... and the nature of such connections is non-disclosed confidential business information.

A "Tier-1 provider" might best be thought of as a very big facilities-based carrier that has a presence in most or all IXs and sells transit services to smaller networks and ISPs.

The ISPs build the access network implement peering or transit connections to other networks, then charge the users for access: a pyramid scheme. The end users end up paying for all.

In addition to access services, the ISP usually provides a Domain Name Server and an e-mail server, and may have many auxiliary products like web hosting and network security offerings.

5.3.4 Resellers

Back in the Flintstones era when dial-up Internet access was first available, telephone companies were a bit slow to react, so for a while, companies like Netcom, MindSpring, Portal, Pipeline, iStar and others had their day.

These organizations were *resellers,* leasing high-capacity router to router and router to IX circuits from carriers, and reselling the access to dial-up users under per-minute or per-month billing plans.

The carriers eventually began competing with resellers, who for the most part went out of business, selling their customers to carriers. For example, Netcom is now part of Earthlink, which is majority owned by Sprint.

For the most part, it is business units of the companies that own the cables coming into your home: the LEC and the cable TV company, along with cellular companies that are the dominant ISPs today.

If you do choose to use a reseller-type ISP, particularly for a business or organization, questions regarding customer service, capacity and availability should be asked.

Another is redundancy - do they have a single point of failure? Do they have multiple connections to different Tier-1 providers? What capacity are those connections?

5.4 Domain Name System

Just as it is necessary to know someone's phone number to call them, it is necessary to know the actual numeric IP address of the destination to be able to send a packet to it.

IP addresses are binary digits. And the IP address of a device can change. How does a person find out what a device's IP address is? How can they remember it?

5.4.1 DNS Servers

For servers, the solution is a human-readable handle for the IP address called a *domain name*. The Domain Name System (DNS) is the structure supporting human-friendly aliases or domain names for servers, like teracomtraining.com.

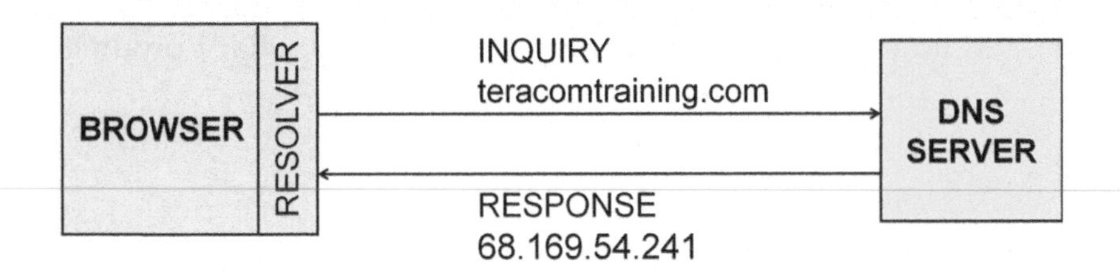

FIGURE 46 ADDRESS RESOLUTION OF A DOMAIN NAME

DNS servers maintain information about each domain, including its IP address in a *zone file* for each domain. DNS servers are analogous to old-fashioned telephone books.

Instead of the human having to enter the numeric IP address of the desired server like 68.169.54.241 in a web browser address bar, the human enters a human-friendly domain name representing the server's address like www.teracomtraining.com.

Resolution of this human-friendly form of the address to the binary IP address is performed by the machines.

When the human enters teracomtraining.com in the browser and hits enter, the IP address corresponding to that domain name is determined by a small program in the browser called a *resolver*.

First, it will check the local cache to see if the IP address for that domain has previously been requested, and if it is past its "best before" date.

As illustrated in Figure 47, if the local cache doesn't have the answer, the resolver will query a Domain Name Server for the domain's A record, which contains the binary IP address corresponding to the domain name.

This process is called *address resolution.*

5.4.2 Domain Zone Files

The Domain Name Server has a database that lists all of the domains it knows about and records with the IP address of the corresponding server, along with many other types of records.

SOA	teracomtraining.com	
A	@	68.169.54.241
A	www	68.169.54.241
MX	@	mx01-spamexperts.com
MX	@	mx02-spamexperts.com
TXT	newsletter._domainkey	p=MIGfMA0GCSqGSIb3D...
SRV	sips	@ 68.169.54.241:5061
SRV	sip	@ 68.169.54.241:5060

FIGURE 47 DOMAIN ZONE FILE

The database is organized into *zone files,* one for each domain, like the example in Figure 47.

The Start of Authority (SOA) record indicates the domain this zone file is for; in the example, teracomtraining.com

"A" records list the IP address corresponding to the domain name in dotted-decimal format. A records also list the IP address of subdomains. In the example of Figure 47, the domain (represented with the @ symbol) has an A record, and the subdomain www also has an A record.

This means that both www.teracomtraining.com and teracomtraining.com will work in a browser address bar, and resolve to the same IP address.

The requested record is returned as the response to the DNS resolution, whence the browser can determine and cache the numeric IP address and start sending packets to that IP address.

DNS servers are open for anyone to use. The IP address of Google's DNS server is 8.8.8.8. A device learns the IP address of an available DNS server as part of the DHCP protocol (Section 16.4).

5.4.3 SIP Records in DNS

DNS will become even more important going forward as the public telephone network transitions to VoIP and SIP.

"Telephone numbers" will eventually be replaced with SIP Uniform Resource Indicators (URIs), which often take the same form as an email address. customerservice@teracomtraining.com will be the "telephone number" for Teracom customer service. It is already the email address; it will become the voice and videoconference address as well in the future.

This requires DNS to look up the actual numeric IP address of the teracomtraining.com SIP server in the DNS zone file for teracomtraining.com to set up the phone call.

As illustrated in Figure 47, SIP servers are listed using the Server resource (SRV) record type. There are two SIP SRV records in the zone file, one for sips.teracomtraining.com and a second for sip.teracomtraining.com.

Each specifies the IP address of the server machine, and a *port number*. The port number indicates the particular computer program to communicate with, depending if encrypted communications (SIPS) or unencrypted communications (SIP) is desired.

5.5 Web Clients

The introduction of client-server computing over the Internet, coupled with the Domain Name System (DNS), resulted in the "World Wide Web".

5.5.1 Browsers

Browsers are the most popular kind of client used on the web. A browser is a software program or *app* that runs on a computer. It provides an extensive set of software functions for acquiring and displaying media.

For the web page developer, the browser provides a platform to render HTML5 code, which encompasses a video player, audio player as well as the usual text and images.

5.5.2 Apps

Browsers are apps specialized in displaying web pages. There are many, many other kinds of apps.

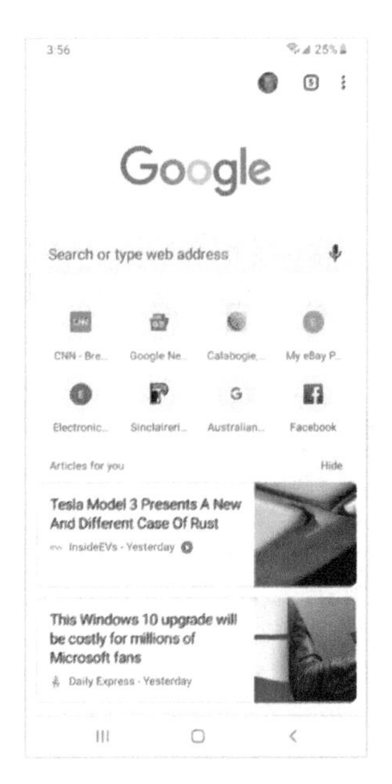

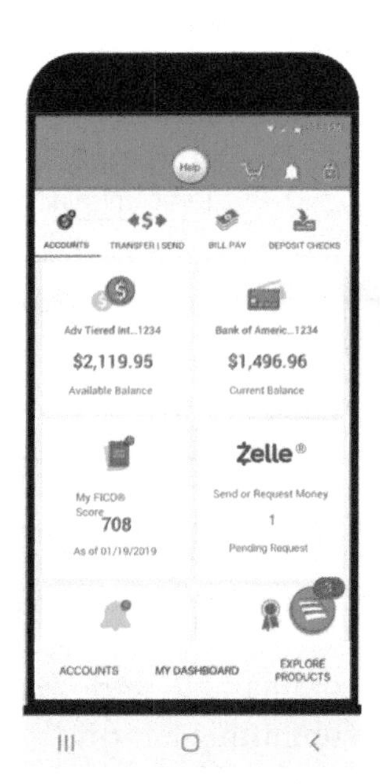

FIGURE 48 BROWSER APP AND BANKING APP

Apps generally run on the operating system, like Android, iOS, or Windows. This allows specific capabilities to be programmed into the client, like a video player app that understands a proprietary coding format, and a Voice over IP client app like Skype that connects only to Skype apps on other phones.

5.5.3 IoT Apps

Apps do not necessarily have human-machine interfaces that are often (if ever) viewed; an IoT app could be collecting readings from a dog collar's GPS chip and transmitting them over the cellular network to a server, which allows the dog owner to view the GPS location as a red dot on a map in a browser.

The dog collar app is not running on a cellphone, but on a tiny computer on the dog collar with an operating system like Unix or android, a cellular or Low-Power Wide-Area radio, with no display.

5.6 Web Servers

5.6.1 HTTP

The protocol for asking a server for a file is Hypertext Transfer Protocol (HTTP). HTTP is a very simple protocol with only a few message types.

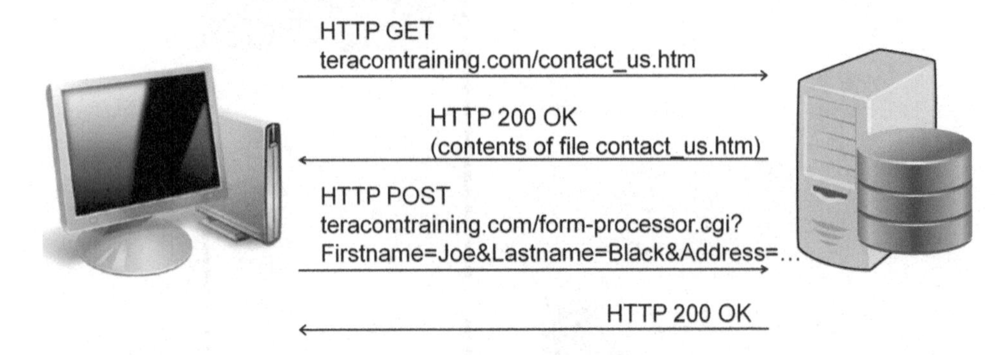

FIGURE 49 HTTP: GETTING A WEB PAGE AND SUBMITTING A FORM

"GET" is the command to retrieve a file. "POST" is used to send data to the server when submitting a form.

5.6.2 HTTPS

Typing https:// at the beginning of an URL in the address box of the browser – or clicking on an HREF with this attribute – tells the browser to use encryption.

The browser and server exchange an encryption key, then use it to encrypt the content of the packets being exchanged.

The Secure Socket Layer (SSL) protocol, now replaced with Transport Layer Security (TLS), causes the contents of the packets to be encrypted by the client for transmission to the server and vice-versa

This would prevent anyone who gets a copy of the message you are transmitting from understanding it. The messages exchanged using HTTP usually contain your username and password when submitting a form to log in to a server.

This information must be encrypted before transmission. The rule in the security business is: if it has not been encrypted, it has been released to the public.

5.6.3 HTML

The language used in web pages is HyperText Markup Language (HTML).

```
<html>
<head>
<title>telecommunications training seminars | telecom certification</title>
</head>
<body>
<a href="http://www.teracomtraining.com">
    <img src="/images/header-468.gif"
    alt="Teracom telecommunications training courses and seminars"></a>   ...
  <a href="https://www.teracomtraining.com/register_for_public.htm"
   class="insetboxlink">Register for a seminar</a><br>
```

FIGURE 50 HTML FOR WEB PAGES

HTML is a plain text language with embedded formatting commands. It has to be interpreted or compiled to create the user screen.

5.6.4 HREFs and URLs

HTML files contain hypertext references (HREFs), which specify the name and location of other files, in a standard format called a Uniform Resource Locator (URL).

The URL includes the protocol to be used to transfer the file, the network address or domain name of the server it is on, the file path, name and type.

5.7 Web Services and Cloud Computing

Making a web server available to the public requires a computer running an operating system and web server application, RAM and hard drives, an Internet connection and an IP address.

5.7.1 Web Server and Back End

As illustrated in Figure 51, any sophisticated website requires a back-end process that does database operations, generates dynamically-produced web pages, renders content, interacts with banking systems and other computing tasks.

Making it easy for people to find the server requires registering a domain name and maintaining a zone file in the DNS that relates the domain name to the IP address of the server.

The domain name can be advertised and submitted to search engines like Google. Google displays a link to the domain name.

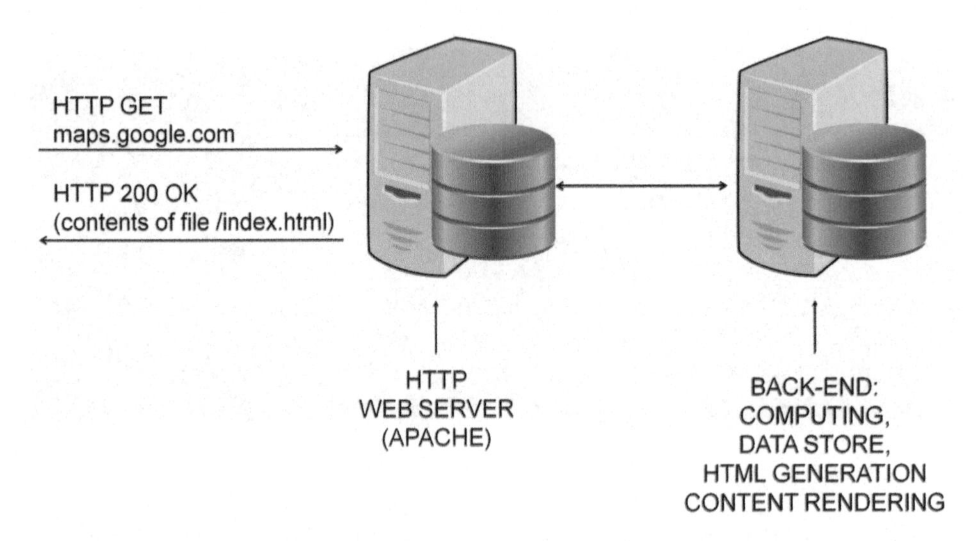

FIGURE 51 WEB SERVER AND BACK-END PROCESSING

When clicked, the user's browser makes a DNS inquiry to get the numeric IP address of the server, then commences sending packets to that IP address.

5.7.2 Doing it Yourself & Dynamic DNS

It would be possible for an individual to accomplish this using a residential Internet connection.

Since residential connections generally come with a dynamic IP address that changes, not a static IP address that does not change, it is necessary to use a service called dynamic DNS to update the DNS zone file each time the IP address that comes with the residential Internet connection changes; or pay an extra monthly fee for a static IP address that does not change.

The individual is responsible for all maintenance of the hardware, uninterruptible power, installing software, all of the security updates, environment and physical security of the computer.

The immediate bandwidth problem (if anyone ever visits the site) is that residential Internet is usually asymmetrical, a lot of downloading capacity and not much uploading capacity, since residences are usually consumers of information; yet this individual's server is serving up pages.

5.7.3 Web Hosting

For most situations, a much better solution is *web hosting*, paying another company to run the server application on their hardware in their secure facility, and get full remote access to it.

To scale up the number of simultaneous transactions past what one computer and/or web server can handle, it is necessary to coordinate multiple instances of the server and its databases.

5.7.4 Virtualization and Cloud Computing

Virtualization accomplishes this function, and is the technology underpinning the term *cloud computing*.

Virtualization is software between the service that is being paid for (web server service) and the hardware it runs on. It enables the end user to modify resources like processing power, memory, disk space and hardware port speed on a browser-based control panel.

Virtualization also allows fault tolerance with multiple instances of the server, and geographic diversity, where the machines actually being used are in various countries, close to the users.

5.7.5 Amazon AWS, Microsoft Azure

There are small web hosting operations, bigger ones like Dreamhost and goDaddy, and then there is Amazon, with Microsoft coming close behind. Amazon is the biggest web hosting company on the planet, making fully half of its profit from Amazon Web Services (AWS).

Amazon built AWS for itself, then turned it into a business; AWS is the web hosting and cloud computing company for Lyft, Pinterest and Apple, and no doubt millions of banks, electrical utilities, insurance companies and other organizations that need reliable computing and storage.

Amazon Elastic Compute Cloud (EC2) and Amazon Simple Storage Service (S3) are two service offerings from AWS. There are dozens of others, including content distribution and security services.

Microsoft's Azure is second in market share.

5.8 Data Centers

A data center is a large building full of racks of computers, hard drives and backups. It is estimated that they will account for 10% of world's electricity consumption in the not-too-distant future.

FIGURE 52 A DATA CENTER. COOLING ON THE LEFT, COMPUTING ON THE RIGHT

Companies like Amazon Web Services and Google have their own private data centers, and use computers and software they design to best meet their needs.

5.8.1 Commercial Multi-Tenant Data Centers

In a commercial data center, one company constructs the building, power supply and cooling. Tenants rent floorspace in Halls that comes with power and cooling, then install racks of computers and hard drives as illustrated in Figure 53, and connect them to the outside world via fiber cables.

Power being an expensive part, in many data centers, it is the other way around: tenants pay for power, which comes with floorspace and cooling.

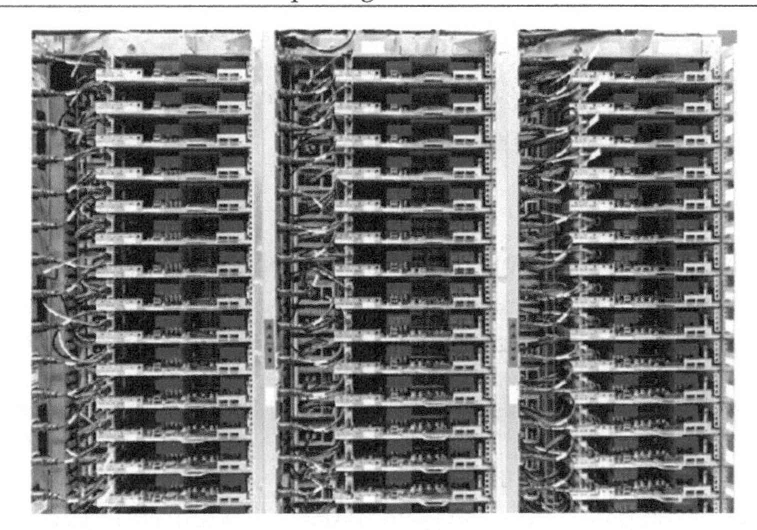

FIGURE 53 DATA CENTERS: THOUSANDS OF RACKS OF COMPUTERS AND HARD DRIVES

With virtualization software, the tenant of the data center has the capability of providing scalable cloud computing with web server front end.

This could be done by the IT department of an organization for their own benefit; or of course, the tenant can sell cloud computing and web server services to its customers.

5.8.2 Collocation

Collocation means physically situating equipment beside another company's equipment. In data centers, it means tenants installing equipment in another company's building.

Collocation is also used when discussing getting physical access to a very high bandwidth upstream Internet connection.

5.8.3 Heat and Electricity

The biggest challenge in data centers is getting rid of excess heat. The computers, with their processors and RAM generate enormous amounts of heat that must be transferred out of the data center to avoid catastrophic temperatures inside the building.

Water is a popular medium to carry heat from inside to outside.

Second is power cost and reliability. Some data centers are located in mountainous areas next to hydro-electric generation plants that provide reliable low-cost power.

Besides a reliable electrical utility's power, other choices for power including solar, wind and batteries; with natural gas, diesel, coal-fired generating stations bringing up the rear.

5.8.4 Connections to Internet Exchanges

Optical Ethernet connections to Internet Exchanges are required to connect the data center to the Internet. Large tenants will have their own redundant connections. A service offering of the data center or a third party can be capacity on a shared physical fiber.

For extreme requirements, for example, a specification of a maximum of 20 ms delay between data centers to implement seamless fault tolerance, the proximity of the data center and length of fiber required to connect it is a factor.

For less extreme requirements, the length of fiber to the data center is moot, allowing the construction of data centers in remote locations where electricity and cooling are economical.

Internet Exchanges are covered in detail in Section 13.1.

5.9 Internet VPNs

The term Virtual Private Network (VPN) is used to mean several different things in telecommunications.

"Network" means connectivity to other locations.

"Private" means dedicated point-to-point circuits for the exclusive use of one customer implement the network.

"Virtual" means that is not true. There are not dedicated point-to-point connections; in reality, many users' packets are interspersed on the network links, but different users can neither see each other's traffic nor each other's sites.

An IP VPN is one kind of VPN. It uses a set of protocols called IPsec to implement secure point-to-point communications using encryption.

An IP VPN is implemented by deploying Virtual Private Network (VPN) hardware at different locations, and connecting it in pairs via the Internet.

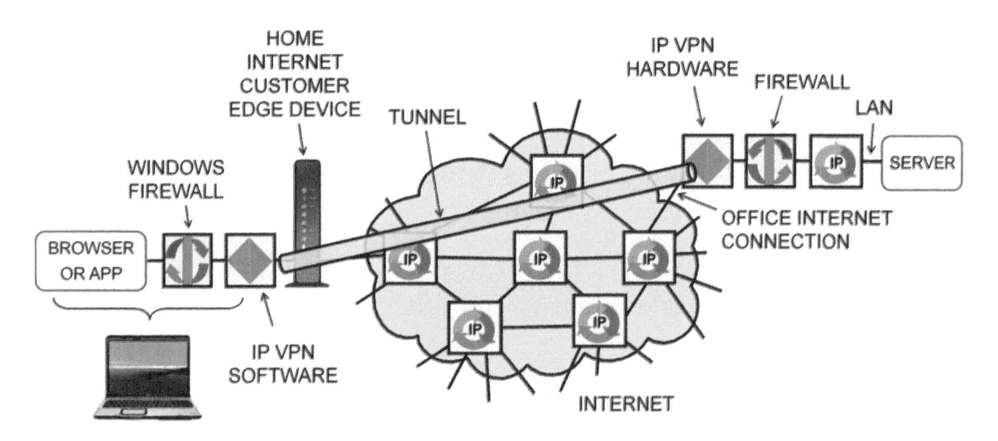

FIGURE 54 WORKING-FROM-HOME IP VPN TUNNEL

Figure 54 illustrates the case of working from home and connecting to servers at work using an Internet IP VPN.

Software on the user's laptop serves as the "VPN hardware" at home. At work, specialized VPN hardware is deployed for performance and security reasons.

5.9.1 Tunnels Implemented with Encryption

On startup, the two pieces of VPN hardware in a pair authenticate each other over the IP network using public key encryption, then exchange a private encryption key for bulk encryption of subsequent transmitted information.

Any IP packets leaving the near-end secure network and destined for the far-end secure network are intercepted by the near-end VPN hardware, encrypted with the private key, then packaged or *encapsulated* inside another IP packet that is addressed to the far-end VPN hardware and transmitted across the non-secure Internet.

The far-end VPN hardware receives these packets and decrypts the contents, extracting the original packet and routing it on its way in the far-end private secure system.

This encrypt-transmit-decrypt process is referred to as *tunneling,* and the point-to-point communications called a *tunnel.*

Strictly speaking, "network" means implementing multiple tunnels between different locations and the need to make a choice which tunnel to take. In the example of Figure 54, the tunnel is a single link and therefore not a network, yet is popularly called "a VPN".

5.9.2 IP VPN vs. MPLS VPN

An MPLS VPN (Section 17.7) is different than an IP VPN. An MPLS VPN is tunnels set up by the carrier internal to their network using MPLS labels, running on carefully-controlled core bandwidth that is not the Internet, and comes with performance guarantees.

MPLS VPN service is widely used by government and industry to connect buildings of an organization that are in different locations.

An IP VPN is set up by the user across the Internet, and typically does not come with a performance guarantee.

5.9.3 Country-Spoofing VPNs

Country-spoofing services are marketed as "VPNs" in advertisements, though they are completely different than IP VPNs and MPLS VPNs.

Most streaming video providers restrict the country to which they will provide service because of restrictions in content distribution licenses.

For example, Netflix USA will only stream to the USA. A resident of Canada cannot directly watch Netflix USA's programming. They can only watch that of Netflix Canada, which is not as good as Netflix USA.

This restriction is implemented by the Netflix USA server only serving IP addresses assigned to an ISP geographically located in the USA.

IP addresses can be geo-located at least to the country level, since they are obtained from an upstream ISP, who obtained them from a regulatory authority that makes public the ISP's country and sometimes city.

Different organizations collect information about ISPs and the addresses they are assigned from regulators and assemble it into a database that can be queried by IP address and return a country code.

A country-spoofing service is one that locates a Network Address Translation (Section 16.9) or *NAT* server in the desired country, which receives packets from the user, changes the **source** IP address field to one from the desired country, and relays the packet to the destination server. It also relays packets in the other direction, from server to user.

Figure 55 illustrates an example of a user in Montreal employing a country-spoofing NAT service in New York City to watch Netflix USA. The packets arriving at the Netflix server in their Syracuse NY data center appear to come from New York USA, not from Montreal Canada.

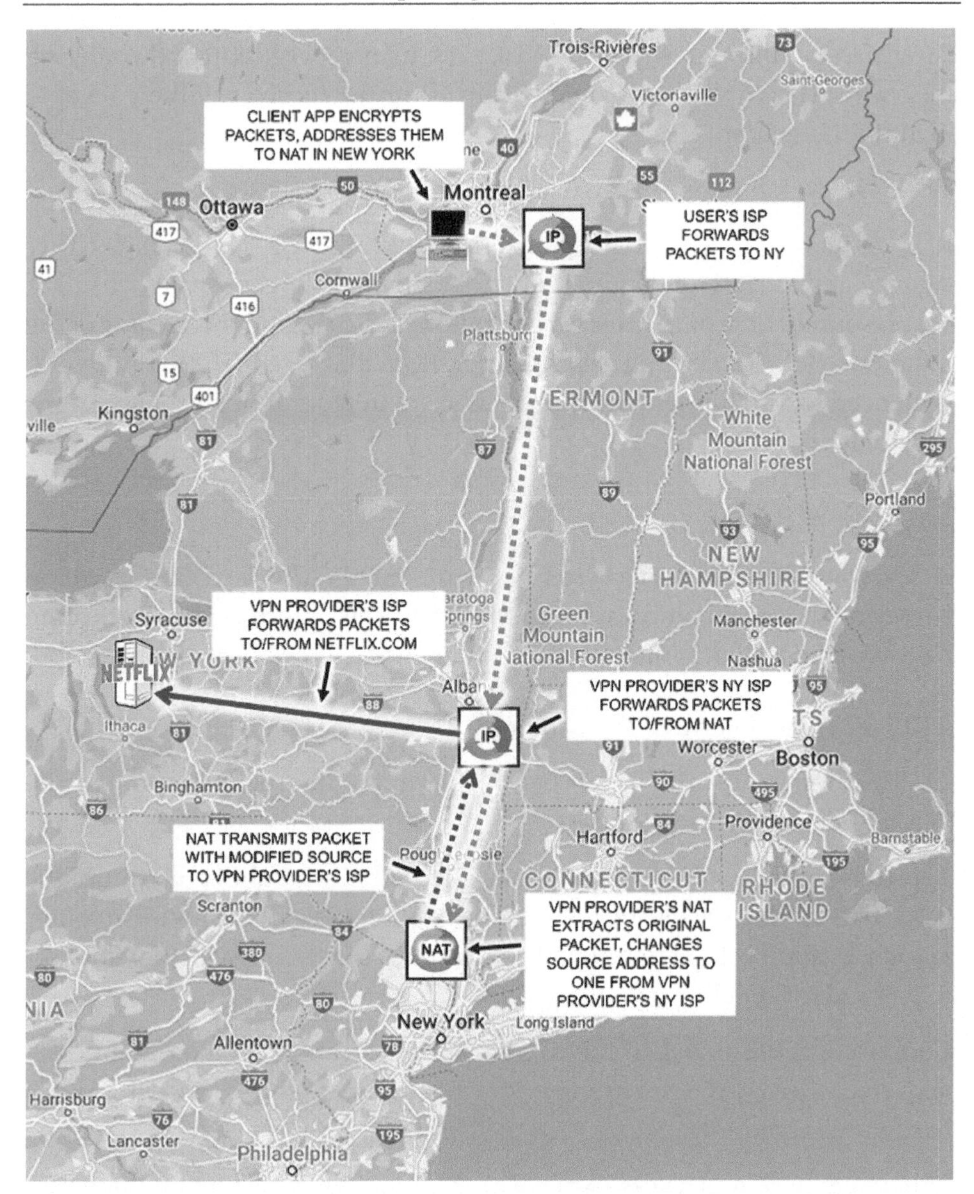

FIGURE 55 COUNTRY-SPOOFING VPN SERVICE

5.9.4 Anonymizer VPNs

Yet another use of the term "VPN" is an anonymizer service.

To implement an anonymizer service, the service provider sets up Network Address Translators in many countries, and chains together IP VPN tunnels between them as described in the previous discussion of country-spoofing VPNs.

After passing through several NATs, each using encryption and changing the source IP address, it becomes difficult for the receiver of the packet to identify the sender.

TOR, The Onion Router, is an example implementation.

It is certainly possible for organizations like the US National Security Agency and the FBI to identify the sender should they choose to focus their vast resources on a particular investigation, as Ross Ulbricht, creator of the infamous Silk Road drug selling operation that used TOR found out: sentenced to two life sentences plus 40 years in federal prison without the possibility of parole.

5.10 SD-WAN

A Software-Defined Wide Area Network (SD-WAN) is communication between an organization's locations over the Internet. Each location requires a broadband Internet connection and SD-WAN hardware.

5.10.1 Pairwise Internet VPN Tunnels

IP VPN tunnels are established pairwise over the Internet between the SD-WAN hardware at each location.

Typically, each location would be assigned a unique subnet, i.e., a unique block of IP addresses (Section 16.3). The routing table in the premise router at each location is then populated with entries that map particular subnets to particular tunnels.

Communications between the organization's locations is then seamless – and free – over the Internet, once the Internet access and SD-WAN hardware is paid for.

This strategy is much less expensive than having a carrier implement a WAN using MPLS on their network between those locations.

5.10.2 No Guarantees

The difference between the two is guaranteed performance.

With an SD-WAN, communications are taking place over the Internet, meaning that while it is less expensive, there are no end-to-end performance guarantees, since the circuits over which the traffic actually moves are unknown and not controlled by the end user.

With an MPLS VPN (Section 17.7), communications are taking place over the carrier's carefully-controlled network, enabling the carrier to sell a performance guarantee called a Class of Service to the end user. Because it requires managed capacity on the carrier's network, a carrier MPLS VPN is more expensive than an Internet SD-WAN.

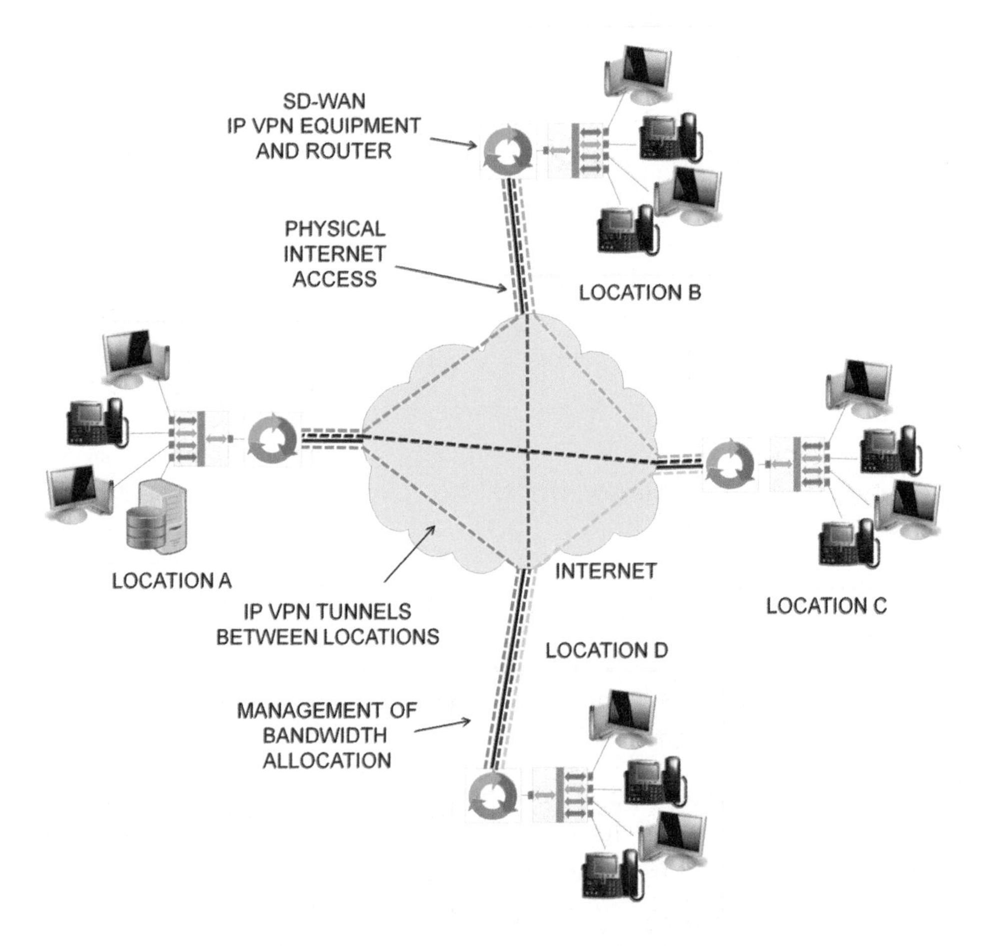

FIGURE 56 SD-WAN: SOFTWARE-DEFINED INTERNET WAN

5.10.3 Bandwidth Management

Each Internet access circuit in the example illustrated will have three tunnels implemented, connections to the other three locations.

SD-WAN hardware may allow the management of bandwidth allocation between the three tunnels, and presumably, non-SD-WAN Internet traffic, that are all sharing the same Internet access.

Whether this is of any practical use depends on the applications using the SD-WAN and Internet access, and the bandwidth of the Internet access.

If there are applications in locations that would tend to use all of the Internet access bandwidth at the expense of other applications and Internet traffic, then the fraction of the bandwidth allocated to the tunnel to that location can be fixed and enforced.

If there are no applications that starve others, and/or the Internet access bandwidth is sufficient to handle all demands, then bandwidth management on the access is not necessary.

5.10.4 Implementation and Standards

Historically, brand-name hardware like enterprise edge routers from Cisco have been used to implement SD-WANs, using Cisco's proprietary IOS communication protocols and messages. This of course requires expensive Cisco equipment everywhere.

An open standard for SD-WAN setup and configuration protocols and messages would allow the use of SD-WAN hardware from any vendor that supports it.

5.10.5 Service Bundling

From a business point of view, an organization can decide to purchase the SD-WAN hardware and operate it themselves, or pay a third party to provide it and maintain it as a service.

There may be cost savings in getting SD-WAN service from the same carrier that supplies the Internet access, or (through a reseller agreement) getting Internet access from the company supplying SD-WAN service.

5.11 Net Neutrality

This section provides a brief overview of all of the things someone could mean when they say "net neutrality".

FIGURE 57 LADY JUSTICE

5.11.1 No Corporate or Government Interference

The Internet Society defines net neutrality with the following statement:

> "Simply put, users should be able to access the Internet content and services they choose without corporate or government interference."

This would be hard to put into practice, as there would be no Internet without corporations and governments. It also suggests all content on the Internet should be free, a quaint idea from the 1980s.

A reasonable interpretation might be that it means an ISP cannot prevent their user from, for example, paying for and accessing Disney+ over their Internet service, instead only allowing access to HBO+.

5.11.2 Relayed Without Regard for Content, Senders, or Receivers

The Internet Society goes on to say:

> "A key element of Internet architecture is that user data is relayed throughout the Internet in the form of standardized packets of information without regard for their content, senders, or receivers.
>
> This nondiscriminatory approach to Internet traffic is a central premise of the Internet's operation. It allows data to easily move across networks without being impeded by the nature of the data.
>
> Fundamentally, this open Internetworking approach is one of the underpinnings that have made the Internet successful."

No discrimination based on content, sender or receiver was a noble idea that could be implemented when the Internet was in its infancy, and traffic was a few ASCII text messages.

However, as soon as there is congestion on network circuits, that is, contention for whose packets will be transmitted and whose will not, management is required to ensure a fair share for all.

In network traffic management systems, content, sender and receiver are three of the main factors taken into account.

Relayed "without regard for content, senders, or receivers" is perhaps best described as a utopian idea that does not mesh well with reality.

5.11.3 Many Different Meanings

There are many participants in the "net neutrality" arena. Some of them are technically knowledgeable about the network, how it is designed and what it does. Some are not.

Network neutrality or "net neutrality" is often used as a broad label in policy and regulatory discussions concerning traffic management.

The term "Net Neutrality" is also used to mean:

- Freedom of expression,
- Competition of service and user choice,
- Impact on innovation,
- Nondiscriminatory traffic management practices, and
- Pricing,

to give some examples.

5.11.4 Criminal Activities

Policing crime on the Internet is of course in direct conflict with the Internet Society's no discrimination stance.

Examples include:

- Pirating: downloading copies of movies or software without paying the copyright holders a licensing fee
- Inciting violence: terrorist indoctrination video entering a sovereign country
- Child pornography and child luring
- Selling illegal goods or services

Government surveillance and storage of Internet traffic in huge data warehouses for later analysis is ever-increasing.

5.11.5 Transparency

Perhaps the most easily defended demand associated with net neutrality is the demand for transparency: a carrier not treat traffic from different sources differently without disclosing the fact to the end user.

5.11.6 Devil in the Details

Other demands are more difficult to implement, as the devil is in the details and the law of unintended consequences quickly comes into play.

For example, it is necessary to prioritize delay-sensitive and loss-sensitive traffic like phone calls and live video streams to be able to guarantee quality and user experience.

That is usually quickly conceded by the "no corporate interference" proponents.

A more difficult issue is that a law barring discrimination based on source and destination would affect the terms of network peering arrangements, to give one example.

5.11.7 No Meters

Some proponents think that net neutrality means that there should be no restrictions on the amount of traffic they can transmit, they should be able to transmit at their access line speed 24/7 without being slowed down.

They refer to the idea as having a "pipe", and are demanding what an engineer would call a *non-blocking* packet network service.

Unfortunately, they are not up to speed on the concept of overbooking the network, giving users more apparent bandwidth for a lower cost. (Section 3.8).

The Internet is not designed to allow all users to transmit constantly at line speed. That would be far more costly to build than the network in place, which is based on statistical multiplexing.

Arguing for non-blocking service is arguing for increasing the cost of Internet access for everyone for no reason.

5.11.8 Zero-Rating

Some people in the developed world (who have Internet and electricity) are in favor of suppressing projects where a business provides Internet access to people who do not already have it in developing countries; but the Internet access is limited, and all web surfing has to be done through a Facebook app, for example. For their own good.

Free restricted Internet is called *zero-rating* by pundits.

6 Telecom Services Overview

6.1 Residential Services

6.1.1 Broadband Internet

Broadband Internet service is the base network service that individual users will have in the future. The reason we need broadband is to watch YouTube videos of guilty dogs and fat hairless cats in HD.

FIGURE 58 BROADBAND FOR RESIDENCES, TO WATCH VIDEOS OF GUILTY DOGS AND FAT HAIRLESS CATS IN HD

A better name for this basic service is *IP dial tone*:

- "Dial tone" meaning the user has network service and the possibility of connecting to anything else on the network, just as with POTS, and
- "IP" meaning the user-network interface is exchanging IP packets at Mega- and Gigabits per second instead of 3 kHz of voltage analogs.

6.1.2 Convergence Achieved

Broadband Internet is the successful implementation of *convergence*: voice, video and data services delivered on the same network connection.

- Anyone with broadband Internet can make outgoing telephone calls to any POTS or mobile line anywhere for a small amount of money.
- Anyone with broadband Internet can watch any television program, movie, TV series or video for a fee, or for free.
- And of course, anyone with broadband Internet can communicate data to any web server, and communicate directly with others as peers for Zoom and phone calls.

6.1.3 PSTN Phone Numbers

At present, the only thing missing from residential broadband Internet service is a PSTN phone number, a network address that can be used to initiate an inbound phone call to the user from anywhere, via the PSTN. Phone numbers are ten digits long in the US and Canada, longer in other countries. A phone number is included with POTS, cellular and phone over cable modem services.

A broadband Internet network service includes an IP address, not a ten-digit telephone number. Without one, it is not possible for anyone to *initiate* an inbound call from the PSTN to the broadband Internet user. The broadband Internet user can make PSTN-terminating phone calls, but not receive PSTN-originating phone calls.

6.1.4 VoIP Service Providers - Internet to PSTN Service

As a workaround, the broadband Internet user can pay a company that connects the Internet to the PSTN, plus a softswitch to relate the user's IP address to a 10-digit PSTN address.

This company could be called a VoIP Service Provider or *VSP* for lack of an official term. The service is sometimes called having a "VoIP phone". A more accurate term might be *Internet phone service.*

When someone calls the 10-digit number from the PSTN, the voice call is terminated on the VSP's gateway, which looks up the broadband Internet user's IP address, then converts the incoming phone call into VoIP and sends the IP packets to the broadband Internet user over the Internet.

This VSP service may become widespread, and even offered as part of broadband Internet service.

In the longer term, the Internet and the PSTN will become the same thing, and 10-digit phone numbers will not be necessary.

SIP URIs, which look like email addresses, will be used as "phone numbers", and resolved by DNS to an IP address to communicate VoIP to a user.

10-digit phone numbers will continue to be valid, but will also be resolved by DNS to an IP address to communicate VoIP to a user.

It will be incumbent on users of rotary dial voiceband analog phones to supply their own gateway to convert that to VoIP for the network interface.

6.1.5 Basic Cable, Streaming, Rentals

The carrier providing Internet service, particularly wired connections, usually also offers video programming. This is an optional, value-added service that is carried on bandwidth separate from the Internet service.

FIGURE 59 TAP, CLICK OR SCAN THIS QR CODE TO WATCH DENVER THE GUILTY DOG AS INTERNET TRAFFIC

In this case, the carrier has signed distribution agreements with content providers to communicate the content to paying customers.

Content includes well-known TV broadcasters like CBS, well-known "basic cable" channels like CNN, niche channels, feature films and archived TV series.

This content may be available from other distributors, or directly from the content producer, over the Internet.

Some of the channels from the carrier may not be available over the Internet.

Some content, like Netflix and YouTube, is only available as Internet traffic.

6.1.6 Dedicated Capacity for TV vs. Internet Traffic

The ISP's video service is delivered over dedicated separate bandwidth, apart from the Internet connection, on the connection to their network. Their video server downloads content to your set-top box over the dedicated bandwidth.

Since it is dedicated capacity, the ISP can guarantee a high-speed connection and thus video quality with no freezing or pixelation.

Any and all other incoming video, like video from streaming service Disney+, is carried as Internet traffic on your ISP service, where throttling and congestion can occur, and there are few if any guarantees.

25 Mb/s Internet service is more than sufficient for two simultaneous streams of High Definition streaming video.

6.2 Business "Data" Services

Business services include Internet service, plus services for reliable communications of business data between locations of an organization, for example, remote access to databases and computing resources, transaction notification and processing, data mirroring and backup.

6.2.1 Internet

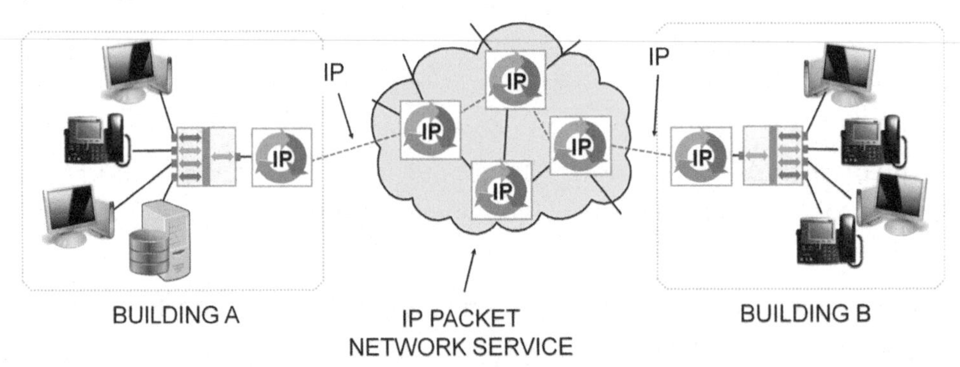

FIGURE 60 BUSINESS DATA SERVICES

Internet service for a business has several differences from residential broadband Internet. The bandwidth can be scaled up to 100 Gb/s, the allocation of bandwidth between uploading and downloading might be symmetrical instead of asymmetrical, the guarantee of packet delivery, uptime and mean time to repair might be stronger, and it will cost more.

6.2.2 Private Network: Dedicated Lines

Communications between an organization's locations, for example, between a bank's data center and all of the bank's branches, could be accomplished with a *private network,* connecting the organization's locations with point-to-point circuits.

Point-to-point service from a carrier is called *private line, leased line, dedicated line* and most accurately, *full-period service.* The point-to-point connections are implemented with private fiber installed for that purpose, leased dark fiber, and legacy T1 circuits.

Implementing many point-to-point circuits in pairs between specific locations, and connecting them to a router at each location implements a *private network.*

While this certainly has application in military, security service and banking, and it is how carriers themselves build their networks, provisioning actual dedicated fiber between locations is hugely expensive and highly inflexible.

6.2.3 Virtual Private Network

Using a carrier's network and shared facilities is a much less expensive and far more flexible choice than private lines.

Implementing a *service* where it *appears* there are dedicated point-to-point circuits between locations, but actually running on shared facilities, is called a *Virtual Private Network* (VPN).

Four different uses of the term "VPN" were listed in Section 5.9. MPLS VPN service is the main service used by business and government.

6.2.4 MPLS VPNs

MPLS VPNs are virtual point-to-point services running on carefully-controlled capacity, not the Internet, for customers like banks and government departments to connect their office buildings and data centers.

Because MPLS VPNs are implemented on carefully-controlled capacity, MPLS VPNs can come with a performance guarantee, called a Class of Service or Service Level.

MPLS VPNs are explained in detail beginning in Section 17.5.

6.2.5 SD-WAN

Software-Defined Wide Area Network means using the Internet to communicate between an organization's locations.

SD-WAN hardware at each location establishes IP VPN tunnels pairwise over the Internet to all of the organization's other locations.

Routing between the organization's locations then becomes a question of what tunnel to take.

SD-WAN hardware might also have software to apportion the shared bandwidth of the Internet access at a location to specific tunnels if there is contention for its use.

SD-WANs are explained in detail in Section 5.10.

6.2.6 Web Services

Another category of business services is web services, which includes web hosting, back-end data center database and cloud computing services, user authentication services, domain registration and DNS services to name a few.

The products and services tabs on Amazon Web Services main page at aws.amazon.com provides a comprehensive list of web services currently available on the market.

There are many other types and providers of web services.

Sophisticated DNS zone file management services are used by sites with high traffic volume, to allow load balancing, fault tolerance and least-cost delivery from one of many data centers around the globe.

This is usually in conjunction with Content Delivery Network service, covered below in Section 6.5.

Web Services are covered in more detail in Section 5.7.

6.3 Business Voice Services

Organizations have many telephones in their buildings, each of which needs a dial tone and a way to connect to the public telephone network.

6.3.1 SIP Trunking: VoIP Between Locations & Dial-Out

SIP Trunking service provides native VoIP connectivity between an organization's locations, plus a gateway to the PSTN.

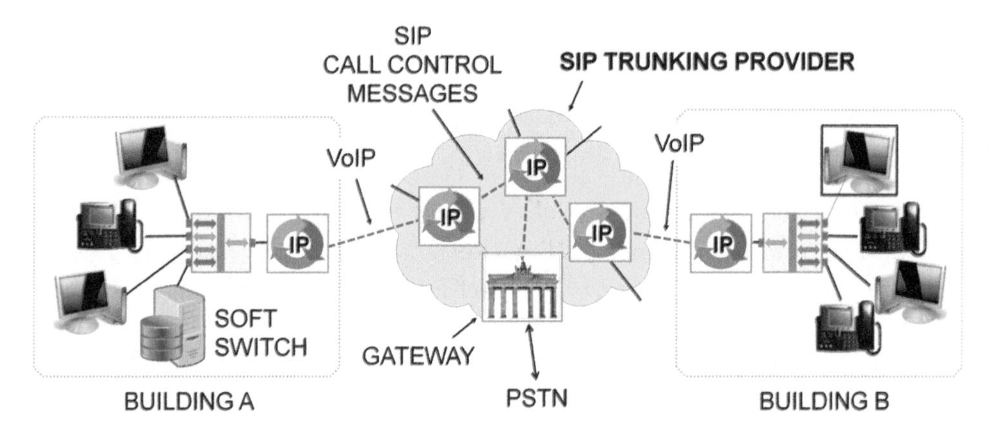

FIGURE 61 SIP TRUNKING

The difference between SIP trunking and sending the same voice packets over the Internet is that SIP trunking comes with a quality guarantee. The Internet does not.

In the case of VoIP phone calls between two locations of the same business, the IP packets containing the voice are carried natively, i.e., without being converted to something else, between the two locations, with guaranteed transmission characteristics like maximum delay and packet loss suitable for voice communications.

The SIP trunking provider also provides VoIP communications to a gateway, where they implement the connection to and from the PSTN.

6.3.2 Connecting VoIP to Ma Bell

In the case of VoIP phone calls from a business VoIP system to the PSTN, for example, phoning home, the carrier providing the SIP trunking service provisions a gateway reachable over their private IP network, and pays to connect its trunk side to the PSTN at a toll center like a LEC.

The gateway will convert the VoIP phone call to the legacy trunk format, DS0 channels. These channels are carried over Tandem Access Trunks to a toll switch in the Toll Center building, where they can be connected to other carriers.

This is covered in detail in Section 13.3.

In the future, agreements for connections to LECs in the form of VoIP and SIP instead of the legacy DS0 channels and ISDN will be formalized in tariffs and this conversion will not be required.

6.3.3 SIP Trunking: Lower Cost than PBX Trunks

The marketing department decided to call this business VoIP service "SIP trunking", the successor to PBX trunking.

It would be more accurate to call the service "VoIP trunking" since its function is to transport voice in IP packets, and SIP is merely the call setup protocol; however, it does carry the small SIP call setup messages, so "SIP trunking" is not incorrect, and does sound better.

In most cases, the SIP trunking carrier provides IP packet transport and the PSTN gateway as services.

In some cases, the SIP trunking carrier provides the IP packet transport, the PSTN gateway plus the SIP softswitch as services.

SIP trunking service is generally much less expensive than the legacy ISDN PRI PBX trunks method, since the pricing model is per PBX trunk, whereas only one IP connection is needed for VoIP.

Carriers may try to present SIP trunking pricing characterized in "PBX trunk equivalents" even though only one circuit is provisioned.

6.3.4 Centrex

Another solution is *Centrex*, the generic term for multi-line business telephone service from the phone company.

As illustrated in Figure 62, with Centrex, the phone company provides the telephones with telephone lines connected directly to a Central Office.

Every telephone gets an extension number which is part of its "outside" phone number. Via software on the CO switch, custom dialing plans can be implemented, for example, only dialing 4 digits and the phone in the office next door rings.

The phone company provides all the equipment and maintenance.

While this provides a very reliable telephony infrastructure, and requires little expertise on the organization's part, it is also expensive, as the billing model is per line per month.

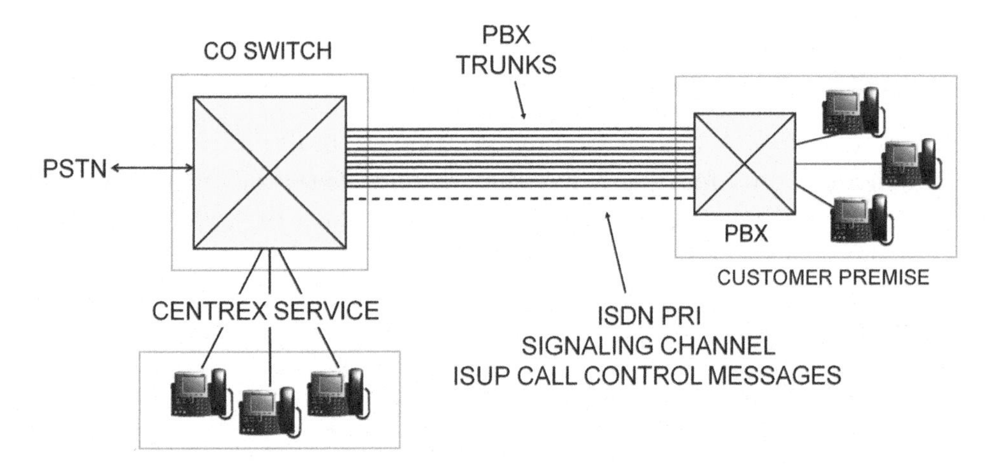

FIGURE 62 CENTREX, PBX TRUNKS AND PRI SERVICE

Providing Centrex to large organizations is typically one of a telephone company's core revenue sources.

6.3.5 Private Branch Exchange & PBX Trunks

An alternative is for the organization to go into the local phone business inside their building, with a private switch called a *Private Branch Exchange (PBX)* in their building. The PBX is connected to the PSTN with *PBX trunks* to a Local Exchange Carrier, at the nearest Central Office building.

The PBX trunks connect the business customer premise to the local phone company's CO, which would then provide switched access to an IXC, to connect to a far-end LEC and its far-end customer to complete a telephone call.

The pricing model is per PBX and phone inside the building, and per PBX trunk outside. One PBX trunk is required per simultaneous external phone call.

6.3.6 ISDN PRI

In medium and large installations, the PBX trunks were implemented with ISDN Primary Rate Interface *(PRI)* technology, which carries PBX trunks digitized at the 64 kb/s DS0 rate on channels, plus two-way call control messages between the PBX and the LEC. The ISDN User Part (ISUP) specifies the message types.

This is being replaced with VoIP and SIP.

6.3.7 Tie Lines and Voice VPNs

In days past, *tie lines* were services that appeared to the customer to be dedicated lines that directly linked the customer's PBXs in different cities.

A sophisticated system of tie lines connecting multiple locations of an organization plus four- and five-digit *dialing plans* was called a Virtual Private Network (VPN). This is a fourth use of the term VPN, in addition to the three of Section 6.2.

6.4 Wholesale Services

Wholesale is the term given to high-capacity communication services purchased by carriers.

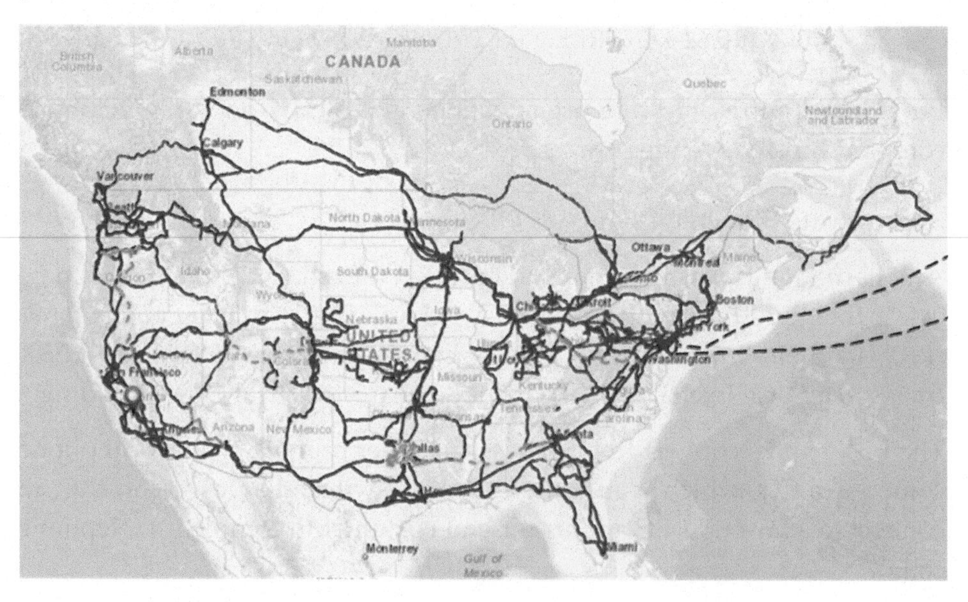

FIGURE 63 ZAYO'S PHYSICAL FIBER INFRASTRUCTURE

6.4.1 Facilities-Based Carriers

All carriers lease connectivity from other carriers.

A *facilities-based carrier* is one that owns transmission facilities to carry their traffic except for a few routes where they lease capacity.

Illustrated in Figure 63 is the physical network of a medium-size player, Zayo Networks. The biggest are AT&T and Global Crossing.

There are countless private military fiber cables and networks not on maps.

Zayo's 133,000-mile fiber network in North America and Europe includes metro fiber connectivity to thousands of buildings and data centers.

Services include dark fiber, private data networks, wavelengths, Ethernet, dedicated Internet access and data center collocation services.

6.4.2 Value-Adding Resellers

Aggregators and *resellers* do not own transmission facilities, their business is leasing capacity from facilities-based carriers and trying to add value to resell it at a profit.

Adding value can include providing seamless communications across many different carriers, and leasing high-capacity services to carry many lower-capacity services.

6.4.3 Services: Dark Fiber

Services range in scope from physical connections only to guaranteed-quality IP packet communications.

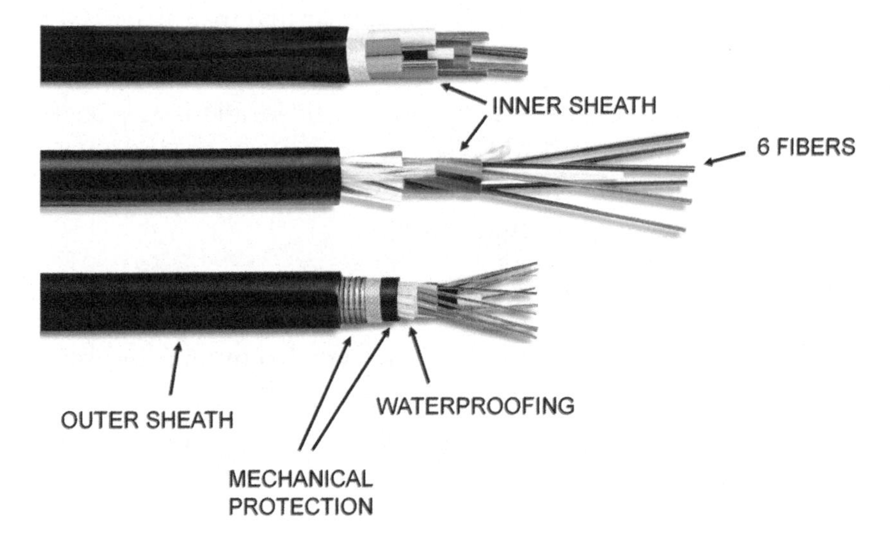

FIGURE 64 CABLES WITH DARK FIBERS

Dark fiber means a very thin tube of glass that can be used as a guide to carry pulses of light up to 80 km (50 mi) at a stretch. Only the glass fiber is leased from the owner. The owner does not provision lasers at each end, hence the term "dark".

The fibers are contained in cables owned, installed and maintained by the carrier. The carrier's wholesale customers can lease specific fibers.

The lessor connects the fiber to their own optical transmission system at each end, to implement point-to-point ultra-high-bandwidth Optical Ethernet communications.

6.4.4 Wavelengths

Wavelengths are fixed portions of the capacity of a fiber identical in concept to Cable TV channels. Frequency-Division Multiplexing, or *Wave-Division Multiplexing* (WDM) is implemented on fibers to increase the overall capacity by transmitting multiple bit streams in parallel. Each bit stream is represented by a laser tuned to emit a particular wavelength of light; light on = "1" and light off = "0".

Leasing one wavelength out of many implies that the landlord is providing the optical transmission equipment, compared to dark fiber where the tenant does. Fiber and wavelengths are covered in detail in Chapter 10.

6.4.5 Carrier Ethernet

Carrier Ethernet is a service that makes it appear that the carrier is a LAN switch. This is also called a Virtual Private LAN Service (VPLS).

This service moves MAC frames between specific locations, just as if they were both plugged into a LAN switch, even though they might be thousands of miles apart. This is more sophisticated than dark fiber or wavelengths, as it supports multiple locations.

6.4.6 IP Services

More complex still are point-to-point IP packet communication services, which make it appear the carrier is an IP router. These services usually come with a Service Level Agreement (SLA) guaranteeing service characteristics like average and peak bitrates, delay and packet loss.

MPLS is a key protocol in implementing the guarantee, internal to the network and not visible to the end-users. This causes the services to be called "MPLS Service", though the service be moving IP packets.

6.4.7 Internet Transit

Wholesale Internet service called *transit* means point-to-point IP packet communication services between a big ISP and customers like a smaller ISP or a data center, with the promise from the big ISP that the customer's packets will be delivered to and received from anywhere in the world.

Large ISPs also rent out blocks of IP addresses from within their bigger block of addresses, and manage the routing of packets to those addresses.

6.4.8 Internet Peering

Another type of wholesale Internet connection is peering, which is connections between different organizations' Autonomous Systems to exchange packets, but not money. The peering is usually accomplished at an Internet Exchange (IX) operated by a third party.

6.4.9 Bit Rates

Carrier capacity lease sizes are large round numbers, following the Optical Ethernet line speeds like 10 Gb/s, 40 Gb/s, 100 Gb/s etc.

Before IP took over, carrier capacity was organized into 64 kb/s DS0 channels. In the US and Canada, carriers leased multiples of DS3-rate (672 DS0 channels, 45 Mb/s) services. In the rest of the world, the Synchronous Digital Hierarchy E3 and E4 rates were common. This is covered in detail in Appendix B.

6.5 Content Delivery Networks

6.5.1 Paying Transit for Data Center to Consumer

In the beginning, Google had one server farm, in one data center. A user had an account with an ISP. Both Google and the ISP had to pay a "Tier 1" Internet backbone connectivity provider for transit services to deliver and/or receive IP packets between Google's server farm and the user's computer.

As things progressed, Google built more and more data centers in various locations. These data centers provide two main functions: mirrors or copies of the data, or content, like YouTube videos; and computing power or "iron" for processing search queries.

The data centers are located

- Near office buildings where the developers work for historical reasons,
- In areas where electricity is cheap such as the interior of Washington State, and
- Near consumers for performance and redundancy reasons.

Providing transit services is a profitable business.

6.5.2 Cutting Out the Middlemen

Paying a carrier for transit services to deliver content from all of these data centers to ISPs and/or to connect the data centers together became a huge cost item on Google's income statement.

The cost could be reduced by eliminating the middlemen, connecting the data centers directly to the ISPs by peering at IXs.

The cost of content propagation to data centers is also reduced by connecting the data centers to each other with dedicated services.

This infrastructure is called a Content Delivery Network (CDN).

Effectively, the content provider, for example YouTube, is bringing the data center-to-ISP function in-house to avoid paying a carrier the "profitable" part of transit services.

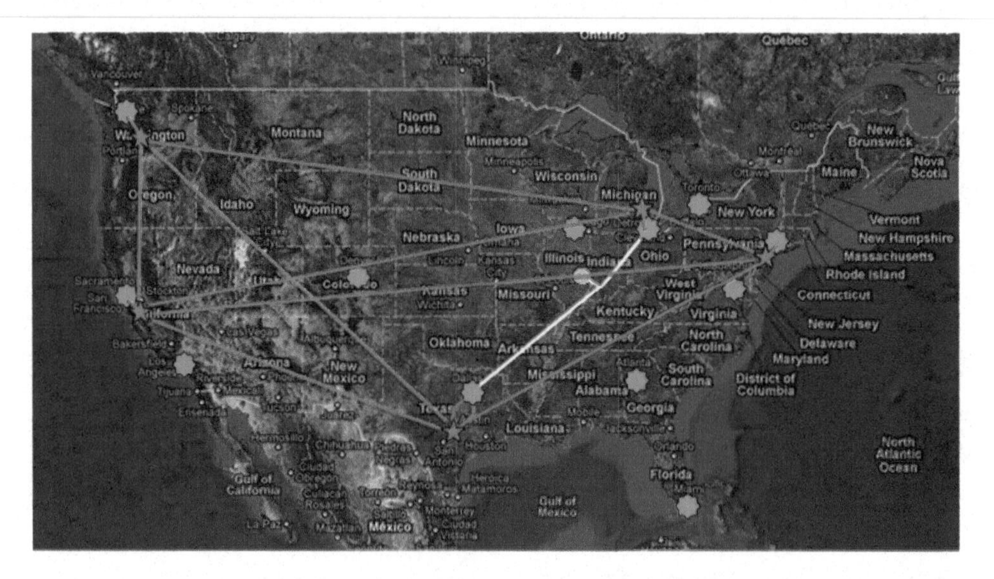

FIGURE 65 YOUTUBE CONTENT DELIVERY

This gives the content provider control over bandwidths and connectivity and a more unified view of their connections to the world compared to a hodge-podge of dissimilar transit agreements with different Tier-1 providers.

In addition, since the CDN has connections to multiple data centers with mirrored content, and multiple connections to an ISP, software can be deployed to determine the optimum data center to use for processing a search query, and the optimum data center to deliver the content from.

This load-balancing function can be performed by the content provider, or provided as a service by a carrier implementing the CDN.

6.5.3 Implementing a Content Delivery Network

All carriers provide the following services, whether or not they are called CDN services:

- Transit services for content providers
- Transit services for ISPs
- Connectivity between content providers' data centers
- Connectivity between content providers' data centers and IXs
- Connectivity between ISPs and IXs

Implementing a CDN is, in its essence, coordinating the design, provisioning and delivery of reliable IP packet communication services for a very big customer account.

Carriers like AT&T design and implement CDNs for their customers, primarily using the AT&T network.

Integrators design and implement CDNs for their customers, sometimes using multiple carriers to optimize the cost.

Big content providers have significant resources allocated to implementing and maintaining all of the aspects of their content distribution.

Google long ago became big enough to continue building its own private CDN with private fiber optic cables, both on land and submarine.

7 Digital Media: Voice, Video, Images, Quantities, Text

7.1 Analog and Digital: What do we really mean?

The words "analog" and "digital" are often used with little regard for their actual meaning. It is useful to review the definitions of these terms to better understand the concept of a digital communication circuit.

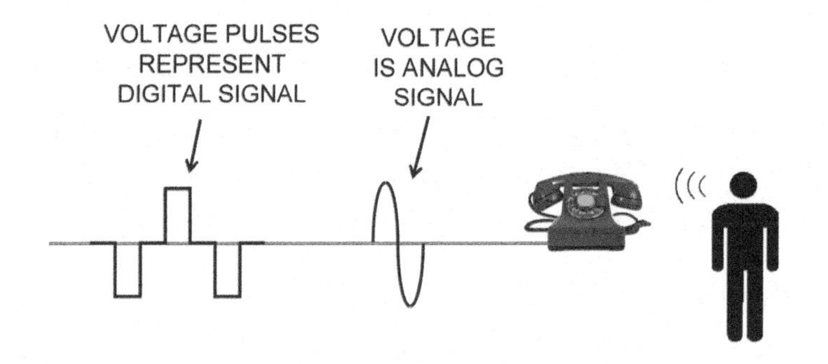

FIGURE 66 ANALOG SIGNALS AND DIGITAL SIGNALS

7.1.1 Analog Signal

"Analog" means "representation". An "analog signal" is a signal that represents another signal.

For example, the voltage on a telephone loop represents the sound pressure waves coming out of the speaker's throat and so is called an analog signal.

7.1.2 Analog Circuit

The term "analog circuit" is inaccurate, but short and catchy-sounding. What is really meant is "circuit capable of carrying an analog signal", and usually, "circuit capable of carrying a voltage that is an analog of the sound pressure waves coming out of a person's throat".

7.1.3 Digital Signal

A "digital signal" is information in a numeric format. The information could have any source. Examples are voice, video, images, text and sensor data. The information is coded into a binary format - 1s and 0s - using some coding scheme.

7.1.4 Digital Circuit

A "digital circuit" is a circuit capable of carrying a digital signal, i.e., a circuit specifically designed to carry numbers from one place to another. In many cases, a digital circuit conveys a digital signal by coding the digital signal into 1s and 0s, then using *pulses* to represent the 1s and 0s.

On a copper-wire digital circuit like a LAN cable, a pulse is changing the voltage on the wires to a fixed amount for a fixed period of time.

On a fiber-optic digital circuit, a pulse is turning a laser on for a fixed period of time. The laser is producing infra-red light at as close to a single pure frequency as possible.

A 1 might be represented by a pulse, and a zero by the absence of a pulse.

Another way of representing a digital signal is to use modems. In this case, some characteristic of a single pure frequency called a *carrier* is changed or modulated in some fixed ways to represent 1s and 0s.

7.1.5 Bandwidth

For the purposes of this course, *bandwidth* means capacity.

The capacity of an analog circuit is measured as the width of the frequency band supported on the circuit, called its *frequency bandwidth*. The units of measurement are changes per second or *Hertz* (Hz).

Wideband and *broadband* are used to describe circuits that support a large frequency bandwidth, like coaxial copper cable ("coax"), which can support 3 GHz of bandwidth.

Companies that offer POTS on twisted pair call the frequency range of their basic service offering the *voiceband*, which is 3 kHz wide.

Companies that operate broadband coax networks refer to the voiceband as *narrowband*.

The capacity of a digital circuit is also called its bandwidth, but is not measured in Hertz, but rather measured in *bits per second* (b/s). A *broadband digital* circuit is one that communicates many bits per second.

7.2 Continuous vs. Discrete Signals

To understand the relationship between analog and digital signals, it is useful to consider the difference between *continuous signals* and *discrete signals*. These are the two types of signals.

7.2.1 Continuous Signals

Continuous signals are signals that vary continuously. A continuous signal can take on any value, or any fraction within given limits.

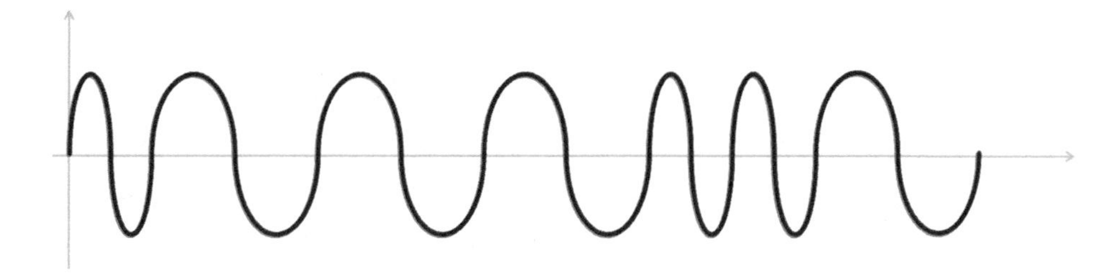

FIGURE 67 CONTINUOUS SIGNAL

Examples of continuous signals are easy to find, since the world is pretty much continuous: length, width, height and time are all continuous.

A good test for a continuous signal is to see if it could possibly take on the value between two other values. If the answer is always "yes", no matter which two values are chosen, then the signal is continuous.

Most *analog* signals are continuous.

A good example of a continuous analog signal is a thermometer. This is a tube of glass with some mercury in it. The height of mercury in the tube is a continuous signal; it can take on any level in the tube.

As the temperature goes up, so does the level of mercury in the tube. As the temperature goes down, so does the level of mercury in the tube.

The height of mercury in the tube is an *analog* of the air temperature.

7.2.2 Discrete Signals

A *discrete* signal is a signal that can take on only specific values.

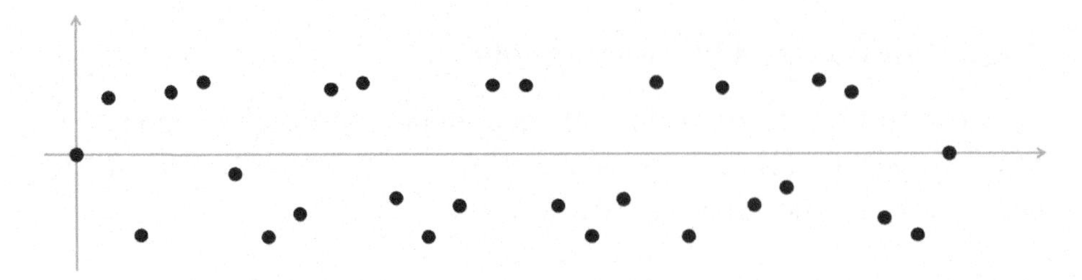

FIGURE 68 DISCRETE SIGNAL

A good example of a discrete signal is the number of people in a room. There can be 14 people in a room, or 15 people in a room. There is no such thing as 14½ people in a room.

Notice that the number of people in a room is discrete in value, but continuous in time. There are always a number of people in the room.

How would we turn this into a signal that is discrete in time? We would have to, on a regular basis, count the number of people in the room and write the answer down on a piece of paper.

This is a good test to see whether a signal is discrete or not: can it be written down on a piece of paper? Or stored on a disk?

Digital signals are discrete.

7.3 Voice Digitization (Analog-Digital Conversion)

We look in detail at the voice digitization process to derive the *coding rate* – the number of bits per second – used in standard digitization of voice.

First, we look at the process that happens on the speaker's line card: converting an analog voice signal to digital. There are three steps: *quantization, sampling* and *coding.*

7.3.1 Quantization

Quantization is the process of changing from a signal that is continuous in **value** to a signal that is discrete in value.

This is accomplished by dividing the possible range of values into a number of "bins" or levels or steps, and assigning a number to each of these levels.

In the example of Figure 69, the range of values is divided into six levels. Then, when asked what the value of the signal is, we say that the signal is "in level #4" rather than trying to measure its exact voltage.

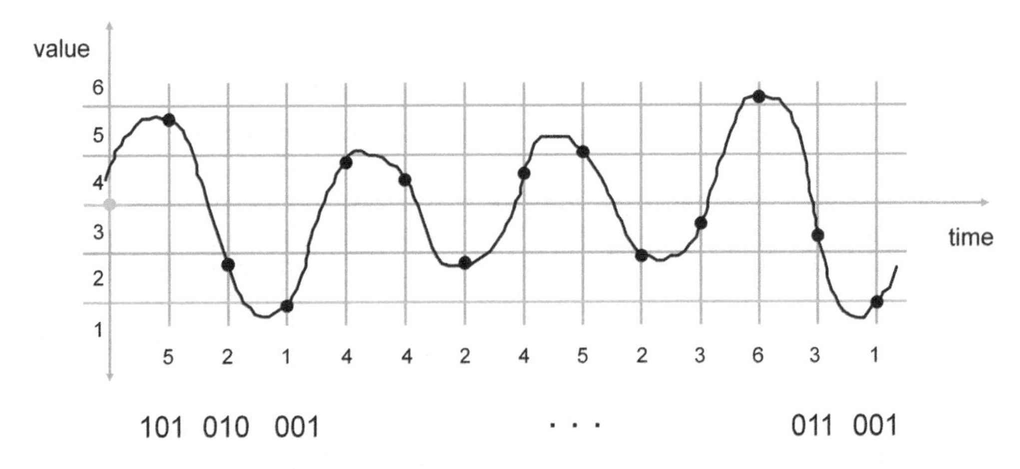

FIGURE 69 QUANTIZATION, SAMPLING AND CODING

Another example of quantization is sugar cubes. Instead of putting some fractional value of a teaspoon of sugar in your coffee, your choice is "one lump or two". The sugar has been **quantized** into uniform lumps.

7.3.2 Sampling

The second step is *sampling*. Sampling is the process of changing the signal from being continuous in time to one that is discrete in **time**.

This must be done on a regular basis, like clockwork.

In the example of Figure 69, the vertical lines indicate the times at which the samples are taken. At each of these times, the signal is sampled, that is, the value of the signal is measured and recorded.

The value of the signal that is recorded is the level it is in at that time.

7.3.3 Coding

The third step is coding. The value of the signal taken at each sample (the level number) must be coded into 1s and 0s so that it can be transmitted over a digital carrier system or stored in a computer.

The objective is to transmit the codes representing the value of each sample to the far-end line card where the reverse process is performed: reconstructing the analog waveform from the received codes.

The whole point in doing this is to move the analog voice signal from the near-end loop to the far-end loop over a digital transmission system without adding any noise – and allowing voice to be carried on the same system as data, video and any other kind of information.

At the sender, sampling, quantization and coding are performed to generate codes representing the value of sequential samples of the analog. These are transmitted to the far end, where the reverse process takes place.

7.4 Voice Reconstruction (Digital → Analog Conversion)

At the far end, the analog is reconstructed to represent speech on the loop leading to the phone. For a residence, this process typically happens on a line card in a switch in a CO or remote. In a VoIP system, this process happens inside the far-end telephone.

7.4.1 Reconstruction

As each code (a binary number) arrives, it is decoded to yield the level or bin number that is the value of the sample, and the voltage on the far-end loop is changed to the value equal to that of the center of the level.

The voltage is changed smoothly from one level to the next.

In the case of a voiceband signal, the voltage would be increased or decreased at a rate so that the result never vibrates more than 3300 times per second, i.e., limited to the voiceband's 3300 Hz limit.

This smoothly-changing voltage is continuous in both time and value – the reconstruction of the original analog.

7.4.2 Quantization Error

However, it is not reconstructed exactly. Due to the fact that only the level number is transmitted – not the exact position within the level – the reconstructed voltage is always set to the center of the level.

The difference between the center of the level and where the signal actually was is a small error introduced into the reconstructed signal, and is called the *quantization error.* The size of the quantization error is directly related to the size of the levels.

To make the quantization error smaller on average, more levels can be defined, to make the levels smaller.

The telephone company uses enough levels so that a human can't hear the quantization error noise.

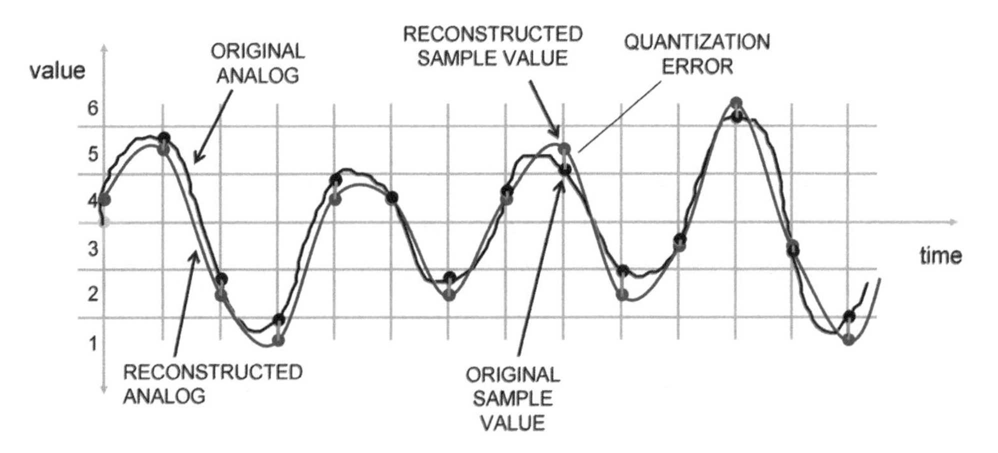

FIGURE 70 QUANTIZATION ERROR

7.4.3 Aliasing Error

If samples are not taken frequently enough, then not enough information will be transmitted so that when the "dots" are connected at the far end, the reconstructed signal is faithful to the original signal.

This is called an *aliasing error*. It can be pictured by removing 2/3 of the samples in Figure 69, then connecting the remaining dots... the result is close to a flat line and does not resemble the original analog.

Harry Nyquist, who obtained a Ph.D. from Yale in 1917 and worked his entire career for AT&T and Bell Labs, discovered that it is necessary to take samples more than twice as often as the frequency bandwidth of the signal to avoid aliasing errors.

This theorem was published in his 1928 paper "Certain Topics in Telegraph Transmission Theory", and is known today as the *Nyquist sampling theorem*. It determines the number of samples per second.

7.5 Voice Digitization: 64 kb/s G.711 Standard

Three steps in voice digitization are *quantization, sampling* and *coding.*

7.5.1 256 Quantization Levels

The telephone system quantizes the voice signal to 256 levels. This number is chosen to reduce the quantization error, which would be heard as noise after the signal is reconstructed, so that a person can't hear it on the line. The diagram shows level numbers 127 and 128 around zero volts.

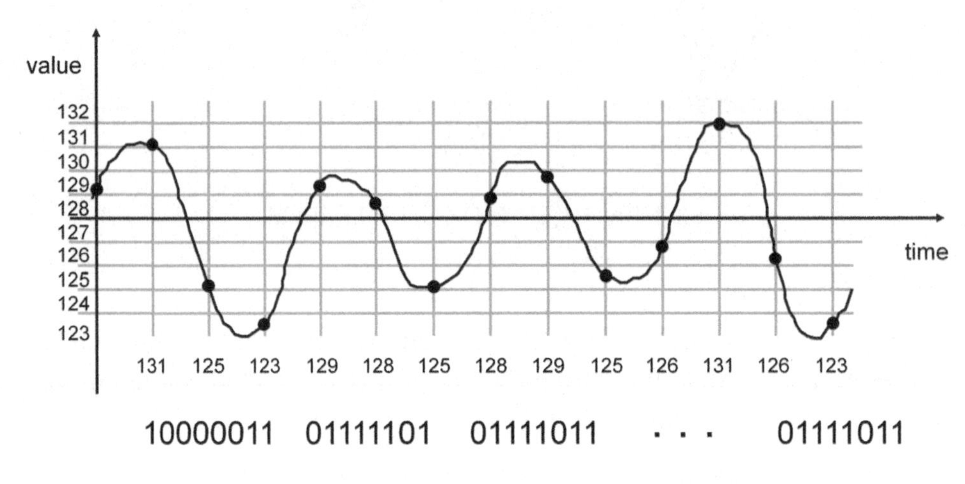

FIGURE 71 64 KB/S VOICE DIGITIZATION

7.5.2 8,000 Samples per Second

The second step is *sampling*. Since this is a voiceband signal, the frequency bandwidth is 3000 Hz, and so the sampling rate must be at least 6001 times per second, following Dr. Nyquist's sampling theorem.

To ensure that there are no *aliasing errors,* the telephone system samples more often: 8,000 samples per second.

7.5.3 8-bit Coding

The third step is *coding*. Traditional telephony uses 8 bits (1 byte) to code the value of each sample.

This technique of using 8 bits per sample is sometimes called Pulse Code Modulation (PCM), a term originally used to describe the entire voice digitization process.

7.5.4 64 kb/s G.711 Codec Standard

To determine the number of bits per second required, multiply the number of samples per second (8,000) by the number of bits per sample (8) to get 64,000 bits per second, or **64 kb/s** for short.

This was standardized by the ITU as codec standard G.711.

7.5.5 64 kb/s DS0 Channels

This 64 kb/s rate was called Service Level Zero in the Digital Hierarchy by Bell Labs, and abbreviated *DS0* (Digital Service Level Zero).

In the previous millennium, transmission systems were built to carry streams of voice digitized at 64 kb/s in channels. In this case, a channel is reserved time slots on a transmission system, a fixed number of bits per second, an unvarying fraction of the overall capacity.

Users, each connected to a channel, take turns using the transmission system, for a specific length of time, one after another, in a strict order that repeats over and over, 8,000 times per second.

These channels could be aggregated into higher bit rate channels, notably DS1 at 1.5 Mb/s and DS3 at 45 Mb/s. This is covered in Appendix B.

7.5.6 64 kb/s Packetized Voice

Going forward, voice is carried in systems originally designed for data.

For efficiency, these systems do not divide the transmission capacity into fixed-size channels, where all users take turns one after another in a strict order, but instead make it first-come, first-served, one *packet* at a time.

Every destination on the network must have a *network address*. On a telephone network, it's called the phone number. On a postal network, it's called the mailing address. On packet networks, it is the *IP address*. IP is a standard way of packet addressing that everyone has agreed to use.

Users create *IP packets* by breaking their transmissions into small segments and pasting the IP address of the desired destination on the front.

The user then transmits the packet to a router, which relays it onward to another router, on and on until it reaches the indicated destination.

At the receiver, the segments of data are extracted from the packets and put together to reconstruct the original transmission.

There are more efficient coding schemes for voice, sometimes called *voice compression*, but they are not the first choice on landline systems. Most of the time, G.711 is employed to avoid compatibility problems.

> Regardless of whether the bits are communicated in a channel or in a packet, the bottom line is that a byte, representing the value of the sample, is transmitted 8,000 times per second to communicate digitized voice when following the near-universal G.711 standard on landline systems.

7.5.7 AMR Codec for Cellular

LTE and 5G use the Adaptive Multi-Rate (AMR) codec, which allows operation at one of eight bit rates between 4.75 and 12.2 kb/s, adaptively chosen based on link quality.

Voice is packetized in segments of 20 ms, yielding 160 samples per packet.

7.5.8 μ-law and a-law

For advanced readers: Figure 71 shows level numbers 127 and 128 around zero volts for clarity. In practice, it might be level numbers 0 and -1 around zero volts.

The figure also shows the levels all being of the same size, which is not completely accurate. In practice, the levels will be smaller close to zero volts and become exponentially larger as the value increases.

This technique further reduces quantization noise by taking advantage of the fact that statistically, the voltage will be around zero most of the time, and so making the quantization finer, i.e. smaller steps, around zero volts reduces the quantization noise.

Two standards for this progressive level size for the G.711 64 kb/s codec are *μ-law* used in North America and *a-law* used in the rest of the world.

7.6 Digital Video, H.264 and MPEG-4

7.6.1 Digital Video Cameras

A digital video camera has a lens that focuses received light on a small array of light detectors, also called *photodetectors.*

The light detectors are usually arranged in groups of three: one sensitive to red, one green, and one blue. There is one triad of detectors per pixel.

The value of each light detector is sampled at the *refresh rate,* often between 50 and 100 times per second.

The value of each sample is coded into a byte. 0 corresponds to no light detected, 255 is maximum light detected, with 253 shades between the two.

The number of pixels, that is, the number triads of light detectors in the camera's detector array is called the picture *definition. High-Definition* (HD) pictures are 1280 pixels across x 720 down, a total of 921,600 pixels.

921,600 pixels, with three color bytes per pixel, refreshed 60 times per second is 1.3 Gigabits per second. This is the output of the digitizer.

"A picture is worth a thousand words" is a well-known saying... but compared to 64 kb/s voice, raw HD video is worth over 20,000 words!

This 1.3 Gb/s bit rate is lowered for storage, transmission and display using mathematical compression techniques. Standard methods of compression are called *codecs*. MPEG is a popular choice.

7.6.2 Factors Affecting Video Quality

A number of factors affect the perceived or *subjective* quality of the images on the far-end user's screen.

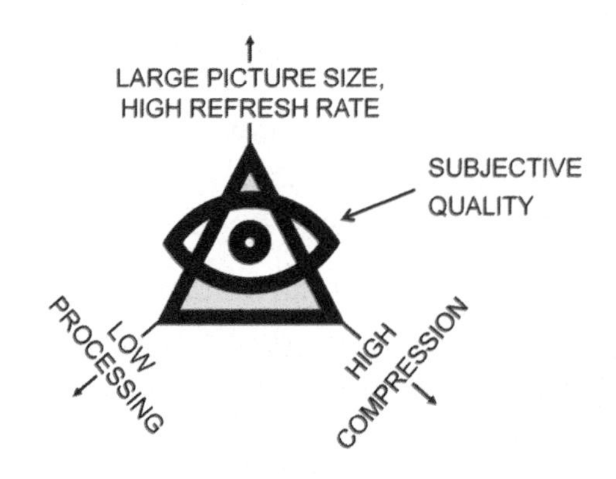

FIGURE 72 DIGITAL VIDEO TRADEOFFS

Aside from network issues like transmission error rate and variability of delay, the main factors are picture definition and refresh rate, the number of bits per second required and the number of processing operations per second that must be performed to implement the compression and decompression in real time.

The objective is to transmit high-definition images using a low number of bits per second while achieving reconstructed picture quality people will be willing to use.

However, these factors are often in conflict: for example, high compression requires intensive processing, and large picture size means a higher number of bits per second. It is one thing to optimize two of the three factors; it is another to optimize all three at the same time.

The reconstructed picture quality is measured by having people watch it and give their subjective rating. It's unclear how objective measurements like signal to noise ratio relate to what a person thinks a picture looks like.

7.6.3 Definition vs. Resolution

Definition, measured in pixels, is the correct terminology for picture size. The term "resolution" refers to the quality of the image after reconstruction. A TV is High Definition (HD), not High Resolution (HR).

However, calling the definition of a display its "resolution" is such a common error that the terms are becoming interchangeable.

7.6.4 Standard Definition, Interlaced and 480i

Standard Definition (SD) in North America is 720 pixels across x 480 pixels down, refreshed 30 times per second.

Human brains have a peak in motion detection at about 15 Hz. Refreshing the picture at 25 or 30 Hz helps minimize the sensation of objects *juddering,* instead appearing to the brain to move smoothly across the screen.

Early developers of television took advantage of the persisting glow of phosphorous on their picture-tube displays to increase the apparent resolution of a display at no cost by doing the refresh in two passes, every odd-numbered line then every even-numbered line.

Each half-picture is called a *field.* Fields are transmitted 60 times per second, leading to the designation *60 Hz interlaced.* Two fields make a *frame.*

SD in North America is abbreviated as 480i.

In the rest of the world, the number of lines per second is the same, but the definition is higher at 720 x 576 since the screen is only refreshed 25 times per second. This is abbreviated 576i.

Videophones and desktop videoconferencing systems in the past supported the Common Interface Format (CIF) at 352 x 258 pixels.

7.6.5 High Definition, Progressive and 720p

When the screen refresh is done in one pass, it is referred to as *progressive* and abbreviated with a p, for example, 480p.

The first step beyond standard definition is High Definition (HD), which is 1280 x 720 progressive and referred to as 720p, and refreshed 50 or 60 times per second.

7.6.6 Full HD 1080 and 2K

Next is 1920 x 1080, interlaced or progressive: 1080i or 1080p.

This is called HD, Full HD, True HD, and in some cases 2K since there are approximately 2000 pixels horizontal definition in the consumer formats.

Advanced readers may want to note that abbreviations with a "K" are also used to describe studio formats, which are slightly different than the consumer formats.

7.6.7 Ultra HD, 4K, 8K, ... 4M

The consumer format 3840 x 2160 is referred to as Ultra HD, Quad Full HD and 4K. After that is 8K with 7680 x 4320 pixels.

Screen definitions will be ever-growing. In the future, displays will have 4,000,000 x 3,000,000 pixels and will occupy entire living room walls. The marketing department will probably call them "4M" displays.

Higher definition (more pixels) can be used to build bigger screens while maintaining resolution (picture quality); higher definition can be used to increase picture resolution while maintaining the physical screen size; or a compromise between the two.

7.6.8 Compression

Compression is required to store and transmit these images. Without compression, 720 x 480 at 30 Hz, with one byte each for red, green and blue is 250 Mb/s. 1280 x 720 at 60 Hz is 1.3 Gb/s. 1920x1080 is 3 Gb/s.

Compression is performed by an algorithm called a coder/decoder or codec, either on a special-purpose integrated circuit chip, or on the shared main processor in a computer.

Video compression is *lossy compression,* meaning the reconstructed image is not exactly the same as the original.

To operate in real time (at playing speed), codecs are usually implemented as highly optimized machine code on custom-built chips containing multiple Digital Signal Processors (DSPs).

7.6.9 MPEG

Standards are required for interoperability. The *Moving Picture Experts Group* (MPEG) and the ITU establish standards in this area.

MPEG-1 was for video on CDs, with the video coded at 1.15 Mb/s.

This was replaced with MPEG-2, which offers a wide range of coding and compression options, grouped in *profiles*. Each profile supports a certain picture size, definition, refresh rate and image quality, and results in a different average bit rate, typically 1 to 3 Mb/s.

MPEG-2 is used as the basis for SD video stored on Digital Versatile Disks (DVDs) and transmitted via cable, satellite and IPTV video services.

7.6.10 MPEG-4 and H.264

Part 10 of the MPEG-4 standard specifies the use of the H.264 codec standardized by the ITU. H.264 provides the same quality of reconstructed signal as MPEG-2 for 1/3 the bit rate with better error tolerance.

H.264 is used for coding HD video for Blu-ray DVDs, and HD channels delivered by cable, satellite, Internet and IPTV video services. HD video is typically coded at 6 Mb/s for broadcast. Much lower rates are used for Internet video.

7.7 Digital Images: JPEGs and GIFs

The color of a pixel is usually represented on a monitor using 24 bits: 8 bits for red, 8 for green and 8 for blue. This means that there are 2^8 or 256 quantization levels for each of the three primary colors.

While there are certainly systems that employ finer quantization, 3 bytes per pixel is typical.

Saving an image as a bitmap saves the three bytes per pixel with no compression, yielding the largest file size.

According to Google, size does matter. They discovered long load times on smartphones was the number one cause of users abandoning a site after clicking on it. Google develops and gives away free automatic software for web servers called Google Pagespeed that compresses the images and serves the compressed version instead of the image the web page developer specified, to speed up page loading.

There are different strategies for compression, often characterized as *lossy* or *lossless compression.*

FIGURE 73 THE MONA LISA

7.7.1 Lossless Compression: PNG

Lossless compression means compressing a bitmap file, and being able to recreate it exactly, bit for bit, like a data file when it is decompressed.

Portable Network Graphic (PNG) implements lossless compression.

7.7.2 Lossy Compression: JPEG and GIF

Lossy compression means that when the image is compressed, some information is discarded, and when decompressed, the result is not bit-for-bit the same.

This kind of compression takes advantage of the way human brains process images to spend fewer bits on parts of the image that the brain doesn't fully look at.

First, it turns out that the brain pays much more attention to luminance (black and white) than it does to the chrominance (color). One way of achieving compression is to convert the image from RGB to a luminance and chrominance representation, then spending half as many bits coding the chrominance values.

It also turns out that humans have edge detectors in their brains, and the brain pays more attention to edges than it does to uniform backgrounds. If it is possible to lower the coding rate of the uniform background areas of a picture, without it *subjectively* looking any worse when reconstructed, compression is achieved.

Edges are high-frequency transitions in the spatial realm, so most compression strategies convert the image to be represented by the frequency at each pixel, allowing fewer bits to be spent coding the low frequencies.

FIGURE 74 EDGES: THE MONA LISA TRANSFORMED TO THE SPATIAL FREQUENCY DOMAIN

The Joint Photographic Experts Group (JPEG) produced a worldwide standard in which compression is achieved by converting the RGB representation to luminance and chrominance, reducing the sampling rate of the chrominance, then transforming to the frequency domain with Discrete Cosine Transform, then performing run-length coding and finally Huffman coding of the DCT values.

Files that have been compressed using this method usually have a .jpg extension.

The Compuserve Graphical Interchange Format (GIF) has one advantage over JPEG compression in that it supports transparency: there is a code for "nothing", making it easier to superimpose images. JPEG forces blank areas to be reconstructed as white, meaning no transparency.

7.8 Digital Images in Email: MIME

The basic email sending program used on most mail servers, sendmail, was developed by UNIX computer programmers for transfers of text files using the UNIX-to-UNIX Copy Protocol (UUCP), which historically only supported 7-bit printable ASCII characters.

What happens if you want to include in an email an image, or a computer program, which is coded using all eight bits for data in each byte?

It was necessary to re-package the 8-bit bytes (which people called *binary*) into groups that look like printable characters (called *text*), email them as if they were text, and then perform the reverse process at the receiver.

7.8.1 UUENCODE, Quoted-Printable and Base-64 Encoding

In the beginning, users did this using a utility program called UUENCODE Now this function is automated: the client e-mail program automatically transforms the message into "text" and tags it with information so that the receiver can perform the reverse transformation.

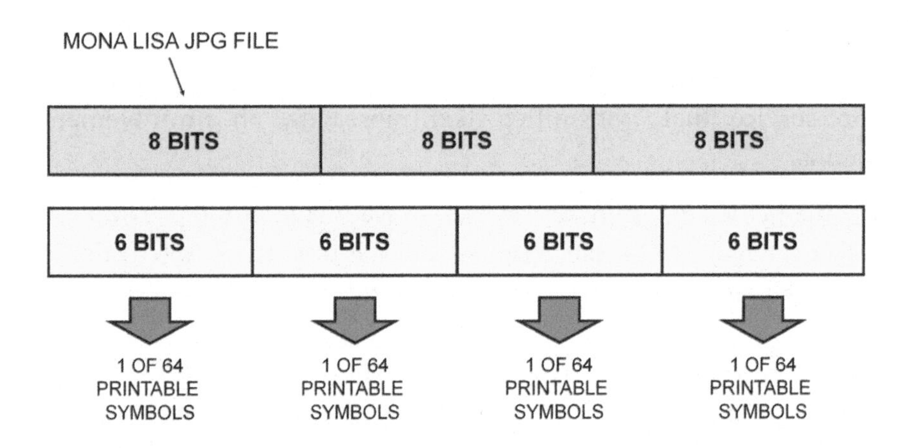

FIGURE 75 MIME FOR DIGITAL IMAGES IN EMAIL MESSAGES

This is described in detail in RFC2045: Multipurpose Internet Mail Extensions (MIME). The Content-Type header field specifies the nature of the data in the body of an entity by giving media type and subtype identifiers like "text" and "jpg".

Quoted-printable encoding and base-64 encoding are popular transforms.

Quoted-Printable Content-Transfer-Encoding is intended to represent data that largely consists of octets that correspond to printable characters in US-ASCII.

Octets with decimal values of 33 through 60 inclusive, and 62 through 126, inclusive, may be represented as the US-ASCII characters that correspond to those octets.

Any octet, except a CR or LF may be represented by an "=" followed by a two-digit hexadecimal representation of the octet's value.

Base-64 encoding is illustrated in Figure 75. A 64-character subset of US-ASCII is used, enabling 6 bits to be represented per printable character.

This subset has the important property that it is represented identically in all versions of ASCII including its international codeset variations.

The tragic part of this is that to be backwards-compatible with 1970s-era computer technology, email programs like Outlook automatically repackage email attachments this way so that they appear to be giant telex messages.

This causes coding and decoding processing delays, plus 33% overhead in transmission. For those who were born after telex was discontinued, telex was a service that transmitted text messages, an improvement on telegrams.

Following is the beginning of the Mona Lisa image of the previous lesson base-64 encoded. The whole image is coded into 121,327 ASCII characters, which is 38 pages of text printed with 11-point Arial font.

data:image/jpeg;base64,/9j/4RXRRXhpZgAASUkqAAgAAAAHABI
BAwABAAAAAQAAABoBBQABAAAAYgAAABsBBQABAAAAagAAACgBAwABA
AAAAgAAADEBAgAiAAAAcgAAADIBAgAUAAAAlAAAAGmHBAABAAAAqAA
AANQAAACA/AoAECcAAID8CgAQJwAAQWRvYmUgUGhvdG9zaG9wIENDI
DIwMTUgKFdpbmRvd3MpADIwMTk6MDk6MjYgMTk6MjQ6MTkAAwABoAM
AAQAAAP//AAACoAQAAQAAAEkCAAADoAQAAQAAACwCAAAAAAAAAAAGA
AMBAwABAAAABgAAABoBBQABAAAAIgEAABsBBQABAAAAKgEAACgBAwA
BAAAAAgAAAAECBAABAAAAMgEAAAICBAABAAAAlxQAAAAAAAABIAAAAA
QAAAEgAAAABAAAA/9j/7QAMQWRvYmVfQ00AAv/uAA5BZG9iZQBkgAA
AAAH/2wCEAAwICAgJCAwJCQwRCwoLERUPDAwPFRgTExUTExgRDAwMD

7.9 Digital Quantities: Number Systems

The purpose of this section is to lead to understanding the fundamentals of binary numbers, and explain *hexadecimal,* used to represent LAN interface MAC addresses and IPv6 addresses.

To explain binary and hexadecimal, we begin with decimal.

All of these are *number systems,* which are the coding step to represent *quantities*. Humans currently use the decimal number system; computers and communication systems use the binary number system.

Decimal is a number system based on **tens**... presumably because most people have **ten** fingers. There are ten symbols in the decimal number system: 0, 1, 2, 3, 4, 5, 6, 7, 8 and 9. Quantities are represented as powers of **ten**.

When expressing quantities in the decimal number system, we use a shorthand notation to indicate how many of which powers of ten are needed to make up the quantity.

10^3 10^2 10^1 10^0

$$
\begin{aligned}
1\ 9\ 6\ 7_D &= \mathbf{7} \times 10^0 &&= \mathbf{7} \times 1_D \\
&+ \mathbf{6} \times 10^1 &&+ \mathbf{6} \times 10_D \\
&+ \mathbf{9} \times 10^2 &&+ \mathbf{9} \times 100_D \\
&+ \mathbf{1} \times 10^3 &&+ \mathbf{1} \times 1000_D
\end{aligned}
$$

FIGURE 76 DECIMAL NUMBERING SYSTEM EXAMPLE

For example, when we write the number "1967", what we mean is

$1 \times \mathbf{10}^3 + 9 \times \mathbf{10}^2 + 6 \times \mathbf{10}^1 + 7 \times \mathbf{10}^0$. This could also be written as

$(1 \times \mathbf{1000}) + (9 \times \mathbf{100}) + (6 \times \mathbf{10}) + (7 \times \mathbf{1})$.

The digits 1, 9, 6, and 7 indicate, for the appropriate power of 10, how many of that power of ten go into making up the quantity 1967.

In other words, the numbers 1, 9, 6 and 7 are placeholders in the shorthand notation, indicating how many of the powers of ten in that place go in to making up the quantity.

7.10 Digital Quantities: Binary

Communication systems and computers use binary numbers to represent information, because computers only have two fingers: on and off.

Computers are built using transistors, and the way that a transistor is used in a digital computer is like a switch: open or closed. All of the processing part of your computer is built out of tiny switches. All of the live memory, the RAM, is built out of tiny switches.

To represent the state of the RAM, i.e., what number is in there, it is most efficient to use a numbering system with two states... binary.

The binary number system is the same as the decimal number system, except it is based on **two** instead of ten. You could even call it base 2 arithmetic if you wanted to. There are **two** symbols in the binary number system: 0 and 1. Quantities are represented as powers of **two**.

$$
\begin{array}{llll}
\overset{2^7\,2^6\,2^5\,2^4\,2^3\,2^2\,2^1\,2^0}{11001001_B} = & \mathbf{1} \times 2^0 & = & \mathbf{1} \times 1_D \\
 & + \mathbf{0} \times 2^1 & & + \mathbf{0} \times 2_D \\
 & + \mathbf{0} \times 2^2 & & + \mathbf{0} \times 4_D \\
 & + \mathbf{1} \times 2^3 & & + \mathbf{1} \times 8_D \\
 & + \mathbf{0} \times 2^4 & & + \mathbf{0} \times 16_D \\
 & + \mathbf{0} \times 2^5 & & + \mathbf{0} \times 32_D \\
 & + \mathbf{1} \times 2^6 & & + \mathbf{1} \times 64_D \\
 & + \mathbf{1} \times 2^7 & & + \mathbf{1} \times 128_D
\end{array}
$$

FIGURE 77 BINARY NUMBERING SYSTEM EXAMPLE

When expressing quantities in the binary number system, we use a shorthand notation to indicate how many of which powers of two are needed to make up the quantity.

For example, when we write the number "11001001", what we mean is

$1 \times \mathbf{2^7} + 1 \times \mathbf{2^6} + 0 \times \mathbf{2^5} + 0 \times \mathbf{2^4} + 1 \times \mathbf{2^3} + 0 \times \mathbf{2^2} + 0 \times \mathbf{2^1} + 1 \times \mathbf{2^0}$.

This could also be written as

(1 x **128**) + (1 x **64**) + (0 x **32**) + (0 x **16**) + (1 x **8**) + (0 x **4**) + (0 x **2**) + (1 x **1**).

The binary digits 1, 1, 0, 0, 1, 0, 0 and 1 indicate, for the appropriate power of 2, how many of that power of two are in the quantity.

In other words, the numbers 1, 1, 0, 0, 1, 0, 0 and 1 are placeholders in the shorthand notation, indicating how many of the powers of two in that place go in to making up the quantity.

Compare this to decimal numbers, and it becomes apparent that the concept of binary and decimal are the same – only the base is different. Decimal is based on ten and binary is based on two.

7.11 Digital Quantities: Hexadecimal

The hexadecimal number system is the same as the decimal and binary number systems, except it is based on **sixteen** instead of ten or two. Hexadecimal could even be called base-16 arithmetic.

There are **sixteen** symbols in the hexadecimal number system:

0, 1, 2, 3, 4, 5, 6, 7, 8, 9, A, B, C, D, E, F.

The letters A, B, C, D, E and F are used as symbols to represent the quantities 10, 11, 12, 13, 14 and 15 respectively, since symbols can have only one character.

$$\begin{array}{ccc} 16^2 & 16^1 & 16^0 \\ 7 & C & 9_H \end{array} = \begin{array}{r} 9 \times 16^0 \\ + C \times 16^1 \\ + 7 \times 16^2 \end{array} = \begin{array}{rcr} 9_D & \times & 1_D \\ 12_D & \times & 16_D \\ 7_D & \times & 256_D \end{array}$$

FIGURE 78 HEXADECIMAL NUMBERING SYSTEM EXAMPLE

Quantities are represented in hexadecimal as powers of **sixteen**. Just as in decimal and binary, when expressing quantities in hexadecimal, we use a shorthand notation. This indicates how many of which powers of sixteen are needed to make up the quantity.

For example, when we write the number "$7C9_H$", what we mean is

$7 \times \mathbf{16}^2 + 12 \times \mathbf{16}^1 + 9 \times \mathbf{16}^0$.

This could also be written as

(7 x **256**) + (12 x **16**) + (9 x **1**).

The hexadecimal symbols 7, C and 9 indicate, for the appropriate power of 16, how many of that power of sixteen are in the quantity.

In other words, the numbers 7, C and 9 are placeholders in the shorthand notation, indicating how many of the powers of sixteen in that place go in to making up the quantity.

Compare this to decimal numbers, and it will become clear that the concept of binary, decimal and hexadecimal are all the same – only the base is different. Hexadecimal is based on sixteen; decimal is based on ten and binary is based on two.

7.11.1 Common Use for Hexadecimal

Not many people have sixteen fingers; so why would we bother with a numbering system based on 16s?

Hexadecimal (or *hex* for short) is perhaps most often used in practice as a **short form for binary numbers**.

Hex	Nibble	Decimal
0	0000	0
1	0001	1
2	0010	2
3	0011	3
4	0100	4
5	0101	5
6	0110	6
7	0111	7
8	1000	8
9	1001	9
A	1010	10
B	1011	11
C	1100	12
D	1101	13
E	1110	14
F	1111	15

FIGURE 79 HEXADECIMAL AS A SHORT FORM FOR BINARY NUMBERS

Long binary numbers written on screens or on paper are cumbersome and largely incomprehensible to humans. How successful would attempting to communicate 011111001001 to another person be? How easily would they remember this number?

A method more useful for human comprehension would be to organize bits in groups and represent values of the groups with symbols.

Since the world has more or less standardized on the octet, i.e., groups of 8 bits, a numbering system that is efficient for groups of 8 bits would be indicated. However, defining symbols for all values that could be represented by 8 bits would require $2^8 = 256$ symbols (like Chinese), too many to be easily comprehended unless learned from birth.

Using decimal to represent the value of a byte is possible, and is used in IP version 4's dotted-decimal notation, but is awkward and difficult. It requires a computer program to do the conversion in one direction.

The solution is to divide a byte in half; into two groups of 4 bits.

To represent any pattern four bits long requires $2^4 = 16$ symbols. Therefore, a byte can be represented with two of these symbols, instead of eight bits. Since $\mathbf{16}^1 = \mathbf{2}^4$, conversion from binary to hex is simple: the binary number is segmented into groups of four bits, and each set of four bits is converted individually to their hex equivalent.

Consider the binary representation of 1993_D: 011111001001.

For readability, instead of commas we use semicolons: 0111:1100:1001.

The hexadecimal equivalent of this number is $7C9_H$.

The hex version uses only 1/4 the symbols, and so is easier to write and to pronounce.

7.12 Digital Text

7.12.1 ASCII

Figure 80 depicts the American Standard Code for Information Interchange (ASCII, pronounced ASK-EE). This is a standard method of representing keystrokes, i.e., text, using binary digits. The official standard for ASCII is ANSI X3.4-1986, which defines a 7-bit code.

For example, the ASCII code for a capital "A" is, at the top of the row 100 and at the left of the column 0001 in Figure 80 to make 1000001.

Bits 4,3,2,1 \ Bits 7,6,5		000	001	010	011	100	101	110	111
		0	1	2	3	4	5	6	7
0000	0	NUL	DLE	SP	0	@	P	,	p
0001	1	SOH	DC1	!	1	A	Q	a	q
0010	2	STX	DC2	"	2	B	R	b	r
0011	3	ETX	DC3	#	3	C	S	c	s
0100	4	EOT	DC4	$	4	D	T	d	t
0101	5	ENQ	NAK	%	5	E	U	e	u
0110	6	ACK	SYN	&	6	F	V	f	v
0111	7	BEL	ETB	'	7	G	W	g	w
1000	8	BS	CAN	(	8	H	X	h	x
1001	9	HT	EM	)	9	I	Y	i	y
1010	A_H	LF	SUB	*	:	J	Z	j	z
1011	B_H	VT	ESC	+	;	K	[	k	{
1100	C_H	FF	FS	,	<	L	\	l	|
1101	D_H	CR	GS	-	=	M	]	m	}
1110	E_H	SO	RS	.	>	N	^	n	~
1111	F_H	SI	US	/	?	O	-	o	DEL

FIGURE 80 ASCII CODE SET

There are a number of codes other than the alphabet and punctuation. These are control codes, for example BEL (7) would tell the receiver to ring its bell, CR (13) is carriage return, FF (12) is form feed and so forth.

ASCII was primarily defined for teletypes, i.e., printed English characters, which (including both uppercase and lowercase) can easily be supported within a limit of 128 codes, and so specified codes that were 7 bits long.

This saved bandwidth compared to 8-bit codes, and allowed the appendage of a parity bit for error detection to make up an 8-bit byte.

However, as computers standardized on octets, more languages with different characters were required to be supported, and parity checking was abandoned, sets of 8-bit codes were required.

IBM used an 8-bit code set called the Extended Binary Coded Decimal Interchange Code (EBCDIC) code set on mainframes. With respect to this chart, an EBCDIC chart would be upside down, backwards, twice as big and none of the characters would be the same. Other than that, EBCDIC is exactly the same thing as ASCII - it is a standard way of coding keystrokes into bytes.

Various ad-hoc extended ASCII code sets were defined, particularly in Microsoft's Disk Operating System (DOS), adding an extra bit, doubling the size of the table. The characters in the extended ASCII table are Greek letters, box drawing characters, és and so forth.

However, there were dozens of variations to choose from, so ¬ sometimes came out as î, if for example a computer and printer did not agree on what extended ASCII variation is in use.

The International Organization for Standardization (ISO) and their friends at the International Electrotechnical Commission (IEC) eventually published a standard code set ISO/IEC 8859-1 (Part 1 of ISO/IEC 8859), an 8-bit code set for the Latin alphabet. The original definition of this code set did not include the control codes, and was not used in practice, but did form the basis for two 8-bit code sets that are now in wide use: ISO-8859-1 and Windows-1252.

ISO-8859-1, defined in RFC1345, is a superset of the original ISO/IEC 8859-1 to include control codes and is at present widely used for plain text web pages and email. It is the default encoding for MIME type "text".

Windows-1252 is almost the same, but substitutes a number of printable special characters like left double quotes where ISO-8859-1 has control codes. These special characters are rendered as a question mark or hollow box when displayed on a web page by a browser, since they are undefined characters in ISO-8859-1, which the browser uses by default.

7.12.2 Unicode

Unicode and its Unicode Transformation Format (UTF) may end up becoming universal standard codes for character sets.

Unicode defines a codespace of 1,114,112 codes in the range 0 to 10FFFFH and methods of representing them, called transformation formats. The most popular is UTF-8, which allows one to four bytes to represent a character.

It is normal to reference a Unicode code by writing "U+" followed by its value in hex. Often, double-byte or four hex characters are used, for example U+005A for Z and U+548C for 和.

In HTML, characters may be expressed as &# followed by the decimal value of the code and a semicolon. For example, Z would be rendered as Z. Or, characters may be expressed as &#x followed by the hex value of the code and a semicolon, for example, Z.

The characters allowed in URLs (web addresses) may be represented as % followed by the hex value of the code (for example, %5A).

8 Fundamentals of Voice over IP

8.1 The Big Picture

Figure 81 provides a framework for the chapter, and an illustration of VoIP components and technologies and how they are organized. The key components are detailed in the following pages.

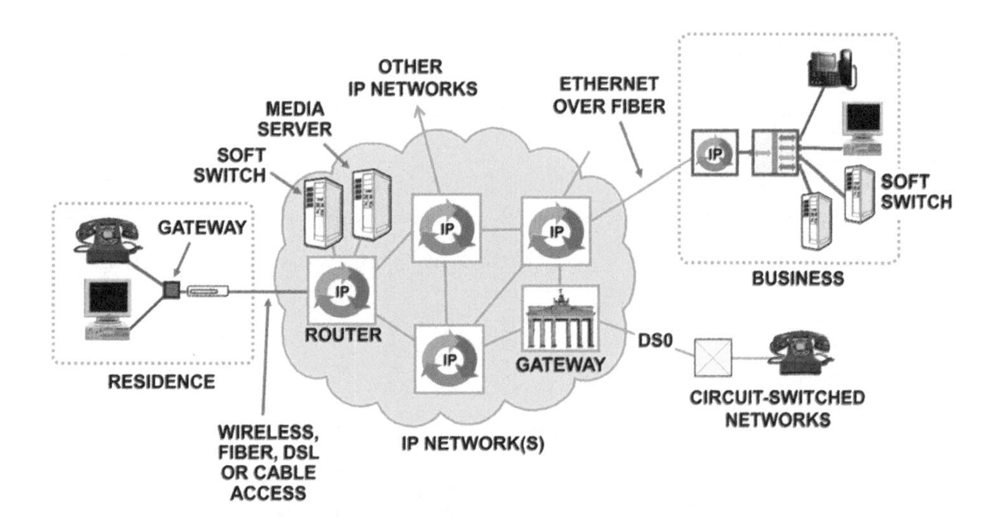

FIGURE 81 VOIP SYSTEM COMPONENTS

To begin understanding Voice over IP, we start with the top-level system architecture and principal components of an IP telecommunications network.

The principal components required for Voice over IP are terminals, soft switches, media servers and gateways. These are detailed in the following pages.

At the heart of Figure 81 is one or more IP networks, constructed of high-capacity point-to-point circuits connecting routers, to which users are provided access.

For residential customers, Passive Optical Networks are being deployed by telephone companies for fiber to the home. For many people, modems over the existing "last mile" copper infrastructure, either DSL on twisted pair or cable modems on Cable TV, is the technology for "high-speed" or broadband access to the IP network and will be for some time.

An adapter that performs analog-digital conversion, packetization and many of the functions of a PC is used to connect a regular POTS telephone to the IP network. This adapter can be a standalone device, or built into the same box containing the modem or fiber terminal.

Wireless access is also a common option, as all 4G and above cell phones communicate "voice minutes" using Voice over IP, as well as being able to communicate VoIP packets to and from the Internet via a "data plan".

Fiber is more common for government and business access, using IEEE 802.3 Optical Ethernet standards for the access circuit.

Devices are connected within the building using Layer 2 (LAN) switches, and a customer-premise router will act as a point of control for traffic between the customer's in-building network and the IP telecommunications network.

8.2 Business VoIP Phones

IP telephones or *terminals* are computers with telephone functions. They may be dedicated-purpose devices that look very much like traditional electronic business telephone sets, or they may be a software application running on a PC, or something in-between.

The "computer" functions of a terminal include LAN and IP functions. At the IP level, the terminal requires software implementing IP and must be able to run the Dynamic Host Configuration Protocol (DHCP) to be assigned an IP address.

The terminal also has to support Ethernet LAN protocols. This means that an IP phone has to have at least one Ethernet jack on it to accept a Category 5, 5e or 6 patch cable to connect to a wall jack and circuitry to handle frames, MAC addresses and other LAN-related functions.

Many IP phones have a second LAN jack that would allow a user to plug their PC into the phone. This would allow the provisioning of only one LAN access cable to the work area, and facilitates the use of power on the LAN cable known as Power over Ethernet (PoE).

FIGURE 82 BUSINESS VOIP PHONES

The "telephone" functions include the microphone and speaker to convert between sound pressure waves and voltage analogs thereof, and analog-digital conversion.

Software implementing protocols required for VoIP must also be present, including a coder/decoder to format the digitized speech into segments, and the RTP and UDP protocols for transmission and interaction with the IP software. Signaling required for call setup and teardown using standards such as SIP is also required.

Terminals may include many optional features such as displays, keyboards, conference bridges, selectable coding rates, local call processing and more. As a minimum, a terminal must support the G.711 standard codec (64 kb/s) and call setup using SIP.

If voice Quality of Service (QoS) is implemented using one of a number of techniques, the terminal may be responsible for tagging packets or frames containing segments of coded speech – setting some bits in a protocol header – to indicate the content of the packet to network devices so that the packets can receive the appropriate priority handling in the network.

8.3 Voice in IP Packets

Figure 83 provides a very high-level block diagram view of the processes involved in actually communicating speech in IP packets from one person to another.

Starting on the left, commands from the speaking person's brain cause the lungs, diaphragm, vocal cords, tongue, jaw and lips to work together to form sounds.

A microphone is positioned in front of the mouth and acts as a transducer, creating a fluctuating voltage which is an *analog* or representation of the sound pressure waves coming out of the speaker's throat.

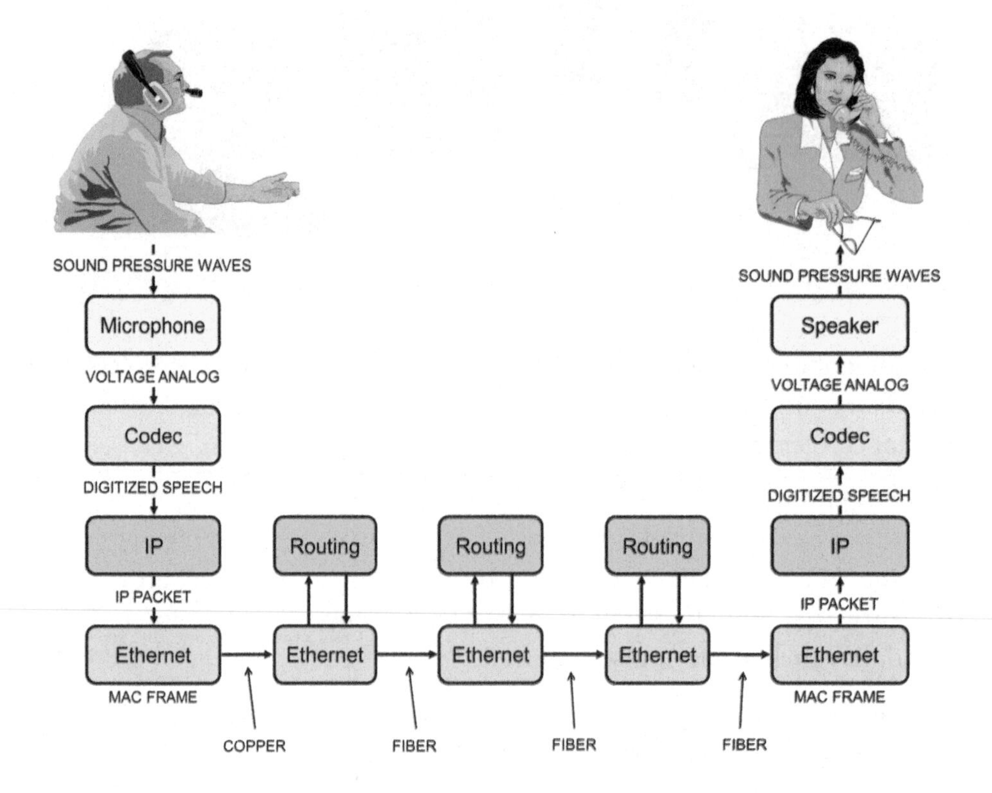

FIGURE 83 VOICE IN IP PACKETS

This voltage analog is fed into a codec, which digitizes the speech then codes (compresses) using a standard technique.

Approximately 20 milliseconds (ms) of coded speech is taken as a segment and has a time stamp, sequence number and error detection mechanism added. This is placed or *encapsulated* as the payload inside an IP packet. The IP address of the source and destination is added in an IP header before the payload.

IP packets are carried from the source to the destination over a sequence of links. The links are connected with routers, which relay the packets from one link to the next.

Link control functions such as access control and framing are performed using the IEEE 802.3 Ethernet and 802.2 MAC standards. At the lowest level, the links are implemented with physical media such as Category 6 cable or Optical Ethernet fiber. They might also be implemented wirelessly.

At the destination, the bits are extracted from the IP packet and fed into a codec, which re-creates the analog voltage.

This in turn is fed into a speaker which re-creates the sound pressure waves, which travel down the ear canal to the inner ear, where small hairs vibrate and cause neurons to fire, resulting in information being processed by the listener's brain.

8.4 Call Managers / Soft Switches / SIP Servers

A *soft switch* is a set of functions implemented on a server. Cisco's term for a softswitch is a Call Manager, which is a more accurate name.

A *switch*, in its simplest form, is a device that causes communications to happen from one point to one other particular point, often when there are multiple "other" points to choose from.

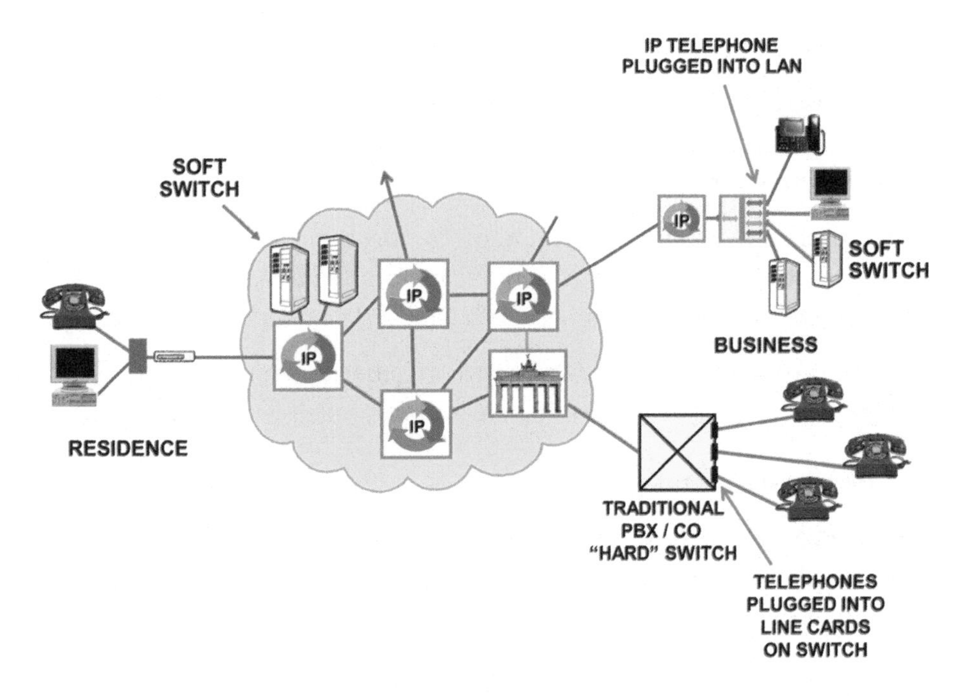

FIGURE 84 SOFTSWITCH / CALL MANAGER / CALL SERVER

As illustrated in Figure 84, a traditional Central Office (CO) telephone switch might be called a "hard" switch, since it has physical line cards that terminate loops and trunks. The switching software running on the computer that is the CO switch directs traffic to be moved between a line card and a trunk or between two line cards during a phone call.

The term *soft switch* is used to mean a computer running switching software that does not have telephone line cards – the communications are instead directed to the correct destination by routers routing packets, a software function.

The essential function of a soft switch is to enable the caller and called party to learn each other's IP address, so that they can subsequently directly exchange packets containing digitized voice, implementing a phone call.

An IP telephone is physically connected to the network with a LAN cable plugged into a LAN switch, which is plugged into a router.

Some traditional line card functions, such as powering the telephone, are performed by the LAN switch. Other traditional line card functions like analog-digital conversion and busy signals are performed by the terminal.

Soft switches are deployed by carriers to provide network-based services, and by end-user business customers to do it themselves.

The Session Initiation Protocol (SIP) standard uses the terms *proxy* and *back-to-back user agent* to refer to the entity performing softswitch functions.

Terms used by product manufacturers for their products that might implement these and other functions include call manager, call server, VoIP switch, communication server, hosted PBX and cloud service.

Regardless of what it is called, there are in general two main functions that may be performed by a soft switch: terminal control and call control.

8.4.1 Terminal Control

Terminal control includes registration, admission and status. Registration means authenticating a telephone (or telephone client software) and associating the telephone and its IP address with a user in a directory.

Admission means controlling whether that telephone is permitted to make or receive calls. Status is keeping track of the current status of the telephone, client software and/or user as an input to processing call requests.

8.4.2 Call Control

Call control can involve a number of different functions, including address resolution, call routing, call signaling and call accounting.

Address resolution means determining the numeric IP address of the called telephone.

Call routing in an IP telephone network is the same as routing of any kind of packet: determining the IP address of the next hop, usually a router, based on the destination IP address and the available networks and services. In the network core, MPLS labels are used instead of IP addresses.

Call signaling includes initiating and terminating the phone call by exchanging control messages with a far-end device. This could include negotiation of the multimedia attributes and coding protocol for the communications. The soft switch may also use the signaling and other information to generate Call Detail Records as an input to a call accounting system.

In addition to these basic functions, specific products may include hundreds of other functions like 6-party conference, call transfer, call waiting, group hunt, ACD agent, malicious call trace, ring again and attendant recall.

8.5 Media Servers: Video Servers

Media means content: voice, video, text, graphics, music... and any other information coded and formatted for communication to a person.

Server means a computer that typically stores the content and serves it up to users (who have *client* software) over the network.

The two types of media server most common in the converged IP network world are video servers and integrated messaging servers.

Video servers are the source of the packetized video delivered to users. Video services include "basic cable", video on demand, pay per view and Personal Video Recorder (PVR) functionality.

8.5.1 Basic Cable and PVRs

Basic cable means delivering programming from companies like CBS, NBC, CBC and CNN. Personal Video Recorders (PVRs) are devices that record video on a hard drive and provide a user interface for selecting what to record and play back. Just as voice mail is a network-based service performing the functions of an answering machine, network-based PVR service allows users to "record" and replay programming with no customer-premise recording device.

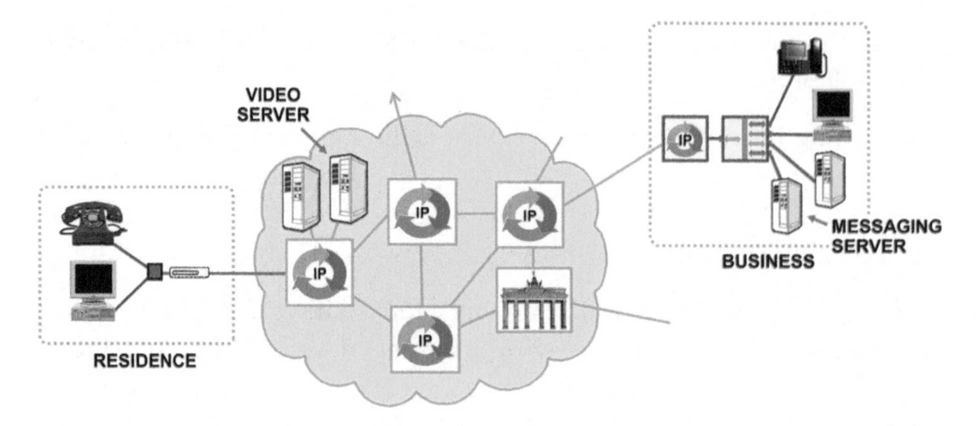

FIGURE 85 VIDEO SERVERS

8.5.2 Video on Demand

Video on demand means that a specific title plays for the user at a time of their choosing. This can take a number of forms. One is movie rentals – no more "old fashioned" video stores! Another is YouTube, where the content is free, supported by advertising. A third business model is Netflix, with video on demand from a library based on a monthly subscription. Pay-per-view could include live feed of boxing events or rock concerts. Anti-piracy and anti-replay technologies are implemented in some cases.

8.5.3 Content Delivery Networks

With the topic of video over IP comes the question of content delivery… to move huge volumes of IP packets with video to the customer.

In the beginning, content providers like YouTube paid top-level Internet service providers to carry the video traffic between YouTube servers, and from the YouTube servers to the customer's Internet Service Provider.

To save money and allow better management, YouTube built their own content delivery network for this purpose with a combination of private line fiber and carrier services.

The customers' ISPs still complained that they had to carry the video on their networks and the content provider was not paying for it. This argument is dubious since the customers are paying the ISP for that.

To lessen this problem, Netflix created a *Netflix appliance*, a video server that caches all of Netflix's content, and gives them to ISPs to deploy close to customers. This way, the ISP only has to get the video once from Netflix.

8.5.4 Integrated Messaging

An integrated messaging server is a computer that can record or receive, store and play voice, email, text and fax messages, and provide a unified interface to the user for all types of messages. A sophisticated system would allow a user to access, retrieve and reply to messages by any method; for example, a user could see a list of voice mail messages on what looks like a web page, and selectively listen to messages through their telephone under control of the web-style interface... or listen to messages through their computer's speaker... or read voicemail messages as text on the screen.

8.5.5 More Media Servers

The list does not end there. If one considers graphics and text to be media, then web servers are a type of media server. Music servers that provide streamed programming, songs streamed and played on demand, and songs to download to transfer to a portable device are media servers.

In the future, countless applications will employ media servers to receive, store and play content to users.

8.6 Gateways

Gateways are devices that convert from one set of communication standards to another. For voice, the two main sets of standards to convert are IP telecommunications and what might be called "traditional" telecommunications.

Traditional telecommunications codes speech into continuous 64 kb/s streams that are carried in DS0 transmission channels. The call signaling employs a set of messages called the ISDN User Part (ISUP) over ISDN PRI or SS7 connections.

In IP telecommunications, speech can be coded using any number of methods but is always carried in IP packets. Call signaling is mostly the exchange of call setup messages following the SIP standard.

The task of converting between sets of communications standards is divided into two main parts: media conversion and signaling conversion.

8.6.1 Media Gateways

Media conversion means converting between the format in which voice, video, fax and other types of content ("media") are represented on one system and the format used on another system.

For example, voice might be converted between the coding method, bit rate and IP packet format of a VoIP system and streaming 64 kb/s DS0 channel format. Media conversion could equally involve converting voice coded using one standard to a different standard, even though both are carried in IP packets.

Media gateways are processor-intensive, and need to be able to convert in real time without introducing any delay into voice conversations.

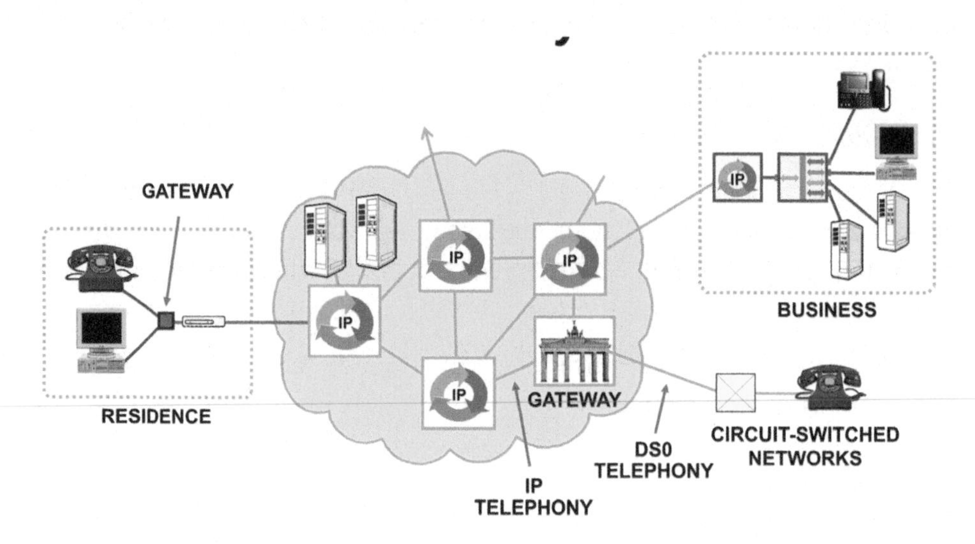

FIGURE 86 GATEWAYS

8.6.2 Signaling Gateways

Signaling conversion means translating between connection setup protocols used on a VoIP system and those used on other networks. For example, the signaling gateway might support SIP call setup messages on the VoIP side, and convert these to ISDN User Part (ISUP) call setup messages for the other side.

Another example of signaling conversion would be conversion between Lightweight Directory Access Protocol (LDAP) email server protocols and Post Office Protocol (POP) email server protocols as part of a unified messaging infrastructure.

Gateways are deployed in Central Offices by carriers to connect older DS0-based networks and CO switches to newer IP-based networks. Gateways can also be deployed at customer locations for the same purpose.

Gateways come in many forms and flavors depending on the equipment vendor implementation, the network application and the required network configuration. In some cases, the signaling conversion function will be physically separate from the media conversion function.

8.7 Voice over IP over LANs and WANs

Another critical part of VoIP is the physical network over which the IP packets are communicated. As illustrated in Figure 87, we differentiate between network infrastructure inside a building, and networks and services to connect buildings.

In-building, the physical network infrastructure is IEEE 802.3 standard Ethernet Local Area Network (LAN) technology. The main components of a LAN are Ethernet hardware interfaces, often integrated in a computer or telephone; software to implement the link control, physical cabling and Ethernet LAN switches.

Gigabit Ethernet (1,000 Mb/s) is becoming standard practice, as is TIA-568 Category 6 twisted pair cabling. The link control (standard 802.2) uses frames with Media Access Control (MAC) addressing to communicate between stations on the LAN. LAN switches provide low-latency connections between stations, and can also provide power on the LAN cable from a wiring closet to power terminals via the LAN.

The network and services used to connect buildings are often referred to as a Wide Area Network (WAN). WAN services from carriers have two main components: access and network connection.

The access is a (hopefully short) physical connection from the customer's building to the carrier's physical network. Technologies including fiber, DSL, cable modems, T1, and wireless are used for this "last mile".

The network connection is a service provided by the network to relay IP packets from one access circuit to another. In the past, technologies like Frame Relay and ATM have been used for this purpose. Now, "native" IP services, that handle IP packets directly, are the norm.

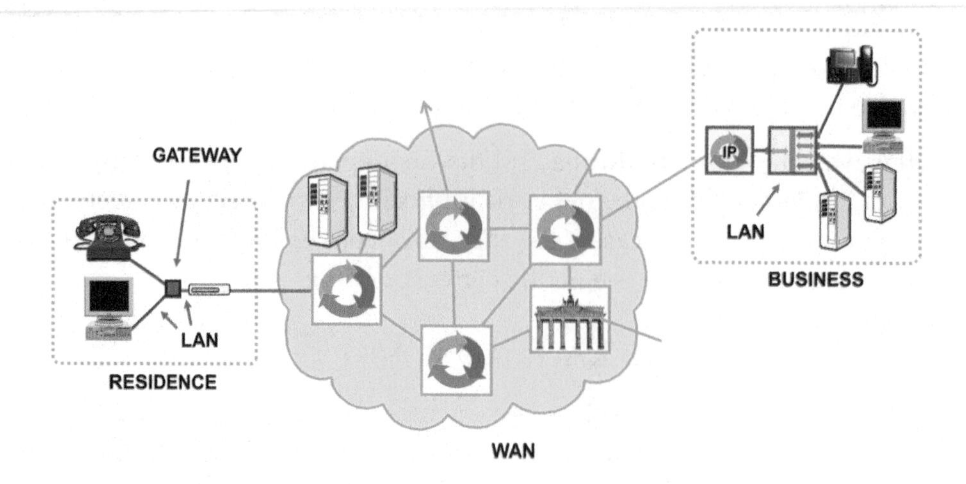

FIGURE 87 VOICE OVER IP OVER LANS AND WANS

One example of a native IP network service is the Internet. However, there are no performance guarantees on the Internet – no guarantee that a packet will be delivered, when or how often that might happen.

IP network services from carriers with Service Level Agreements (SLAs) guaranteeing a Class of Service (CoS), which is transmission characteristics like packet loss, delay and jitter are available.

For connections between an organization's locations, these are often called MPLS services, as MPLS is the underlying technology used to implement the prioritization to realize the Class of Service.

SIP trunking is a native IP service that comes with a CoS suitable for voice, connecting an organization's locations plus a gateway to the PSTN.

8.8 Key VoIP Standards

VoIP technology brings together telecommunications protocols standardized by the International Telecommunications Union (ITU), protocols related to IP defined by the Internet Engineering Task Force (IETF), standards for LANs published by the Institute for Electrical and Electronic Engineers (IEEE) and standards for cabling published by the Telecommunications Industry Association (TIA).

A number of the key functions, including the Session Initiation Protocol (SIP) for call setup and management, Session Description Protocol (SDP) for call characteristics, The Real-Time Transport Protocol (RTP) to re-synchronize segments of speech received in packets at the far end, the User

Datagram Protocol (UDP) for sequencing the segments and directing them to the correct "port" on the far end, and of course the Internet Protocol (IP) are defined in IETF Request for Comment (RFC) documents.

The ITU Telecommunications Standardization Sector (ITU-T) developed the G.700 series of standards for coding voice, as well as H.323, an older umbrella standard for multimedia over IP that is not much used anymore.

VoIP terminals and other system components are connected to LANs, defined in the 802 series of standards by the IEEE, which define the MAC addresses and how frames are communicated between stations.

802.3 Ethernet LAN technology: 10BASE-T, 100BASE-T and Gigabit Ethernet run on cables certified to conform to Category 5, 5e or 6 as defined by the TIA. On copper cables, Power over Ethernet (PoE) can be used to power telephones.

Ethernet on fiber: Optical Ethernet, and Ethernet over radio: 802.11 Wi-Fi are widely-used standards.

8.9 Where All of This is Headed: Broadband IP Dial Tone

Where is all of this headed? Broadband IP dial tone.

Dial tone means you have the ability to transmit and receive IP packets to and from any other point on the planet.

Compare this to traditional "dial tone": the ability to transmit and receive signals with the 3 kHz voiceband to and from any other point on the planet.

Broadband means at a high enough bit rate to watch YouTube videos of guilty dogs in 4K definition. The IP packets can of course contain any content: a phone call, videos, web pages.

Just as with traditional telephone service, your ISP can provide the switching service and value-added services like voicemail and email and gateways to other networks – or it can be provided by a competitive carrier, or you can do it yourself. All possibilities exist.

Changing from voiceband POTS to IP packet communication as the basic network service results in the separation of the basic communication function vs. the call control function and value added-services.

This greatly enhances the possibility of competition for providing the control and value-added aspects: the softswitch function, voicemail, video content and more.

To implement network quality guarantees, called Classes of Service, suitable for live phone calls and streaming video, *MPLS* is used to manage IP traffic in the core of a carrier's network.

IP packet communications has been called "the Internet" up to the present, a different product than telephone service.

As broadband Internet access becomes ubiquitous, whether wireless or via copper or fiber, it will become the only kind of telecommunications service most people pay for. Most people will not pay for a second service called POTS or "voice minutes". They will have a "data-only" wireless plan.

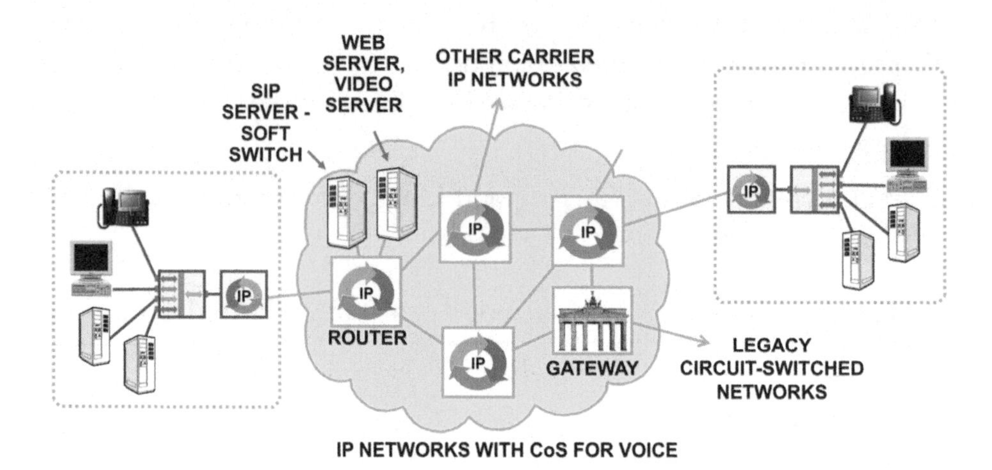

FIGURE 88 BROADBAND IP DIAL TONE

Once there is a widely-implemented standard way of setting up Internet phone calls, a person using Skype will be able to call a person using Google Hangouts or any other client software. The "number" they dial might be a ten-digit phone number, or it might look like an email address.

At this point, the differentiation between the Internet and the telephone network will disappear

"Telephone service" becomes a helper application to allow the caller to know the called party's IP address is, so they can send them voice packets.

The "Web" becomes a helper application to allow a client to find out what the address of a server is, so they can send data packets, receive packets with video or web pages.

All packets are carried over the Internet, which has absorbed the PSTN and become the IP Packet-Switched Telecommunications Network (IP-PSTN).

9 Wireless

9.1 Radio

Wireless generally means the use of radio, which is electromagnetic waves at frequencies measured in Gigahertz (GHz), that is, vibrating 10^9 or a billion times per second.

Wireless could also mean using electromagnetic energy vibrating on the order of 10^{14}, hundreds of trillion times per second (this is called light); but one of the problems in wireless communications is obstacles.

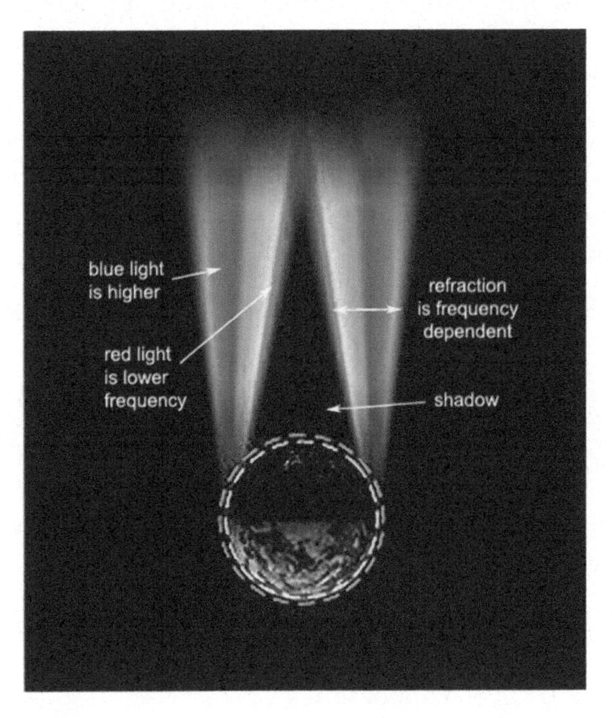

FIGURE 89 LOWER FREQUENCIES RESULT IN SHORTER SHADOWS

It turns out that the higher the frequency, the longer distance it takes for energy to *refract* or bend around an object. Light does refract around objects – this is how we can tell there are planets around other suns – but the length of the shadowed area behind the object is too long for use on a terrestrial scale.

If the frequency of the energy is reduced, the length of the shadow behind an obstacle shortens. In addition, lower-frequency energy can penetrate through objects like walls and clouds more easily (the reason why fog horns are very low frequency).

For these reasons, wireless communication systems tend to use energy at Gigahertz frequencies, two or three hundred thousand times lower than light, and call it *radio*. Accordingly, this chapter is on communications centered at Gigahertz frequencies, in frequency bands with widths measured in Megahertz (MHz), which is 10^6 or millions of Hertz wide.

Radio is used in many different kinds of systems with different applications, including everything from demagogues broadcasting angry rants on talk radio shows using analog AM, to mobile cellular systems for telephone calls, messaging, and mobile Internet, *trunked radio* for police communications, fixed wireless to remote residences, short-range wireless Wi-Fi LANs, geosynchronous communication satellites, Low Earth Orbit satellites and more.

Video and music broadcast, speech communications and point-to-point microwave links were the biggest applications for radio in the past. Mobile phone and Internet is a big business in the present. In the future, wireless will be ubiquitous.

To represent speech on broadcast radio, a single pure frequency called a *carrier frequency* is chosen, and the amplitude (volume) of the carrier frequency is varied in a continuous fashion as an analog of the sound coming out of the speaker's mouth. The frequency of the carrier could also be varied continuously as an analog of the sound.

These are called Amplitude Modulation (AM) and Frequency Modulation (FM) respectively.

Representing 1s and 0s is a more complex task. Since radio bands do not include zero Hertz, sometimes called DC, pulses cannot be used to represent 1s and 0s as on copper wires.

Instead, it is necessary to use techniques similar to those used in telephone line modems to represent the 1s and 0s, such as shifting back and forth between specific amplitudes, shifting between different frequencies, shifting between different phases, or combinations thereof.

9.2 Spectrum

9.2.1 The Need for Regulation

Every country has the sovereign right to manage energy at radio frequencies in its territory. When the airspace through which the radio waves travel is public property (which is most of the time), and there is contention for its use (which is all of the time), regulation is required to allow the rational use of the shared resource.

In the United States, the Federal Communications Commission (FCC) regulates the wireless spectrum for use by anyone other than the Federal government. Industry Canada, which includes the former Department of Communications, regulates the airwaves in Canada.

Canada, Mexico and the USA coordinate radio frequency usage due to geographic proximity. Likewise, the International Telecommunication Union (ITU) in Geneva facilitates coordination between other countries, particularly outside of North America.

9.2.2 Radio Spectrum

The range of radio frequencies, called the radio *spectrum,* is divided into blocks of frequencies allocated for different services. *Allocations* are divided into *allotments,* which are bands of frequencies assigned to specific users or to the public.

A *license* to emit radio-frequency energy at the specified frequencies in a specified area is issued by government to record the assignment of the allotment and the conditions for its use.

Allotments for television broadcast, also called Over the Air (OTA), were organized as frequency channels, each notionally 6 MHz wide.

AM Radio	0.5 – 1.7	MHz
CB Radio	26 – 27	MHz
TV Channels 2-6	54 – 88	MHz
FM Radio	88 – 108	MHz
TV Channels 7-13	174 – 216	MHz
Maritime	457	MHz
600 MHz band	614 – 698	MHz
700 MHz band	698 – 806	MHz
800 MHz Cellular NA (1G →)	806 – 894	MHz
900 MHz ISM Band NA	902 – 928	MHz
900 MHz GSM Band	880 – 960	MHz
1800 MHz GSM Band	1710 – 1880	MHz
1900 PCS NA (2G →)	1850 – 1990	MHz
4G, 5G: many between	1800 – 2400	MHz
2.4 GHz ISM Band NA Wi-Fi, Bluetooth, phones microwave ovens	2.4	GHz
2.5 GHz Mobile & Fixed	2500 – 2690	MHz
3.5 GHz Fixed Wireless	3550 – 3700	MHz
3.7 GHz C-Band for 5G	3700 – 4000	MHz
5 GHz U-NII bands NA Wi-Fi, point-to-point	5.15 – 5.925 5.925 – 7.125	GHz
Millimeter-wave 5G	28, 30, 47	GHz

FIGURE 90 SPECTRUM AND BANDS

9.2.3 Capacity vs. Performance Tradeoff

It turns out that the capacity, the widths of the bands, is very good at high frequencies. However, transmission characteristics, like penetration through walls, are not so good. Lower frequencies experience much better transmission characteristics but have lower capacity.

The sweet spot for good capacity with long range and good in-building penetration appears to be within the 1800 MHz – 3000 MHz range with current technologies.

9.2.4 Two-Way Radio: FDD or TDD

Television is one-way communications. Phone calls and Internet access of course require two-way simultaneous communications, sometimes called *full duplex*.

Using different frequencies for the base station's transmitter and the mobile's transmitter to allow two-way simultaneous communications is called *Frequency-Division Duplexing* (FDD).

The alternative is to assign the same frequency to the base station and mobile and have them alternate transmitting in time, called Time-Division Duplexing (TDD), also known as half-duplex.

9.2.5 Frequency Bands

Figure 90 lists popular bands currently in use, beginning with old-school analog AM radio, and CB radio.

Analog TV channels 2-13, interrupted by an allocation for analog FM radio, follow, beginning at 54 MHz. Channels 2-13 were called Very High Frequency or *VHF channels*. Channels 14-69 were Ultra-High Frequency or *UHF channels*.

UHF channels are now repurposed to mobile communications, beginning with channel 38 and the 600 MHz band.

9.2.6 600 MHz Band

The band 614 – 698 MHz, formerly UHF channels 38-51, is now referred to as "The 600 MHz band".

The 600 MHz band was re-allocated to create seven paired 5 MHz blocks: 5 MHz for the base station for downloading, and 5 MHz for the handsets for uploading, abbreviated as 5+5.

5 MHz is relatively low capacity; not enough to provide broadband to many people at the same time.

However, radio in the 600 MHz band enjoys transmission characteristics like range and in-building penetration that are far superior to systems at higher frequencies.

This makes the 600 MHz spectrum ideal for Internet of Things (IoT) applications like trackers and monitors, where the data rate requirement is low and wide coverage and excellent in-building penetration is desired.

9.2.7 700 MHz Band

The band 698 – 806 MHz is "The 700 MHz band".

After being repurposed from UHF channels 52 – 69, the 700 MHz band was organized into the Lower 700 MHz band and the Upper 700 MHz band.

The Lower 700 MHz band was reallocated to 6 MHz blocks for mobile communications. The Upper 700 MHz band was reallocated to blocks from 1+1 to 11+11 MHz wide, plus a 12+12 MHz allocation in for Public Safety.

The 9/11 terrorist attack on the US revealed room for significant improvement in communications. The NYC police private radio system could not communicate with the fire department or the FBI. Even calling each other on cellphones was difficult due to demand overload.

As a major upgrade, the US congress funded FirstNet, a nationwide network exclusively for First Responder and Public Safety (i.e., Government) use. It was built by the FirstNet Authority, who started with $7 billion in funding from the US Treasury in February 2012.

9.2.8 800, 900, 1800 and 1900 MHz bands

The 800-MHz band was initially allocated for the first generation of mobile cellular in North America, ending up with two 10+10 blocks plus two 5+5 blocks within 824-894 MHz.

In the rest of the world (called "Europe" in the telecom business), the 900-MHz and 1800-MHz bands were allocated for second generation mobile called GSM: The Global System for Mobile Communications.

In North America, the 1900 MHz, or 1.9 GHz band, which actually encompasses 1850 MHz to 1990 MHz, was initially allocated for the second generation of cellular, called Personal Communications Services as depicted in Figure 91 with 15+15, 10+10 and 5+5 blocks.

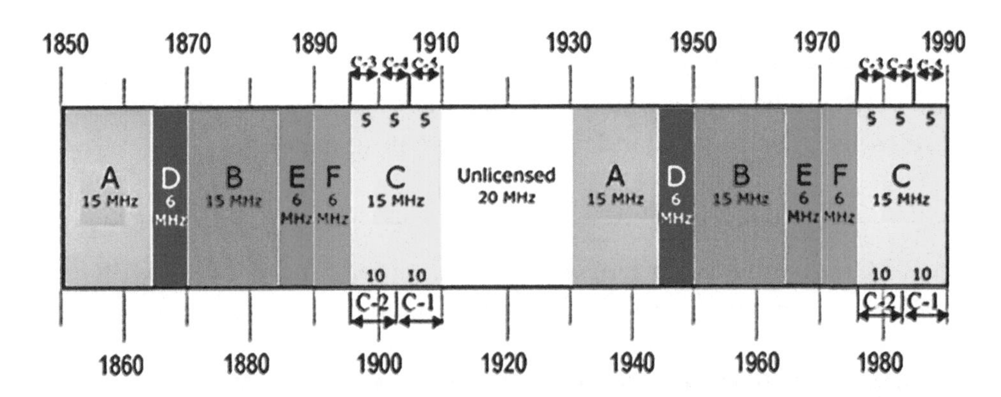

FIGURE 91 1.9 GHZ "PCS BAND"

New bands are allocated, and older bands are re-allocated to new generations of systems on an ongoing basis. Bands allocated for 3rd-generation CDMA and 4th-generation LTE will run 5G in the future.

9.2.9 2.4 GHz and 5 GHz Unlicensed Bands

The 900-MHz band is an *unlicensed band* in North America, which essentially means it is allocated to the public, and a license to emit energy at that frequency is not required. These bands are also called Industrial, Scientific, and Medical (ISM) bands. The 900-MHz band was used for cordless phones and baby monitors in North America.

FRIGIDAIRE
HOUSEHOLD MICROWAVE OVEN
ELECTROLUX HOME PRODUCTS
INC CHARLOTTE, NC 28262
UL LISTED 54NJ

MODEL NO.	SERIAL NO.	MANUFACTURED
FFMV1645TB	KG82001262	May, 2018
Vac/Hz: 120/60 Rated Input: 1550W Rated Output: 1000W Microwave Frequency: 2450MHz	For Customer Assistance Call Support at 1-800-374-4432 www.frigidaire.com	FCC ID : VG8EM044KYY MADE IN CHINA

THIS PRODUCT COMPLIES WITH DHHS RULES 21 CFR SUBCHAPTER J. Not for built-in installation.
PN:16070000B08767

FIGURE 92 MICROWAVE OVEN OPERATING IN THE 2.4 GHZ ISM BAND

At 2.4 GHz is another ISM band. This band is used by numerous technologies including Bluetooth, Wi-Fi, cordless phones, private point-to-point transmissions and… microwave ovens, to avoid the need for a license. This makes the 2.4 GHz band noisy.

Bands starting at 5 GHz are part of the Unlicensed National Information Infrastructure (U-NII). U-NII bands between 5 and 6 GHz are used by 802.11a Wireless LANs, and between 6 and 7 GHz are used by 802.11ax.

Energy at 5 and 6 GHz encounters significant impairment with transmission through walls and other obstructions. Unobstructed line-of-sight between the transmitter and receiver is necessary to achieve the peak bit rates specified in the standards and marketing materials.

9.2.10 2.5 GHz Band

The 2.5 GHz band has a promising future in broadband wireless, with two 50 MHz-wide channels and one 17.5 MHz channel available for mobile and fixed applications. The channel width has increased from 10 and 15 MHz in the 800 and 1900 MHz bands to 50 MHz, dramatically increasing the capacity available.

9.2.11 3.5 GHz Band

New spectrum from 3550 to 3700 MHz is beginning to be used in North America for Fixed Broadband Wireless services, providing broadband wireless Internet to residences and businesses as an alternative to cable, DSL, fiber or a cellular data plan.

This service is provided from cell towers, and currently uses LTE, but it is not accessible by mobile phones.

The 150 MHz of spectrum is divided into 10 MHz blocks. In the USA, licensees can aggregate up to four blocks to create 40 MHz channels.

However, as frequency increases, transmission characteristics like penetration through obstacles worsens. 3.5 GHz energy is significantly blocked by near-field interference, usually trees in front of the antenna.

Line of sight between the tower and the antenna at the customer premise is required to achieve broadband bit rates.

9.2.12 3.7 GHz C-Band 5G

The 3.7 GHz band (3700 – 4200 MHz) was allocated to Fixed Satellite Service (FSS) space-to-Earth communications, and Fixed Services (FS) using 20+20 MHz for point-to-point long-haul analog radio telephone trunk carrier systems.

Paired with the 5.925-6.425 GHz band (Earth-to-space), these frequencies are known as the Conventional or C-band in the satellite business.

By repacking the incumbent license holders in the 3.7-4.2 GHz band into the upper 200 MHz of the band, 280 MHz of spectrum between 3.7 and 4

GHz has been cleared and reallocated to mobile and fixed communications for broadband terrestrial 5G applications.

This portion of the C-band is now allocated for mobile and fixed communications (5G) in 20 MHz blocks, which can be aggregated up to 100 MHz channels.

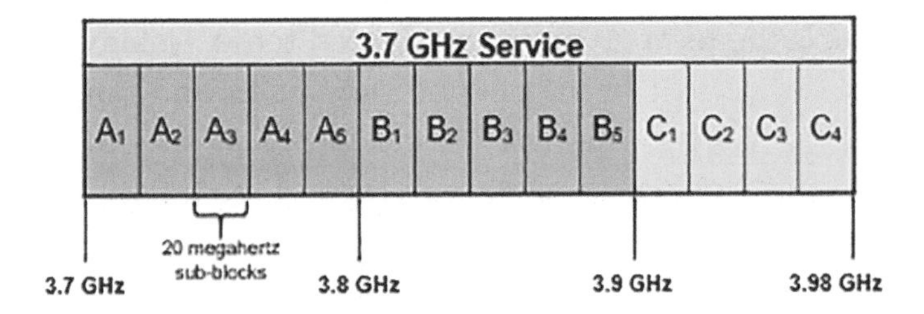

FIGURE 93 3.7 GHZ C-BAND FOR 5G

Interference with radar altimeter systems was a significant cause for concern in the weeks before the rollout of 5G in the C-band.

Not only would the 5G licensees and the altimeters interfere with each other, it is illegal for aircraft to emit radio at those frequencies in the USA.

A number of foreign airlines, who claimed they didn't hear about the reallocation, were still using the now-cleared 3.7 – 4 GHz band, and urgently requested 5G rollout at airports be stopped. It wasn't.

Such aircraft cannot use their radar altimeters in the USA until they are brought into conformance with FCC regulations, preventing landing in zero-visibility conditions. The carriers had to upgrade their altimeters.

9.2.13 Millimeter-Wave Bands

The FCC has completed spectrum auctions in the 24 GHz band, the 28 GHz band, and the upper 37 GHz, 39 GHz, and 47 GHz bands, representing almost 5,000 MHz (5 GHz) of spectrum for 5G… 200 times more than the bandwidth that was allocated for 1G.

The bands around 30 GHz are called *millimeter-wave* bands, since the wavelength of the energy is measured in millimeters. At these frequencies, molecules in the air block energy, a phenomenon called *atmospheric absorption,* which limits the useful range to hundreds of meters.

One application for 5G is data communications for assisted-driving cars, where a central controller assembles cars on the highway into convoys or *platoons* to increase density and reduce energy usage.

Since line-of-sight is required and atmospheric absorption limits range to hundreds of meters, full coverage on a highway at 30 GHz would require radio base stations installed on every second or third streetlight all the way along a highway.

9.3 Mobile Network Components and Operation

Mobile network is the term given to distributed radio systems designed so that many users who may or may not be moving around can share a radio band and communicate amongst themselves and to other people or computers on wired networks like the PSTN and the Internet.

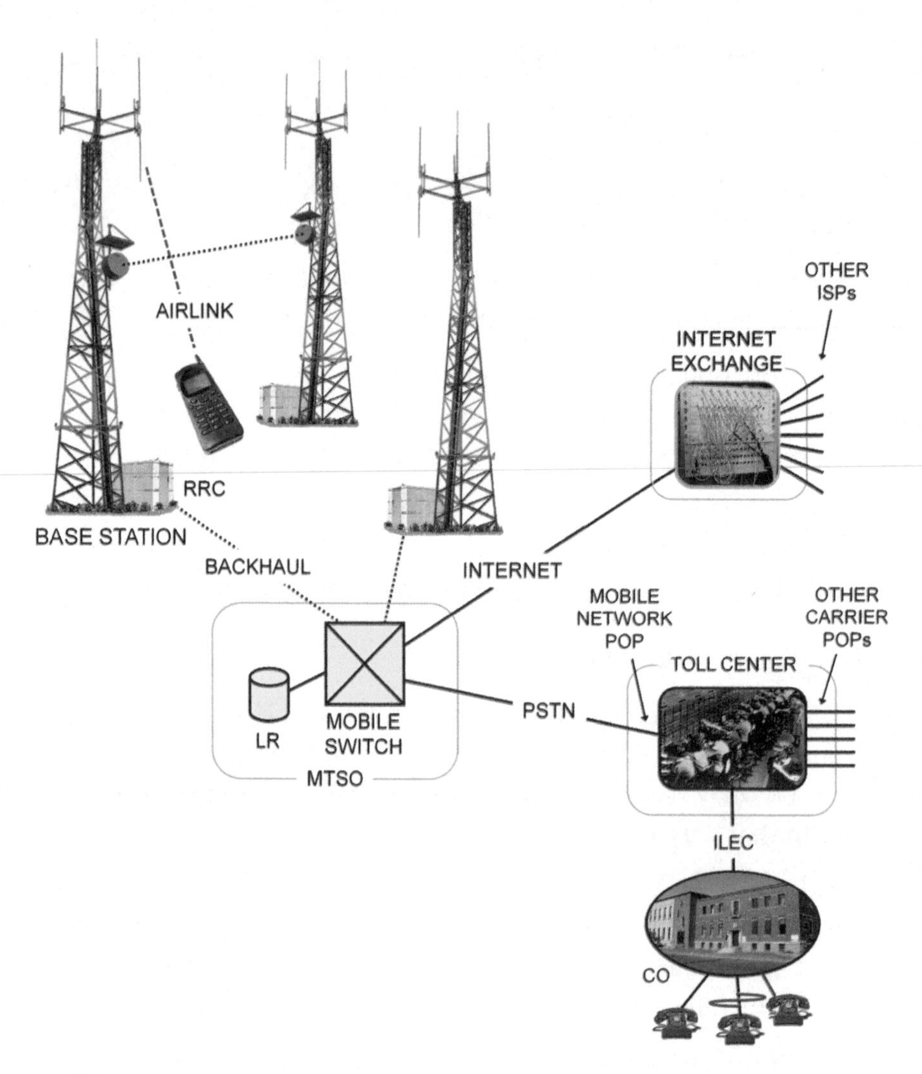

FIGURE 94 MOBILE NETWORK COMPONENTS AND OPERATION

9.3.1 0G: The Mobile Phone System

The first kind of radio systems connected to the PSTN were called *MPS*: the Mobile Phone System. They employed base stations in large metropolitan areas and automobiles fitted with big whip antennas to communicate to base stations in large metropolitan areas.

The caller had to call a "mobile operator" and ask for a particular "mobile number", and would (maybe) be patched through.

The geographical areas where service was available – the *coverage* – was very limited.

There was very little *capacity* – not many people could use the system at the same time.

And ironically, it did not support mobility: once the call was connected; if the person with the mobile radio drove too far away from the base station, the call would be dropped.

9.3.2 Mobility

The definition of *mobility* is having the ability to start a communication session using a terminal communicating with a particular antenna, then move away from the antenna and not lose communications, but rather be handed off to another antenna.

9.3.3 Base Station, Cell, Airlink, Handset and SIM

The cellphone, also called a *mobile,* terminal, smartphone or *handset* contains the components necessary for communication: microphone, speaker, codec, screen and radio and others.

The phone also contains a *Subscriber Identity Module (SIM)* card, with the *International Mobile Subscriber Identity,* one half of an encryption key pair used to authenticate the SIM data, and the *Location Area Identity,* a code from a nearby base station representing the current location of the device.

The phone is connected to the network via an *airlink* to a *base station,* called an evolved NodeB (eNB) in LTE.

A base station includes the Radio Resource Controller (RRC) that schedules communications between devices and the network, allocates capacity, controls the signal power used to communicate, negotiates the power state of each device and more.

The base station enclosure also contains racks of radio transceivers, which produce energy at radio frequency in the form of electricity. The transceivers are connected with thick coaxial cables to antennas that convert the electricity to electromagnetic waves at the same frequencies. A physical support for the antennas such as a tower or building is another element of a base station.

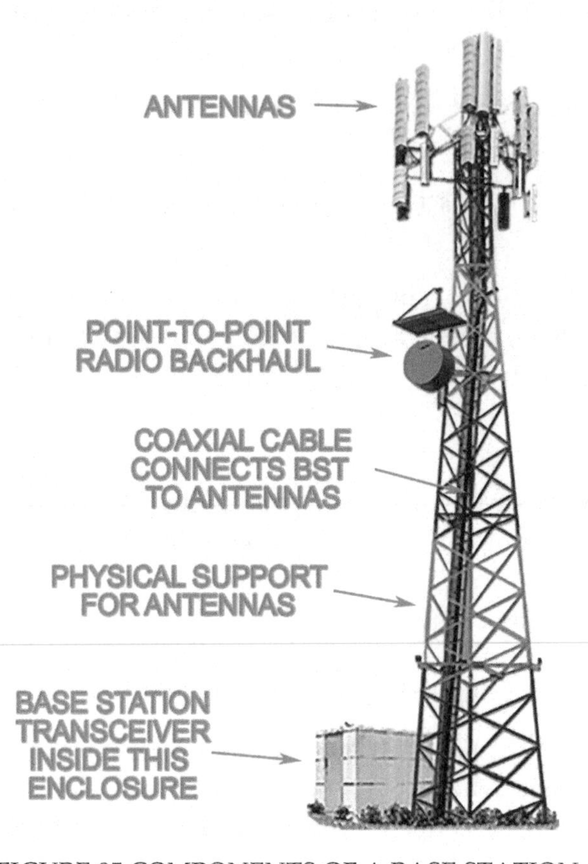

FIGURE 95 COMPONENTS OF A BASE STATION

The area on the ground covered by the base station is a *cell*.

9.3.4 Mobile Switch

The base stations are connected to a mobile switch, housed in a building called a Mobile Telephone Switching Office (MTSO).

In the MTSO, routers direct Internet traffic to an ISP, or to a specific content provider. The routers direct phone calls to a *mobile switch*.

A mobile switch is a telephone switch with additional capability to keep track of users moving between base stations.

In the mobile switch is a database called the *Location Register* (LR). The Location Register is used to keep track of users and where they are (or last were) via radio control channels, *Electronic Serial Numbers* (ESNs) and *Subscriber Information Module* (SIM) cards.

9.3.5 Backhaul

The connection from the base station back to the mobile switch to connect to the network is called the *backhaul*.

In some cases, the base station will be connected to the mobile switch with fiber, particularly if the operator is an affiliate of the phone company, the ILEC.

In other cases, a point-to-point microwave link will be used to connect one base station to another that has a connection to the mobile switch. This method of backhaul is often used by their competitors.

Due to its relatively low capacity, it is difficult to offer broadband mobile Internet via cellular with microwave backhaul. 10 Gb/s or more fiber backhaul to a base station is required to support many broadband users.

The mobile switch is connected to the Public Switched Telephone Network (PSTN), to allow calls to landlines.

Separate connections are made to Internet Exchange (IX) buildings for connections to Tier-1 Internet Service Providers, peer ISPs and networks and to content delivery networks.

9.3.6 Registration and Paging

On power-up, the handset registers with the switch, which records the ID of the base station the handset is using in the LR. When the handset is moved, the value of the Location Area Identity will change, and the handset initiates a location update to the LR.

When there is an incoming call, the mobile switch will page the handset from the base station(s) corresponding to the Location Area Identity value.

If the handset does not answer the page, the network will resend the page on neighboring base stations in the area or in the extreme, all of the base stations on the network.

Once the handset answers the page and the user presses the "talk" button, voice communications take place over the airlink and through the backhaul from the base station to the mobile switch.

For a mobile-to-mobile call, the communications will be routed to a base station. For mobile to landline, the call will be routed to the PSTN.

9.3.7 Handoff

If a user moves during a call, at some point, the user will be *handed off* from one base station to another.

This means that the network will switch to using a different base station to communicate to the handset, and, depending on the technology employed, may involve changing the radio frequency of the handset.

The handoff implements *mobility* – the ability to maintain communications while traveling.

9.4 Cellular Principles

9.4.1 AMPS: The Advanced Mobile Phone System

To meet the requirements of coverage, capacity and mobility, *cellular* radio systems were deployed.

The first generation of cellular, the improvement on MPS, was called the Advanced Mobile Phone System (AMPS).

1G employed Frequency-Division Multiplexing, where the allotment of radio frequencies is divided into narrow radio channels. Since multiple users could access the system, it was referred to as Frequency-Division Multiple Access (FDMA).

Radio frequency bands or spectrum was allocated for this service by the federal government between 800 and 900 MHz in North America, and above 900 MHz in many other countries.

In North America, an allotment was given to an affiliate of the incumbent telephone company in each market (which in a wonderful piece of jargon was called the *wireline cellular*) and an allotment was given to a competitor.

An operator would need a real estate department to find locations where they could construct the base stations, which included, in the first generation, large unsightly towers to support the antennas.

9.4.2 Cells

An operator would divide their block of frequencies into seven groups of frequencies, then at a base station, tune the base station transceiver to use one of the seven groups of frequencies.

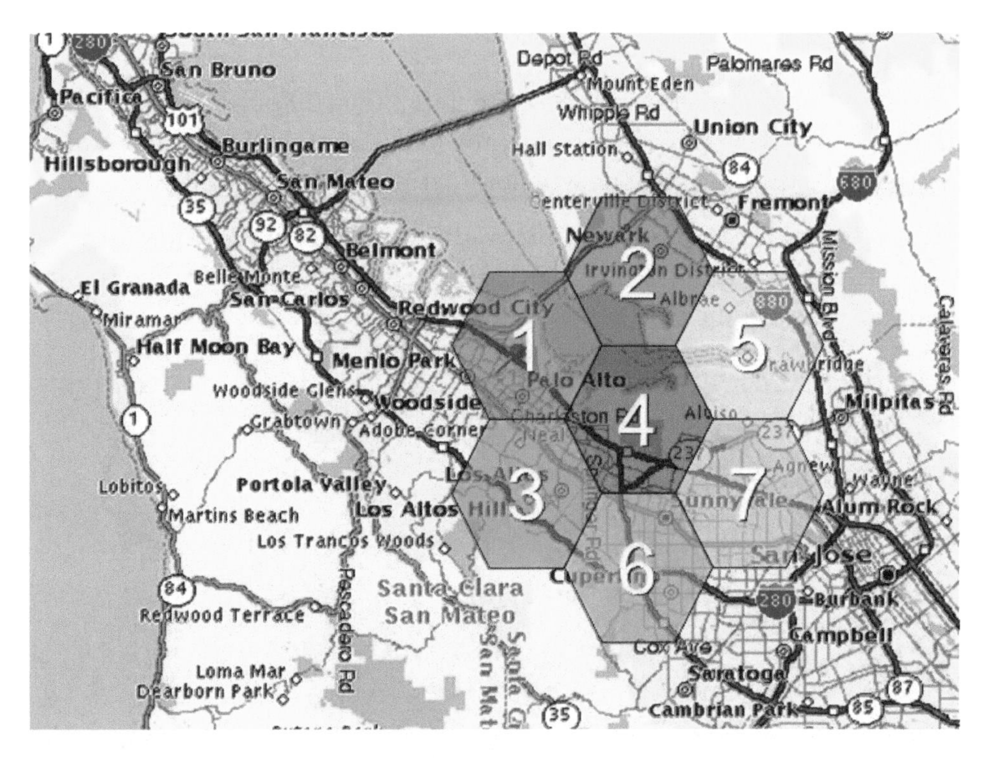

FIGURE 96 RADIO COVERAGE IN CELLS

The radio coverage area around the tower would be something like 3 miles or 5 kilometers in radius. On the ground, this is the cell, for example the area labeled "1" in Figure 96 centered on Menlo Park, California.

Then the operator would find another location six miles or ten kilometers away, and build a second base station, using a second group of frequencies, for example, the cell labeled "2" in Fremont across the Bay.

This pattern is continued to build seven base stations, using all seven groups within the allotment.

At that point, the operator would have coverage in the geographical area illustrated in Figure 96, all the way over to Cupertino CA where Apple is headquartered, and all of their allotted spectrum would be used.

9.4.3 Frequency Re-Use

Then, the operator could find another geographic location, for example, near Woodside, where Neil Young lives on a 1500-acre ranch in the middle of some of the nicest real estate on the planet, and convince Neil to let them build an eighth tower and base station, where they could, in this example, re-use frequency group #7.

Since the second tower is in Woodside, more than 20 miles away, and the radios have a relatively short range, the same frequency group #7 can be re-used, and the base stations will not interfere with each other.

This is the key idea behind a cellular radio system: being able to re-use the same frequency groups over and over again in different geographic locations, to meet the coverage requirement.

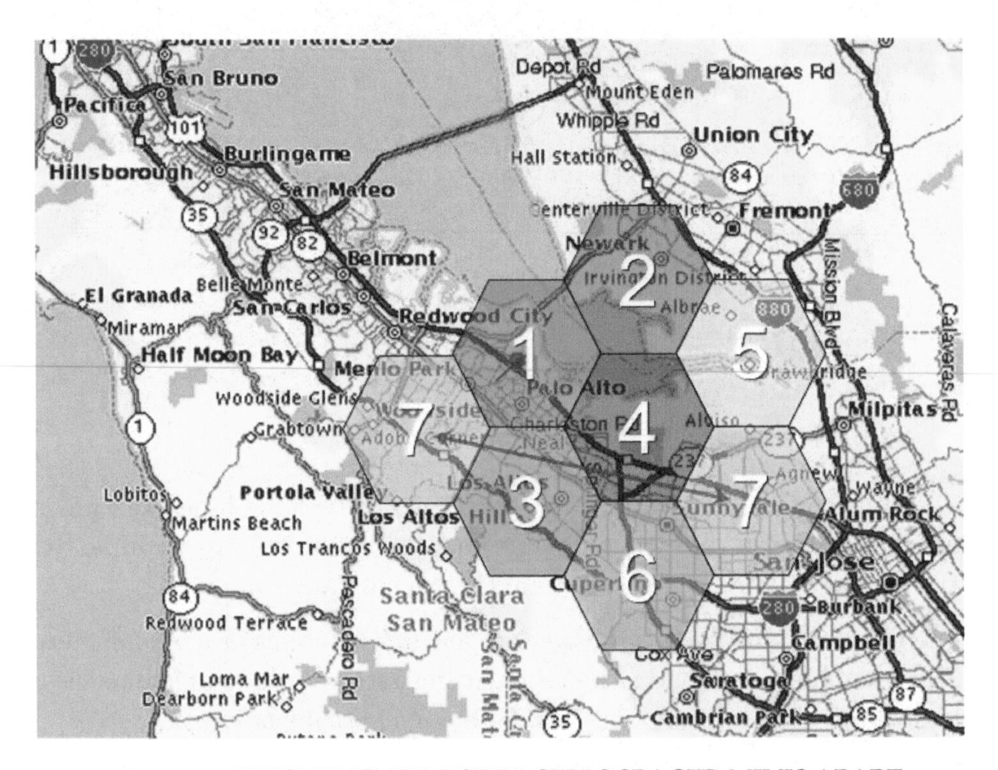

FIGURE 97 FREQUENCY RE-USE IN CELLS SPACED MILES APART

To implement mobility, when a user moves too far away from a base station, they have to be handed off to another base station without having their phone call dropped.

9.4.4 1G: Analog Frequency-Division Multiple Access

To meet the capacity requirement, in AMPS, the group of frequencies used in a cell was divided into 45 sets of 30 kHz radio channels, and each user in a cell is assigned a set of radio channels.

Voice or modem signals were represented on the radio channel using FM radio: continuous frequency modulation of the channel frequency. This analog technique resulted in middling- to poor-quality voice communications... and there was no encryption or coding of the voice, allowing eavesdropping with relatively simple equipment.

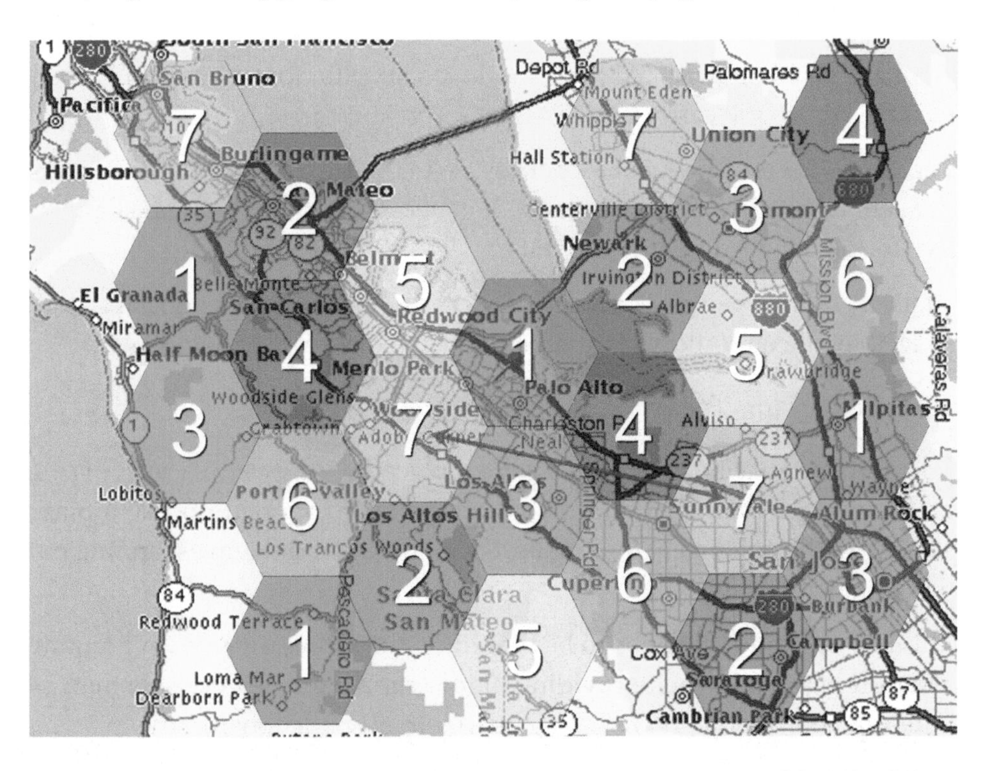

FIGURE 98 FREQUENCY RE-USE IN CELLS TO MEET THE COVERAGE OBJECTIVE

9.4.5 AMPS Handoffs and Dial-Up Modems

AMPS was not so good for data communications using a modem over the analog radio channel due to disconnect during handoff. When a user drove too far away from a base station, they had to be handed off to another.

Two things have to happen: since each cell employs different frequencies, the cellphone has to change which frequency it's operating at; and the network has to hand off the phone call from one base station to another.

This takes about 0.2 seconds, during which the communications will be interrupted or muted. This was a bit annoying for voice; it caused users' modems to disconnect.

Several strategies were attempted to keep the modems from hanging up, but in practice, it was necessary to "pull over" to send a mobile fax.

Unfortunately, even if a user is stationary, they can still change cells: if someone drives in the far side of their cell, they might be bumped to the next (overlapping) cell to make room for the newcomer.

9.4.6 AMPS Capacity

Another problem with AMPS was low capacity meaning a high per-channel cost to the carrier.

There are 45 sets of frequencies per cell, so in theory, there could be 45 users per cell. In practice, it's 40, because it is necessary to keep some channels free for people driving into the cell.

40 users, 3 miles radius… about 1.5 users per square mile… doesn't work so well in Silicon Valley, Manhattan or just about anywhere else.

9.4.7 Sectorization

To improve capacity, round cells were replaced with *sectors*. A sector is the area covered by an antenna that radiates energy in a pattern like a pizza slice, with the antenna at the smallest end and the radio waves fanning out within a narrow angle.

This allows many antennas to be placed on the same tower, each pointing at a different angle, each providing higher capacity to fewer users than an omnidirectional antenna radiating energy in all directions.

9.5 Second Generation: Digital

9.5.1 PCS and GSM

The second generation of cellular technology employed lower power, smaller cells and implemented digital communications.

It was in some cases referred to as Personal Communication Services (PCS), and in many places, the Global System for Mobile communications (GSM).

The advantage of implementing digital communications is better sound quality, better signaling and control capability, and mobile access to the Internet and other networks.

Second-generation cellular (2G) was initially deployed in North America on frequency bands centered around 1.9 GHz, whereas AMPS was deployed on frequency bands centered around 800 MHz.

Handsets were dual-mode, meaning they could support both AMPS at 800 MHz and PCS at 1.9 GHz.

Several different technologies were deployed by different carriers for spectrum-sharing for the second generation.

These included techniques called CDMA or Code Division Multiple Access and TDMA or Time Division Multiple Access (TDMA) in North America.

FIGURE 99 2G BRANDS

In the rest of the world, a 2G TDMA scheme called GSM, the Global System for Mobile communications was widely deployed.

Fundamentals of TDMA and CDMA are covered in upcoming sections.

9.6 PSTN Phone Calls using the Phone App: "Voice Minutes"

9.6.1 The Native Telephone App

One of the many apps a smartphone runs is the native telephone app, i.e., the app included with the Operating System (OS). It is the app that runs when the handset icon on a smartphone is tapped or swiped, and the one that "rings" when someone calls your mobile phone number.

This app routes phone calls to the Public Switched Telephone Network.

Other apps, like Whatsapp, route phone calls over the Internet.

9.6.2 Speech Digitized and Packetized in the Phone

The cellphone contains a microphone, which creates a voltage that is an analog of the strength of the sound pressure waves at the microphone.

This analog signal is fed into a codec or *vocoder* inside the terminal that digitizes the analog waveform, then codes it into a standard format.

The result is 1s and 0s representing the digitized speech. These 1s and 0s are packaged into IP packets.

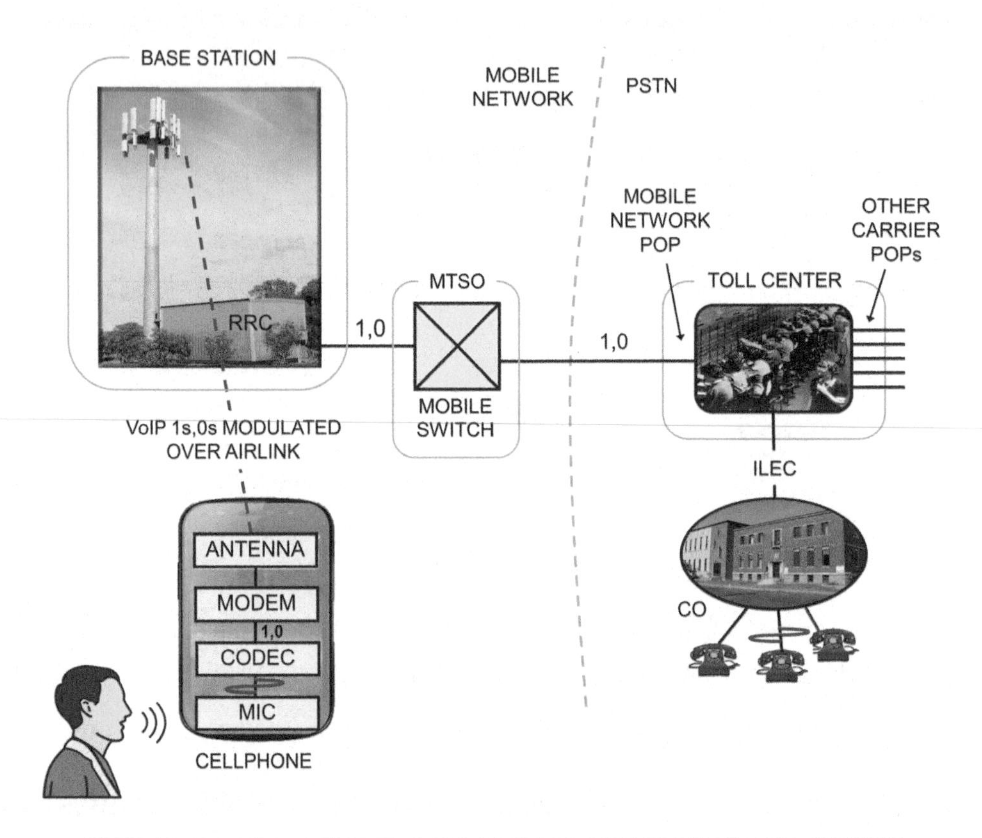

FIGURE 100 PSTN PHONE CALLS USING THE PHONE APP: "VOICE MINUTES"

9.6.3 Radio Frequency Modem

A modem that operates at radio frequencies represents the 1s and 0s that make up the packets using a modulation technique like Quadrature Amplitude Modulation (QAM), on channels specified by the RRC.

9.6.4 Antenna

The signal produced by the modem is in the form of electricity. This electricity is conducted to an antenna in the phone, from which the modem waveform radiates into space in the form of radio waves, or electro-magnetic energy, at the speed of light.

9.6.5 The Base Station

At the base station, an antenna receives a tiny (!) fraction of the power transmitted from the phone, added to all of the other signals received. The antenna converts the radio waves to an electrical signal that is fed into the Radio Resource Controller, which decodes and demultiplexes all of the bit streams to yield IP packets.

9.6.6 Backhaul to Mobile Switch and Call Routing

The VoIP packets with digitized speech are then backhauled or transmitted back to the mobile switch, to be routed to the PSTN for a mobile-to-wireline call, or routed to another base station for a mobile-to-mobile call.

9.6.7 Speech Coding Standards and Bit Rates

The speech is coded at between 4.75 and 12.2 kb/s for transmission over the airlink, far less than the 64 kb/s DS0 rate used in the PSTN.

For mobile-to-wireline calls, the speech has to be converted to the format and rate for interconnection between the mobile network and PSTN.

In the past, the coding and transmission standard for the PSTN was the G.711 codec at 64 kb/s, carried in a DS0 channel, carried in a DS3 in a SONET Optical Carrier (OC) frame.

In new landline systems it is the G.711 codec producing 64 kb/s carried in IP packets in Ethernet frames. In the future, a different coding technique than 64 kb/s G.711, like AMR, may become standard across all networks.

9.6.8 Connection to the PSTN

The connection to the PSTN happens at a building traditionally called a toll center, where all of the carriers connect. Carriers each have a Point of Presence (POP) in the toll center, which is the connection point to their network. Calls are switched between different carriers' POPs. See Section 13.3 for more about POPs and Toll Centers.

Connections in the toll center in most cases are implemented under a dinosaur-era tariff called Tandem Access Trunks, which is circuit-switched DS0 channels and ISUP call control messages. In the future, connections in the toll center will be VoIP and SIP call control messages.

9.7 Mobile Internet: "Data Plan"

The great thing about digital cellular is that we can take advantage of its inherent capability to move bits to communicate not just digitized speech, but 1s and 0s that are representing e-mail messages, web pages, video, music or literally anything else.

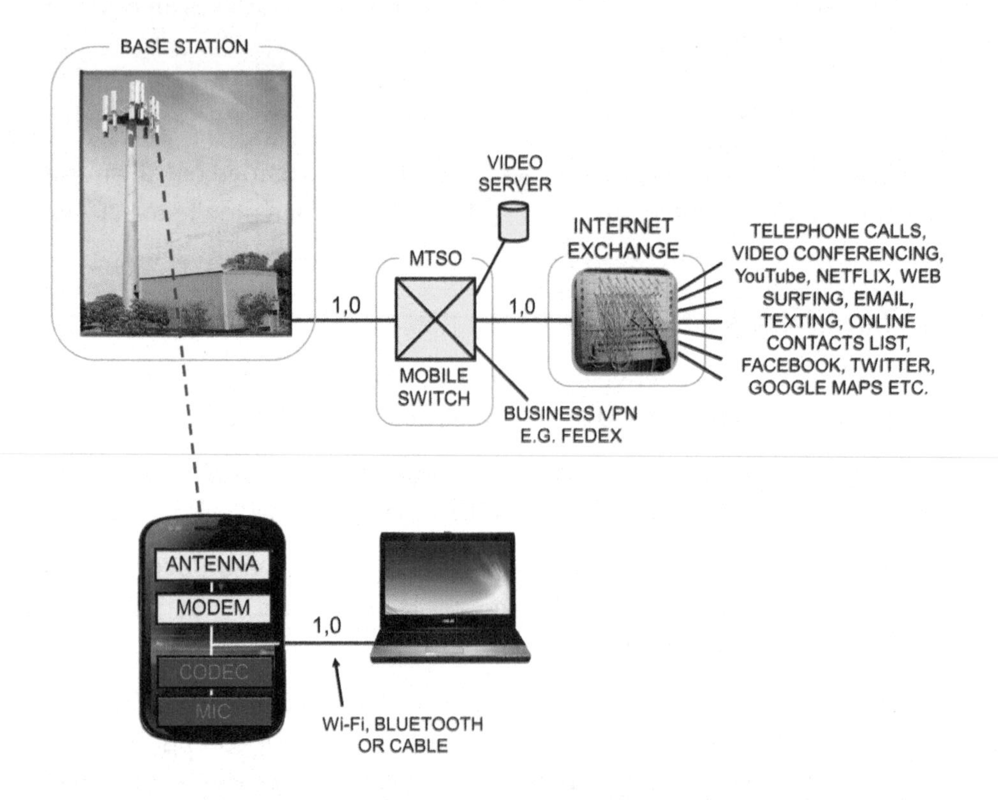

FIGURE 101 MOBILE INTERNET: "DATA PLAN"

In a quaint, old-fashioned use of terminology, all traffic apart from telephone calls and Short Message Service (SMS) texts is called "data" in cellular billing plans.

"Internet traffic" would be a more accurate term than "data", as it includes web pages, email and social media, Skype Internet to PSTN telephone calls, Skype videoconferencing, YouTube videos, Netflix videos, Google maps data and app downloads, along with millions of other applications communicating IP packets to the Internet via the carrier.

As illustrated in Figure 101, the base station sends packets through the mobile switch to the Internet at the Internet Exchange building the mobile carrier selected. At the IX, the mobile carrier has agreements with other IP networks to exchange traffic to get the packet delivered to the far-end device on the Internet.

A smartphone can run all of the applications to generate and receive any of this kind of traffic.

9.7.1 Cellphone as a Tethered Modem

In some cases, it is desired to share the smartphone's Internet connection with other devices.

Figure 101 adds a PC connected to the cellphone of Figure 100, showing how the modem in the cellphone and the modem in the base station can be used to move data from a laptop to the Internet via the cellphone.

When using a laptop connected to a cellphone with a cable as illustrated, the cellphone is said to be acting as a *tethered modem*, i.e., physically tied to the computer.

USB Cable Tethering

One way to implement this is to connect a computer to a data port on a cellphone with a USB cable.

The computer then sees the cellphone like an external modem, very much as if one had plugged a landline modem into the computer.

Wireless Tethering with Bluetooth

One could also connect the computer to the phone with a Bluetooth wireless link, which is a low-budget radio completely separate from the cellular radio, running in the 2.4 GHz ISM band.

9.7.2 Mobile Wi-Fi Hotspot

The most convenient option once set up is to activate the feature in a smartphone that turns it into a Wireless LAN (Wi-Fi) access point, on the 2.4 GHz and/or 5 GHz unlicensed bands.

The computer then connects to the smartphone's Wi-Fi access point just as it would connect to any other Wi-Fi access point. The smartphone internally bridges the Wi-Fi connection to its cellular Internet data connection.

9.7.3 Packet Relay to the Internet

Essentially, the microphone, speaker, screen and keyboard in the phone are ignored; the computer connects to the modem in the cellphone and uses that along with the radio, antenna and battery to communicate IP packets from the computer to the cellular network base station.

From there, the received packets are routed to the mobile switch and relayed to local content servers, the Internet, or to a carrier providing a service with guaranteed quality and security.

9.7.4 Dongles

A very similar story can be implemented with what the marketing department might call a "stick" – a modem, radio and antenna built into a small *dongle* that plugs into a USB port on the computer.

The dongle implements the same capability as a cellphone, but without the speaker, microphone, codec, battery, keyboard and screen.

9.7.5 Smartphone as the Terminal

Of course, a smartphone is a computer; if the user experience can be satisfactory on a small screen, there may be no need to have a separate computer using the smartphone as a bridge.

All smartphone apps communicate to the Internet directly through the smartphone's cellular radio when Wi-Fi is turned off. Or not, if you don't have a "data plan".

Instead of computers connecting to the smartphone, the Wi-Fi radio can be used to connect the smartphone to a Wi-Fi access point with an upstream Internet connection.

Since the Wi-Fi connection to the Internet is free, the smartphone will use it instead of the cellular network Internet connection, which is not.

9.7.6 Billing Plans and Roaming

If a user leaves their carrier's serving area, and uses mobile Internet without a roaming data plan in place, they will be charged the "default" or "casual use" rate, which can be astronomically high, for example, $5 per MB.

That's $5,000 per GB. $22,500 to download a DVD. $125,000 to download the data on a single-layer Blu-ray DVD.

It is not unknown to hear of people getting a bill of $20,000 for using their cellphone for Internet access without a roaming plan then watching YouTube and Netflix video.

One joke had Canadian astronaut Chris Hadfield forgetting to sign up for a roaming plan before leaving for the International Space Station. When he returned to Earth after a year, he was presented with a million-dollar roaming bill.

9.7.7 The Holy Grail of Convergence

The connection to the Internet supports phone calls, television, movies, music, email, web surfing and everything else, on the same connection on the same device: the Holy Grail of convergence people have been talking about for the last 50 years. Steve Jobs achieved it with his iPhone.

9.8 Mobile Operators, MVNOs and Roaming

9.8.1 Mobile Network Operator

Mobile Network Operator (MNO) is the term usually used to refer to a facilities-based carrier, i.e., a company that owns base stations, a mobile switch, backhaul between them, and spectrum licenses, and sells services to the public... and to other carriers.

The MNO implements external links to other carriers for PSTN phone calls and for Internet traffic.

For PSTN phone calls, the MNO implements a fiber optic connection to a building traditionally called a Toll Center or Class 4 switching office. The termination of their fiber in that building is called a POP. It is their physical point of presence in the building.

Many other carriers have POPs in the building, including the ILEC, IXCs, CATV companies, other mobile carriers, and any other company that wants to connect phone calls to a phone on the MNO's network.

The operator of the toll center, usually the ILEC, provides a switch in the Toll Center to switch phone calls from one carrier's POP to a different carrier's POP.

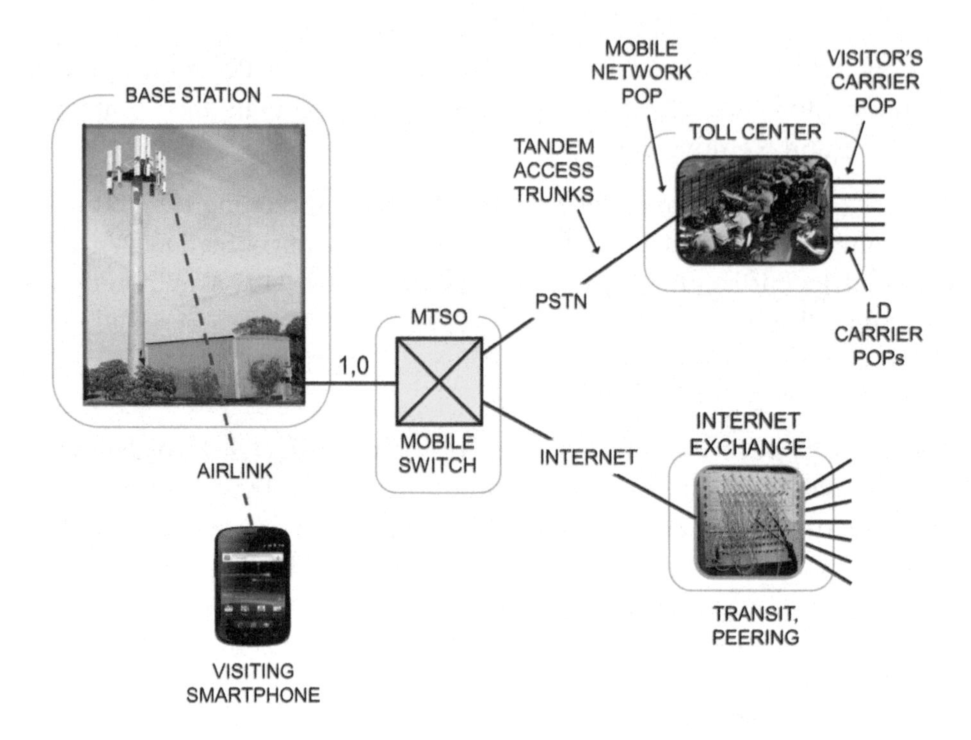

FIGURE 102 MOBILE OPERATORS, MVNOS AND ROAMING

For Internet access, the MNO implements a fiber optic connection to one or more Internet Exchange buildings, where they pay the operator of the IX to route packets to other carriers with whom the MNO has established IP packet transit and peering arrangements.

9.8.2 Mobile Virtual Network Operator

A Mobile Virtual Network Operator (MVNO) is the term used to refer to a non-facilities-based carrier… one that does not own the hardware or spectrum licenses or POPs. Instead, the MVNO enters into a long-term contract with one or more facilities-based carriers to have them supply a "white label" service that the MVNO sells.

Typically, the MVNO will develop a unique branding and sell smartphones and tablets to go along with its service.

When the MVNO deals exclusively with one carrier, the MVNO bill to the customer would be typically generated by the facilities-based carrier as a white-label service. If the MVNO is very large and deals with multiple carriers, the MVNO may operate their own billing system, which is a significant investment.

The facilities-based carrier bill to the MVNO includes a volume-discount rate for IP addresses and Internet traffic, voice-minute airtime and switched access to the POP for PSTN phone calls.

The MVNO also has to pay for connectivity from the POP to other toll centers for "long-distance" connections, and the switched-access charge at the far end.

The rate plan the MVNO pays could be a mix of fixed-rate leases and usage-based billing.

Unless the MNO is obliged to sell capacity to MVNOs through regulations and tariffs, the nature of the plan is confidential business information.

9.8.3 Roaming

Roaming service is very similar to the service provided to MVNOs, in that it is the MNO that is providing the airlink, base stations, backhaul, mobile switch and connections to the PSTN and Internet.

In the case of roaming, the visitor uses their own phone, and billing is usage-based.

Roaming is an important feature for smaller players: they are facilities-based in selected cities, but to offer a national and international service to their customers, they must have roaming agreements in place with MNOs in other locations.

By denying roaming service to smaller or startup carriers, or charging an exorbitant price for roaming, an incumbent carrier can erect a barrier against competition. In many countries, the right to roam and the wholesale cost of roaming is regulated to encourage competition.

9.9 3G: CDMA and HSPA

The third generation of cellular is usually referred to as *3G*.

The main objectives of the third generation were to improve capacity, the number of simultaneous users, and to increase the number of bits per second that can be transmitted over the airlink, for mobile wireless high-speed Internet access and video.

9.9.1 IMT-2000

To try to avoid a repeat of the 2G CDMA vs. TDMA dichotomy, in 2000, a standards committee attempted to define a world standard for 3G called IMT-2000.

They failed.

The result was a "standard" describing five incompatible implementation variations. Like many other technologies, we ended up with one solution for "North America" and a different solution for "Europe".

To support higher bit rates over the airlink, more frequency bandwidth is required.

Out of the five variations in IMT-2000, the two serious ones both specified CDMA as the method for spectrum-sharing – but disagreed on the width of the radio bands and how many bands there should be.

9.9.2 1X or CDMA2000: IMT-MC

Service providers using CDMA for 2G, primarily North American and certain Asian countries, favored a strategy that was basically a software upgrade from 2G, employing existing 1.25 MHz radio carriers and allowing multiple carriers.

This is called IMT-MC or multi-carrier CDMA.

Qualcomm's brand name for this was CDMA2000.

The service provider could purchase licenses for as many bands as desired, and the bands could be variable sizes to meet different countries' radio licensing plans, providing a flexible and scalable capacity.

FIGURE 103 VERIZON 1X 3G TECHNOLOGY

A single 1.25 MHz carrier version referred to as *1X* was widely deployed.

9.9.3 UMTS or W-CDMA: IMT-DS

Service providers using GSM TDMA for second generation, primarily cellular carriers outside North America, favored the deployment of CDMA in a 5 MHz wide band.

This was called IMT-DS, Direct Spread, Wideband CDMA (W-CDMA) and Universal Mobile Telephone Service (*UMTS*).

An incentive for GSM operators was they would be able to re-use some control infrastructure from the second-generation TDMA GSM systems.

However, practical functioning of a multi-user, multi-base station, mobile CDMA network requires among other things constant control of the power on the cellphones, so that the received power at the base station is the same from all the phones; and compensation for time delay differences in identification signals received from different base stations.

In the 1X systems, this was accomplished using techniques patented by Qualcomm (and paying Qualcomm a royalty for every cell phone and every base station transceiver), and the United States government's Global Positioning System respectively.

European operators, with their UMTS, did not favor the notion of paying an American company royalties, and did not favor building a network dependent on the American government's GPS.

Since UMTS required mathematical calculations across a 5 MHz band, compared to 1X's 1.25 MHz band, at the time, the processor in the phone required to perform such calculations drew so much current from the battery that the battery heated up to the point that people burned their hands on the phones during trials.

The GSM/UMTS Europeans embarked on a seven-year-long odyssey attempting to circumvent Qualcomm patents, and avoid using GPS.

After a number of strategies failed, a Euro-GPS called Galileo was created for UMTS; the first satellite was launched in 2005.

This delayed deployment of UMTS until 2007.

1X was deployed and working years earlier.

The tipping point between 2G and 3G in the GSM/UMTS camp was reached in the summer of 2007, when more new activations on these carriers' networks were 3G CDMA (UMTS) instead of 2G TDMA (GSM).

The 2G TDMA technology GSM still had far more users, but like 1G analog, GSM ends up in the dustbin of history.

Note that "GSM" is used erroneously to refer to carriers in the UMTS camp, and the 3G, 4G and 5G phones compatible with their networks.

9.9.4 Data-Optimized Carriers: HSPA and EV-DO

For Internet access and watching video on cellphones, variations of the coding schemes optimizing for the statistical characteristics of "data" were developed and deployed by both camps.

In both cases, these were deployed on carriers (the 1.25 or 5 MHz bands) apart from those used for telephone calls.

Accessing these data carriers required either a "stick", the USB dongle described in an earlier section, or dual radios in a phone, one tuned to the traditional carrier for telephone calls and a second tuned to the data-optimized carrier for watching video.

The 1X camp developed a variation called 1X Evolution Data-Optimized (1XEV-DO), allocating a carrier for data communications and promising 2.4 Mb/s over the airlink in the first incarnation.

Proposals for future revisions of EV-DO promised to support more than 70 Mb/s over the airlink.

In the UMTS camp, the variation was called High Speed Packet Access (HSPA), referring to improvements in the UMTS downlink, often called High Speed Downlink Packet Access (HSDPA) and in the uplink, High Speed Uplink Packet Access (HSUPA) and also Enhanced Dedicated Channel (E-DCH).

Revisions of HSDPA promised download rates of 14.4 Mb/s then 42 Mb/s.

9.9.5 The End of the Standards War

Market forces finally pushed the two camps together.

The fact that there were far more 2G GSM users on the planet meant that for one thing, handset manufacturers produced 2G GSM phones before 2G CDMA phones.

GSM phones were less expensive and had better features.

This trend was continuing into 3G, where UMTS phones would have the same advantage over 1X phones.

Another fact was that Steve Jobs of Apple only permitted carriers operating TDMA systems to have the iPhone, then only permitted carriers with HSPA systems to have the iPhone 3G.

Finally, the 1X camp threw in the towel and decided to go with the UMTS camp's proposal for the fourth generation to level the playing field.

As soon as that decision was made, then the deployment of 1XEV-DO was more or less capped, and some 1X carriers began deploying HSPA instead.

And the fact is, as soon as carriers that were in the 1X camp, like Verizon in the US and Bell and TELUS in Canada gave up and agreed to conform to the 3GPP "European" standards, Steve Jobs allowed the iPhone on their networks.

As the iPhone was at the time the most popular phone, this was a major incentive for the 1X camp.

One of the legacies of Steve Jobs that most people will not be aware of is in additional the iPod, iPhone, and iPad, but a decisive part in ending the mobile spectrum-sharing standards war.

9.10 4G LTE: Mobile Broadband

9.10.1 Universal Terrestrial Radio Access Network Long-Term Evolution

After more than 20 years, a universally-accepted standard for mobile radio was achieved. By inventing the world's most popular consumer electronics device – the iPhone – and only allowing "GSM" carriers to sell it, Steve Jobs almost single-handedly brought the "GSM vs. CDMA" standards war to an end.

FIGURE 104 3GPP AND LTE

To get the iPhone, North American service providers supported the "European" Third Generation Partnership Project (3GPP) standards group's Release 8, known as Universal Terrestrial Radio Access Network (UTRAN) Long Term Evolution (LTE).

LTE is an all-IP network, moving VoIP and Internet traffic over the airlink. LTE's spectrum-sharing method, Orthogonal Frequency Division Multiplexing (OFDM) on the downlink, is a return to first-generation Frequency Division Multiple Access with some major improvements.

One improvement is the definition of a thousand or more channels, called subcarriers in OFDM, and dynamic assignment of one or more channels to users called OFDMA, enabling multiple simultaneous users and the possibility of multiple channels for a single user for parallel downloads and thus high bit rates. On each channel is a modem running QAM or QPSK.

9.10.2 Radio Resource Controller

A device called the Radio Resource Controller (RRC) at the base station manages the allotment of channels, and also controls the wake/sleep state of the phone to conserve battery life.

Service providers can deploy LTE on radio carriers 1.4, 3, 5, 10 or 20 MHz wide. The highest bitrates, on the order of 100 Mb/s are achieved on a 20 MHz carrier. 5 MHz carriers allow overlay on frequency allocations for UMTS.

Cell sizes can range from femtocells measured in the tens of meters (in-home), picocells up to 200 m (office-building), microcells between 200 m and 2 km, which includes cells in urban areas typically 1 km, cells in suburban areas typically 5 km, and macro cells in rural areas 30 km and theoretically up to 100 km in diameter.

Multiple-Input, Multiple-Output (MIMO) antenna designs can increase the bitrate using spatial multiplexing, which is basically gluing several transceivers and antennas together and communicating in parallel.

9.10.3 OFDM

The modulation and spectrum-sharing scheme for LTE is OFDM, which is different than FDMA, TDMA and CDMA.

It's not really necessary to understand the details of how the modulation actually works... but since OFDM is used in most modems, it's worth knowing the basic idea: hundreds of modems working in "parallel" within the available frequency band.

A controller allocates modems to users, sometimes multiple modems in parallel to a single user to give them more bandwidth. The radio waveform resulting from all of them is calculated in a single step.

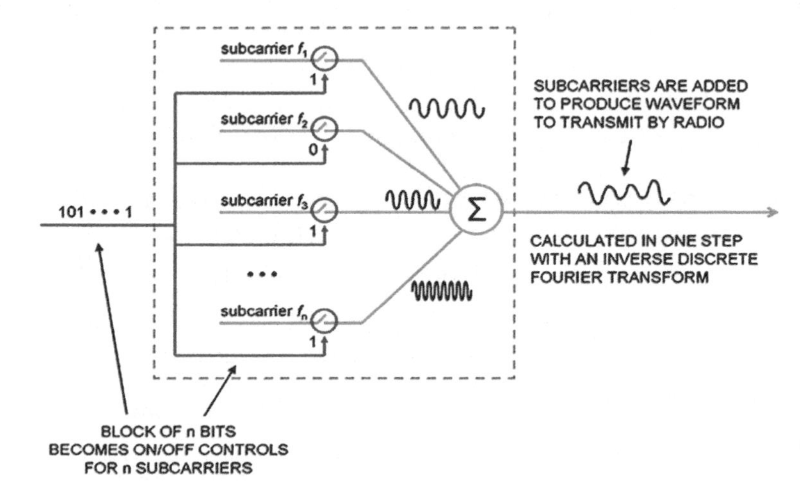

FIGURE 105 SIMPLEST EXAMPLE OF OFDM PARALLEL COMMUNICATION

For those interested in a longer explanation: the idea behind OFDM is the definition of hundreds or thousands of *subcarriers* within the main carrier.

A subcarrier is a single frequency which will be modulated like any other modem carrier signal, spreading energy in a small band around the center.

Essentially, this is taking the idea of a modem and its modulation and implementing it hundreds or thousands of times in narrower frequency bands within the larger frequency band.

The incoming bit stream is then divided up in some way to use as the input bits to modulate each of the subcarriers.

In the simplest implementation, illustrated in Figure 105, the incoming bits are used to turn subcarriers on or off, splitting the incoming bit stream at a rate of 1 bit per subcarrier and implementing binary Amplitude Shift Keying, with one of the amplitudes zero.

In the most complex implementation, the incoming bit stream would be allocated at a rate of 6 bits per subcarrier, used to implement QAM-64 on each of the subcarriers.

The outputs of the subcarrier modulations are all added together to produce a transmittable waveform.

To allow multiple users at the same time, Orthogonal Frequency-Division Multiple Access (OFDMA) is implemented, allowing adaptive assignment of specific subcarriers to particular users.

The beautiful part of OFDM (at least to Engineers) is that the modulation of each subcarrier and adding them all together is calculated in a single step with a digital signal processing operation called an *Inverse Discrete Fourier Transform*. At the receiver, a Discrete Fourier Transform performs the reverse process to yield the original bit stream.

This calculation is performed, and a waveform modulated at the same frequency as the spacing of the subcarriers, which has the result of making the harmonics of all of the subcarriers cancel each other out at the receiver. Mathematicians and systems engineers use sets of orthogonal cyclic polynomials to describe the waveform, hence the term *orthogonal*.

In the LTE standard, the subcarriers are spaced 15 kHz apart, and the baud rate, the rate at which the modulation is changed, is 15,000 times per second.

Prior to modulation, *Forward Error Correction* is implemented, adding redundancy to the bit stream so correct decisions can be made based on maximum likelihood in the presence of impairments like noise and fading.

The bit stream is also shuffled or *interleaved*, re-arranging the order of the bits in time so that burst errors are no longer sequential errors.

In the uplink, LTE uses a pre-coded version of OFDM called Single Carrier Frequency Division Multiple Access (SC-FDMA) to avoid needing an amplifier across the entire carrier, which would increase handset cost and shorten battery life.

9.10.4 3GPP Standards Committees

The 3GPP Technical Report 25.913 contains the detailed requirements specification for LTE. The system architecture, in Technical Specifications 36.300 and 36.401, is simplified to two principal network elements: evolved Network Base stations (eNBs) and Evolved Packet Cores (EPCs). eNBs communicate with EPCs, with each other and with user equipment.

The ITU defined "4G" as supporting at least 1 Gb/s downloading. 3GPP Release 8 LTE does not meet that criterion, and so in a strict standards committee environment, LTE would be called a 3G technology.

An updated version, 3GPP release 10, called LTE-Advanced, supports 1 Gb/s, causing those standards committee members to consider LTE-Advanced to be a 4G standard.

Everyone else has referred to LTE as 4G from the start.

9.10.5 Qualcomm Patents

One of the reasons for the 3G standards war was the requirement to pay American company Qualcomm royalties on patents for several techniques necessary for a mobile CDMA system.

LTE is not CDMA, so those royalties are avoided... but it turns out that Qualcomm filed or has purchased many patents that underpin LTE.

Additionally, since LTE phones will have to be backwards-compatible with 3G CDMA networks, Qualcomm sees "no impact" on patent royalty revenue for the first ten years of LTE development according to COO Sanjay Jha.

9.11 5G NR: Enhanced Mobile Broadband, IoT Communications

FIGURE 106 5G LOGO

The increasingly inaccurately-named Third Generation Partnership Project's Release 15 is the first full specification for a standalone Fifth-Generation mobile communications system.

5G uses a new air interface called New Radio in the standards committees, which has 40% better spectral efficiency (bits/second per Hertz) than LTE.

Further increasing the bit rate that can be delivered to a user, 5G supports massive MIMO, hundreds of parallel streams implemented using spatial multiplexing.

Spectrum for 5G is grouped into above 6 GHz and below 6 GHz.

9.11.1 Below 6 GHz

5G will be deployed on existing spectrum below 6 GHz, eventually replacing LTE.

More interest is currently focused on deployment of 5G mobile communications on new frequency bands, particularly in the 2.5 GHz band which has a favorable balance of available bandwidth and signal propagation through obstacles.

The 3.5 GHz band has significant bandwidth available for sectorized systems, but does not propagate so well through obstacles like walls and trees.

Other new spectrum is at comparatively low frequencies, the 600-, 700- and 800-MHz bands. These bands have the advantage of long reach, i.e., signals can propagate over long distances and inside buildings; but the number of 20- and 40-MHz bands available for sectors is limited.

9.11.2 mmWave

5G in frequencies above 6 GHz is starting with millimeter wave bands, which are 30 – 300 GHz, where the length of the radio wave is 1 – 10 mm in vacuum. At these bands, ultra broad bandwidth can be delivered – hundreds of Gb/s – but the maximum useful airlink range is 150 meters with line-of sight.

Subsequent 3GPP releases will add new features and enhancements. Release 16 is called the 5G System - Phase 2 in the standards committees and promises support advanced applications as well as efficiency enhancements.

9.12 Spectrum-Sharing Roundup: FDMA, TDMA, CDMA, OFDMA

Cellphones transmit and receive signals over shared radio bands. In this section, we'll go over the methods of sharing.

To separate users so that they do not interfere with one another, nor hear each other's conversations, service providers use one of four radio band or *spectrum sharing* methods: Frequency-Division Multiple Access (FDMA), Time-Division Multiple Access (TDMA), Code-Division Multiple Access (CDMA) and Orthogonal Frequency-Division Multiplexing (OFDM) and its multiple-access version.

In this section, we'll explain how FDMA, TDMA, CDMA and OFDM work, and how they were deployed for first and second generation with names like AMPS, TDMA (IS-136), GSM and 2G CDMA (IS-95).

In subsequent sections, we'll take a closer look at CDMA for third generation (1X and UMTS), then 4G and 5G which use OFDM.

9.12.1 FDMA

Frequency-Division Multiple Access (FDMA) is a spectrum-sharing method where a block of spectrum is divided into small frequency bands called *channels*, which are organized into groups.

Notionally, there was one group per cell.

Users are assigned a set of channels when their call begins. There are channels for voice each way and for control signals each way.

To communicate voice, a carrier frequency is centered in a channel and its frequency is varied or modulated continuously in proportion to the voltage coming out of the microphone – which in turn is an **analog** or direct representation of the strength of the sound pressure waves coming out of the talker's mouth. This is referred to as *analog radio.*

The same idea is used for FM radio.

In the mobile system, if the user does not move, and there are not a lot of other users moving around, the user will stay on those radio channels for the duration of the call.

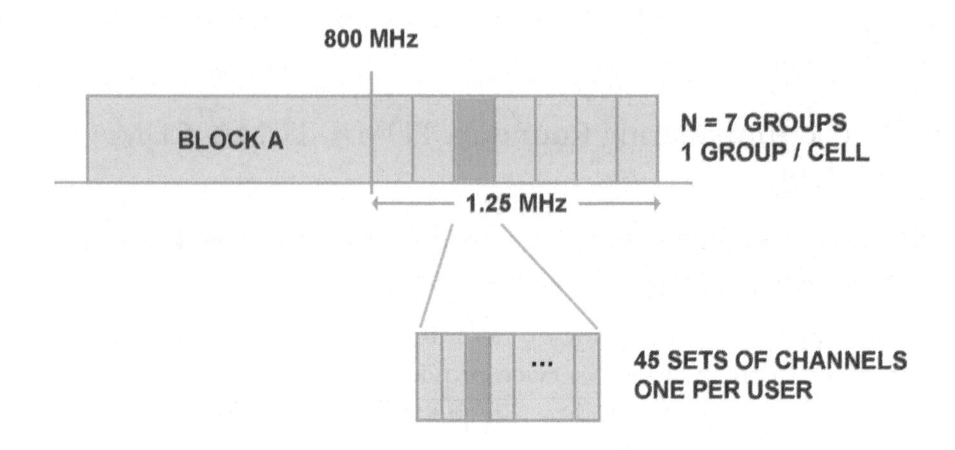

FIGURE 107 FDMA - AMPS

This makes eavesdropping easy, as all that is required is an FM radio scanner that can be tuned to the cellphone frequency – there is no encryption or coding of the voice.

If the user moves, or others do, the mobile switch may hand them off to another cell.

This means that both the base station and the radio frequency channels may change during the call, since each cell uses different groups of channels.

During a handoff, the end-to-end communications path will be interrupted or *muted* for a short period of time while the base station and frequency channel change is made.

This interruption is heard as a "click" during a voice call; it causes modems to disconnect.

FDMA was the method used in first-generation "analog" systems, including AMPS, NMT, and TACS used in various countries.

AMPS

In North America, the Advanced Mobile Phone System (AMPS) was the standard for 1G FDMA, deployed in radio bands 25 MHz wide at 800 MHz (824-849 MHz uplink, 869-895 MHz downlink).

The radio band is divided into 30 kHz channels, with groups of channels allocated to base station transceivers front-ended with antennas mounted on towers, providing radio coverage in an area around the tower: the cell.

The groups of channels are allocated so that they can be re-used geographically far enough away so that they do not interfere.

Organizing the channels into 7 groups is referred to as N=7, and allows coverage of arbitrarily large areas using a honeycomb pattern for the cells as illustrated in Figure 98.

The capacity of the system is limited by the number of channels in the group of frequencies, and is relatively low… 1.5 users per square mile in the previous calculation.

It is possible to *sectorize* a cell to achieve higher capacity, that is, more users per square mile.

Sectorization means using directional antennas with 120 degree or 60-degree beamwidths, instead of an omnidirectional antenna with a 360-degree beam.

Sectorization implements a number of pie-wedge-shaped cells emanating from a single tower.

Typical plans for AMPS were to use 7 groups with 3 sectors for "rural" areas and 4 groups with 6 sectors for "urban" areas.

However, the resulting capacity is still too low, and data communications was very difficult using dial-up modems over the analog radio system.

9.12.2 TDMA

Time-Division Multiple Access (TDMA) means a radio channel shared in time between a number of users.

Users transmit and receive modem signals one after another in a strict order in time slots on a radio channel.

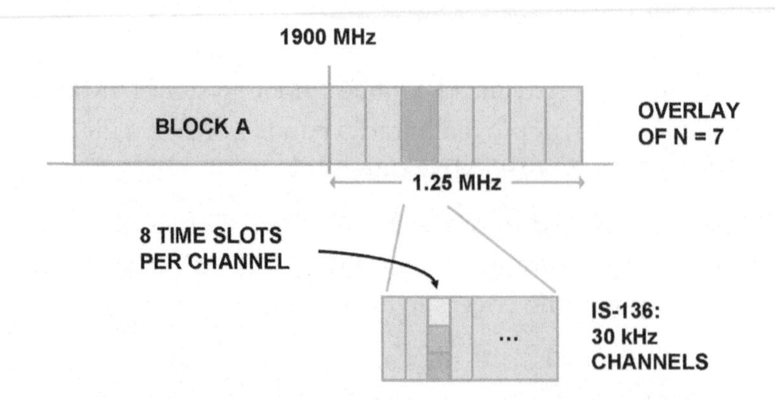

FIGURE 108 TDMA: TIME-SHARING RADIO CHANNELS

TDMA is called digital, since the modems integrated in the handset and base station move 1s and 0s, which could be digitized speech, text messages, web pages, video or anything else.

IS-136

In North America, the IS-136 standard was deployed by some service providers for 2G. This was sometimes called D-AMPS, as it was digital implemented on 30 kHz AMPS channels.

GSM

In the rest of the world, another form of TDMA called the Global System for Mobile Communications (GSM) was deployed for 2G. GSM became the most popular technology.

GSM channels are 200 kHz wide with 16 time slots, meaning seven users per 200 kHz: 7 time slots each way for voice or data and 1 for control.

This was widely deployed, and became the most popular spectrum-sharing technology.

IDEN

The Integrated Digital Enhanced Network (IDEN) from Motorola is an overlay on TDMA that allows group walkie-talkie functions similar to trunked radio systems used on construction sites and at sporting events.

Inefficiency of TDMA

TDMA provided an improvement in capacity, by having users time-share radio channels; but it is not an efficient way to share.

Users are assigned specific time slots and these time slots are reserved for them, whether they have anything to transmit or not.

This is inefficient for telephone calls, where one speaks only half the time, and highly inefficient for data communications like web browsing, where one has nothing to communicate most of the time.

9.12.3 CDMA

Another spectrum-sharing strategy is Code-Division Multiple Access (*CDMA*). CDMA is completely different than FDMA and TDMA.

The allocated spectrum is organized into 1.25 or 5 MHz frequency bands called *carriers*. Users transmit at the same time, in the same cell, on the same carrier.

Each user has a code, which is a binary number 64 bits long. If the user wants to transmit a "0", they send their code. If they want to transmit a "1", they send the mathematical complement of their code.

The codes are such that when codes from multiple users are received, added together, mathematical operations reveal which codes were transmitted.

This is analogous to being at a cocktail party where everyone is speaking at the same time in the same space, but each pair of people are speaking unique languages - and you only understand your partner's language.

You can understand this by relating the words you hear to your vocabulary and finding correlations. Only your partner's words make sense. Everything else sounds like background noise to you.

Spread Spectrum

Anyone with a graduate degree in Electrical Engineering specializing in communications will know that since the frequency bandwidth of a signal is proportional to the bit rate times the signal to noise ratio, and vice-versa, sending a 64-bit-long code instead of a single 1 or 0 has the effect of transmitting at a higher bit rate, which spreads the energy of the transmission across a wider frequency channel than non-coded transmission does.

For this reason, CDMA is referred to as "spread spectrum".

Spectral Efficiency

CDMA systems employ variable-bit-rate codecs and statistical multiplexing, which means that if no noises are coming out of the user's face, not much is transmitted.

The capacity of the system is designed based on the statistics of how often noise does come out of a person's face during a phone call; speaking is treated as bursts of sound.

CDMA is the most *spectrally-efficient* method of those discussed here, meaning that it allows the greatest number of phone calls per Hertz of radio band than any other.

Since the CDMA system was designed to treat speech like bursts of data, it is also inherently efficient for data communications.

Qualcomm, a technology company mainly located in San Diego, patented certain methods of power control and synchronization necessary to make mobile, multi-user, multi-base-station CDMA work, and sells chips or licenses that implement these patented techniques.

CDMA for second generation cellular followed the IS-95 standard, deployed by Verizon and Sprint in the USA and Bell and TELUS in Canada.

CDMA is the spectrum-sharing technique underlying most 3G cellular, including both the 1X and UMTS variations.

9.12.4 OFDM and OFDMA

Orthogonal Frequency-Division Multiplexing (OFDM) is the spectrum-sharing method for the downlink for the fourth generation, called LTE, and for the fifth generation, referred to as "New Radio" in standards committees.

OFDM is a frequency-division technique with significant improvements on 1G analog FDMA.

In an OFDM system, hundreds or thousands of *subcarriers* are defined within the main carrier. A modem representing 1s and 0s operates on each subcarrier.

OFDMA – the Multiple Access implementation of OFDM – allows multiple users to participate in the OFDM downlink and uplink at the same time.

Users are dynamically assigned one or more subcarriers by a radio resource controller, giving each user their own bandwidth to communicate.

Communication of high bit rates is implemented by assigning multiple subcarriers to a single user and splitting their bit stream into multiple parallel streams, one for each subcarrier.

Essentially, OFDM is taking the idea of a modem and its modulation and implementing it hundreds or thousands of times on a subcarrier, i.e., a narrower frequency band within the larger frequency band, then assigning multiple "modems" to each user to employ in parallel.

Other aspects of OFDM – of interest mainly to mathematicians and Engineers – include the fact that the combined output waveform is calculated in a single step; and the baud rate, the rate at which the modulation is changed from one state to another, is the same rate as the subcarrier spacing, which has the effect of cancelling harmonics to allow successful transmission on tightly-spaced subcarriers.

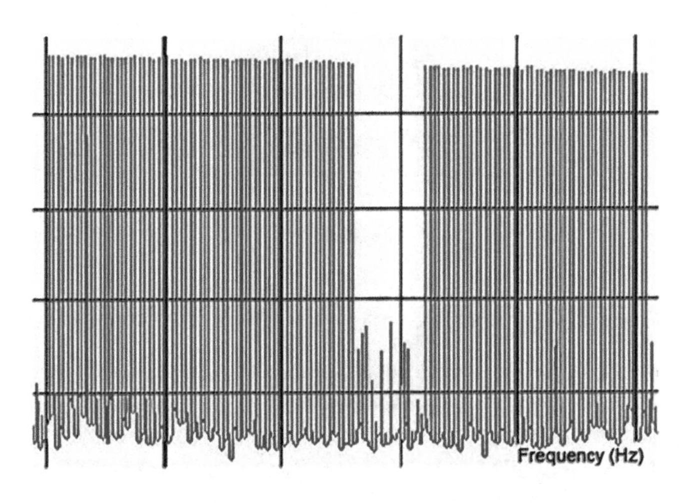

FIGURE 109 OFDM SUBCARRIERS

The uplink adds a Discrete Fourier Transform to the processing so that the frequency components of the modem waveform for a single subcarrier are transmitted instead of the waveform itself. This is called DFT-spread OFDM or Single-Carrier FDMA.

OFDM is used for 4G, 5G, 802.11 wireless LAN Wi-Fi, DSL and cable modems.

9.13 3.5-GHz Fixed Wireless Broadband Home Internet

Once the density of homes is above 500 homes per square kilometer (5 homes per hectare = about 2 homes per acre), the economics of installing fiber infrastructure with a long-term payoff becomes interesting.

Broadband to the home in areas with more sparse population and without suitable cabling infrastructure is usually most cost-effectively delivered wirelessly.

At present, the capacity of mobile LTE networks is such that download speeds fall far below "broadband" on weekend evenings when many people are watching Netflix and videos of hairless cats and guilty dogs.

Over the next decade, millions of people will get broadband to the home via *fixed* wireless.

FIGURE 110 3.5 GHZ FIXED WIRELESS BROADBAND HOME INTERNET

Fixed wireless broadband to the home means a wireless link, with an antenna fixed to the outside of the home, connected (and powered) by a LAN cable to an indoor device, which provides Wi-Fi and LAN jacks.

The 3.5 GHz band allows the allotment of numerous 40 MHz channels. LTE and its Radio Resource Controller is used to share the resulting bandwidth on this shared fixed wireless network. MIMO is used to increase the maximum bitrate available to users.

Offering 25/1 service (25 Mb/s down, 1 Mb/s up) allows the carrier to support between 100 to 150 users per sector, with 10 Gb/s backhaul for each multi-sector base station.

5G with its massive MIMO, enhanced transmission modes, additional spectrum like 2.5 GHz and more sophisticated customer antennas in the future will enable 50/10 as a standard service offering.

For cost reasons, it would be ideal to provide both mobile Internet to cellphone users, and fixed wireless for broadband home Internet from a single radio technology and a single set of base stations.

Due to licensing, spectrum cost and regulations, fixed wireless broadband Internet to the home is deployed as a standalone sectorized radio system.

9.14 Wi-Fi: 802.11 Wireless LANs

In this section, we provide an overview of the 802.11 wireless LAN standards, often called *Wi-Fi*.

We will concentrate on understanding the variations of 802.11, the frequency bands they operate in, bit rates to be expected and practical issues.

Since 802.11 is wireless LANs, there are a number of associated topics: LAN frames, also called MAC frames, MAC addresses, LAN switches, IP addresses, routers and network address translation.

Those topics are covered in other chapters, particularly Chapter 15 and Chapter 16. In this section, we concentrate on the radio aspects.

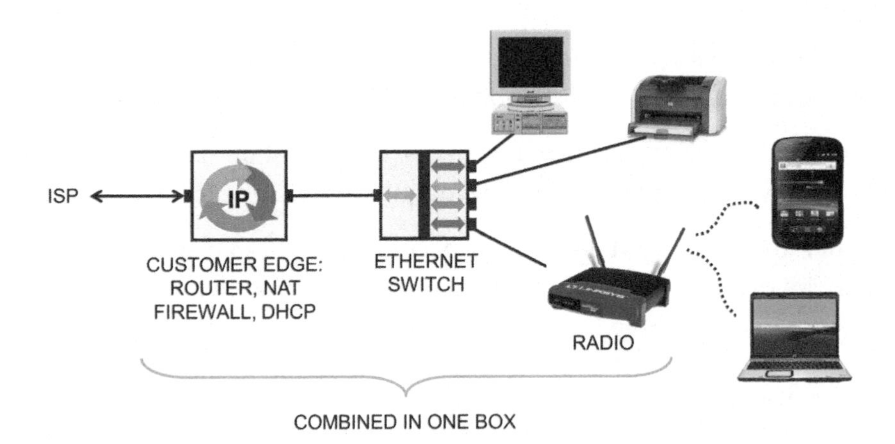

FIGURE 111 WI-FI WIRELESS LAN EQUIPMENT

9.14.1 System Components

Perhaps the most widespread broadband wireless data communication technology today is wireless LAN technology, also referred to as 802.11, Wi-Fi and hotspots.

Wi-Fi is a certification standard for wireless Ethernet LANs conforming to the IEEE 802.11 series of standards, with space as the physical medium instead of cables, operating in unlicensed bands at data rates measured in the tens and hundreds of Mb/s and ranges measured in the tens or hundreds of feet.

A typical set-up employs a radio base station, called an Access Point (AP), and wireless interfaces built into devices including computers, printers, cameras, phones and music players.

The Access Point is usually part of a device called the Customer Edge (CE). The CE also includes an Ethernet switch, routing, Dynamic Host Configuration Protocol (DHCP), Network Address Translator (NAT), firewall and port forwarding, and if supplied by an Internet Service Provider (ISP), will also include a DSL or Cable modem, or for the lucky few, a fiber port.

9.14.2 Service Set ID

The name of a Wi-Fi AP's radio service is defined by the user, and usually broadcast as its Service Set ID (SSID). This shows up on the list of available networks on a wireless device trying to connect.

9.14.3 Unlicensed Radio Bands

All Wi-Fi systems operate in *unlicensed* radio bands, also called Industrial, Scientific and Medical (*ISM*) bands in North America.

"Unlicensed" means that it is not necessary to obtain a license from the national government to emit electromagnetic energy at these frequencies.

At other frequencies, it would be necessary to obtain a license, which usually involves proving that you will not interfere with anyone else at the requested frequencies in a specific geographic area.

9.14.4 Half-Duplex

The first five versions of 802.11 wireless LANs are *half-duplex*: only one device can transmit at a time. This means that the actual throughput will be less than half the advertised rate during two-way communications, and worse with each added station.

Devices transmit only if they have data to send. Otherwise, another device can use the bandwidth.

This technique is called statistical Time-Division Duplexing (TDD). It is more efficient than Bluetooth, with its channelized TDD reserving half of the slots for each direction whether they will be used or not.

Wi-Fi 6, conforming to the 802.11ax standard, is the first to implement full-duplex communications for multiple simultaneous users using OFDMA. This results in a dramatic increase in performance, particularly in areas with a high density of users.

9.14.5 802.11b and g

Standards are published by the 802.11 working group of the Institute of Electrical and Electronic Engineers (IEEE). These include 802.11a, 802.11b, 802.11g, 802.11n and 802.11ac and the latest 802.11ax, also called Wi-Fi 6.

802.11b and g operate in the 2.4 GHz ISM band, offering a maximum of 11 and 54 Mb/s respectively. 802.11b uses Direct Sequence Spread Spectrum like cellular CDMA, and 802.11g implements OFDM like LTE, both allowing multiple devices at the same time in the same space.

However, many other devices including cordless phones, baby monitors and Bluetooth operate in this band, meaning significant performance-lowering interference. It is possible to listen to 802.11b transmissions on an analog 2.4 GHz cordless phone. Microwave ovens operate at 2.4 GHz and can cause radical interference with Wi-Fi communications.

9.14.6 802.11a

802.11a operates in a 5 GHz unlicensed radio band, supporting a maximum of 54 Mb/s using OFDM. The 5 GHz band is relatively free of interference, but the higher frequency also means shorter range and poorer penetration through walls. In practice, line-of-sight between the access point and the terminal are necessary to achieve 54 Mb/s.

9.14.7 Wi-Fi 4: 802.11n

802.11n uses the 2.4 and/or 5 GHz bands, optimizing for power to noise ratio between the bands. 802.11n also supports 20 MHz and/or 40 MHz channels – using more of the wireless spectrum when available to enhance performance, and allows parallel transmission using 1 to 4 radios in Multiple-Input, Multiple-Output (MIMO) systems to achieve very high data rates.

In theory, 802.11n will implement 150 Mb/s with a single antenna... but that would be on the moon, where there are no atoms between the transmitter and receiver and no interference. As soon as there is anything between the transmitter and receiver – like water molecules, plaster, concrete and so forth, and/or interference, the power-to-noise ratio and thus bit rate drops.

9.14.8 Wi-Fi 5: 802.11ac

802.11ac operates at 5 GHz with wider bands and 4x4 MIMO to achieve in practice 500 Mb/s and in theory 3.5 Gb/s.

9.14.9 Wi-Fi 6: 802.11ax

802.11ax uses up to 160 MHz channels, QAM-1024 and 8x8 MIMO to achieve in theory up to 9.6 Gb/s, OFDMA to share bandwidth efficiently between many users, and device sleep/wake control like LTE and 5G. The 6 GHz unlicensed band is used.

9.14.10 VoIP over Wireless LANs

802.11 wireless LANs are mostly used to access the Internet for email, web surfing, YouTube, Netflix and the like. Voice over IP (VoIP) telephone calls over Wi-Fi using, for example, Whatsapp is a growth area.

A cellphone, iPod, or tablet with 802.11 and Whatsapp to connect over the Internet to the telephone network has no need to pay for cellular service to make a phone call whenever the device is in range of an 802.11 network... which may be "most of the time" in the future.

9.14.11 Wi-Fi Security

A major concern with wireless LANs is security.

Network security that can be enacted is Media Access Control (MAC) address filtering: setting the base station to only accept connections from specific, predefined wireless LAN cards.

This protects the network connection from access by unauthorized users – but does not protect legitimate users' transmissions from eavesdropping.

If someone can get physically close enough to receive signals, there is no way to prevent them from eavesdropping on communications, which can include intercepting and re-using usernames and passwords and intercepting and "wikileaking" sensitive information.

This is particularly troublesome in coffee shops, airports and anywhere else the communications are not encrypted, an "open" hotspot.

In 2010, a plugin for the Firefox browser was made available that allowed someone sitting in such a coffee shop to eavesdrop on everyone else's communications – and with one click, to re-use other people's credentials to log in to servers.

This means that secure encryption of communications over the airlink is now mandatory, not optional.

If it can be ensured that the users always, without fail, implement client-server encryption (sometimes called Transport Layer Security... though anyone who has taken an OSI Layers course will know it is Presentation Layer security), by using VPN software for connecting to work, typing https:// for all web surfing, and using encrypted email communications, then there is no need for encryption of the airlink.

However, users cannot be relied upon, so encryption of the communications on the airlink between the access point and terminal must be implemented whenever possible.

Wired Equivalent Privacy (WEP) was the first encryption algorithm for wireless LANs; but its use is not recommended as there are software tools available that can crack it in a matter of minutes.

Wi-Fi Protected Access WPA2 should be implemented on the airlink whenever possible. WPA3 is a minor update to WPA2.

9.15 Communication Satellites

In this last section of this chapter, we will take a quick overview of communication satellites, understanding the basic principles and the advantages and disadvantages of the two main strategies: Geosynchronous Earth Orbit and Low Earth Orbit.

9.15.1 Transponders

Communication satellites are orbital platforms that carry multiple base station transceivers with antennas pointed towards the surface of the earth.

Instead of base station transceiver, the term transmitter/responder or *transponder* is used in the satellite business.

In two-way systems, radio signals are transmitted from the earth to the transponder, which responds with radio signals directed back down to the surface at a different frequency to avoid interfering with the surface transmitter.

There are two basic choices for the orbits of communication satellites: geosynchronous orbit or low earth orbit.

9.15.2 Geosynchronous Orbit

Geosynchronous satellites are parked 22,300 miles (35,680 km) above the surface of the earth above the equator.

At that radius, the orbital speed is the same as the rotational speed of the earth, and hence the satellite appears to stay in the same spot in the sky.

This is *geosynchronous* or *geostationary*, depending on your point of view.

Geosynchronous communication satellites are operated by the International Telecommunications Satellite Organization (Intelsat), the International Marine Satellite Organization (Inmarsat), numerous private companies, government and military.

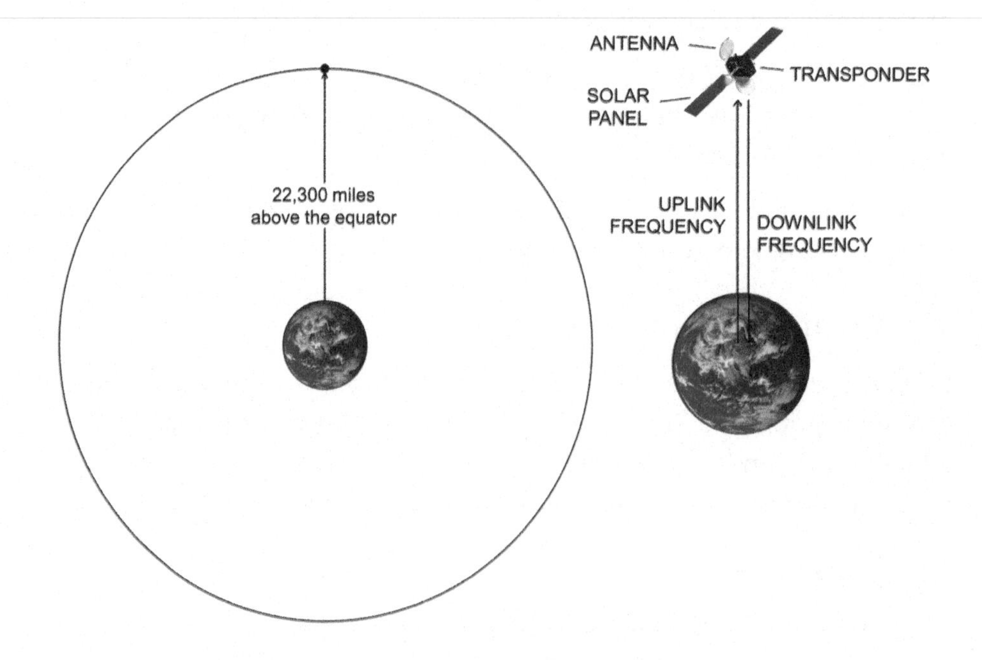

FIGURE 112 GEOSYNCHRONOUS SATELLITES

Each country has the right to spots in geosynchronous earth orbit above their country based on international agreements. The countries can use these spots, or lease them. Occasionally, there are disagreements as to who has the right to a spot and two countries will launch a satellite and park it in close proximity to the other, causing interference.

The three main advantages of geosynchronous satellites are broadcast, broadcast and broadcast: a transponder on a geosynchronous platform is 22,300 miles up in the sky – and if desired, can provide radio coverage to a third of the earth's surface.

The main disadvantage is path delay: the radio waves have to travel 22,300 miles up and 22,300 miles down.

At the speed of light, that takes about 1/4 second each way.

For interactive communications between two points on the earth via a geosynchronous satellite, that means up and down for the inquiry, and up and down for the response, a total of just under one second delay.

If the two locations cannot see the same satellite, then an intermediate ground station must be used, meaning a path delay of about two seconds.

This wreaks havoc with the protocol people use to decide who gets to talk next during a phone call, and the extended delay can cause users to hear echoes that are normally suppressed.

No one likes using use geosynchronous satellites for phone calls; trans-oceanic fiber optic cables are preferred because they are much shorter, meaning that the path delay is negligible; about 25 milliseconds from New York to Paris on fiber, for example.

One-way communications is the natural application for geosynchronous satellites.

Television is the biggest market.

Radio-frequency modems communicate video that has been digitized, coded using MPEG-2 or H.264 (MPEG-4 Part 10) and encrypted, from a Digital Broadcast Center of a satellite TV company up to the transponder, which repeats the modem signal back to the earth.

Access to the Internet is also implemented on geosynchronous satellites for those who do not have DSL, Cable, fiber or cellular service available.

The "upload" path from the customer to the Internet can be over modems over a regular phone line, and the download path via satellite, making the customer premise electronics cheaper and cutting the delay in half.

Two-way satellite communications is also available.

Another business service based on geosynchronous satellites is Very Small Aperture Terminal (VSAT), which means "small dishes" in plain English.

Applications for this two-way wide-area data communication service include emergency backup communications capabilities and nationwide VPN services.

9.15.3 Low Earth Orbit

The path delay problem of geosynchronous platforms can be fixed by bringing the satellite closer in. These types of communication satellites are called Low Earth Orbit (LEO).

LEO be used for voice communications because the path delay is reduced to an acceptable level.

This introduces two different problems: the satellites do not stay in the same position in the sky to an earthbound observer, and the coverage or footprint of the satellite is reduced.

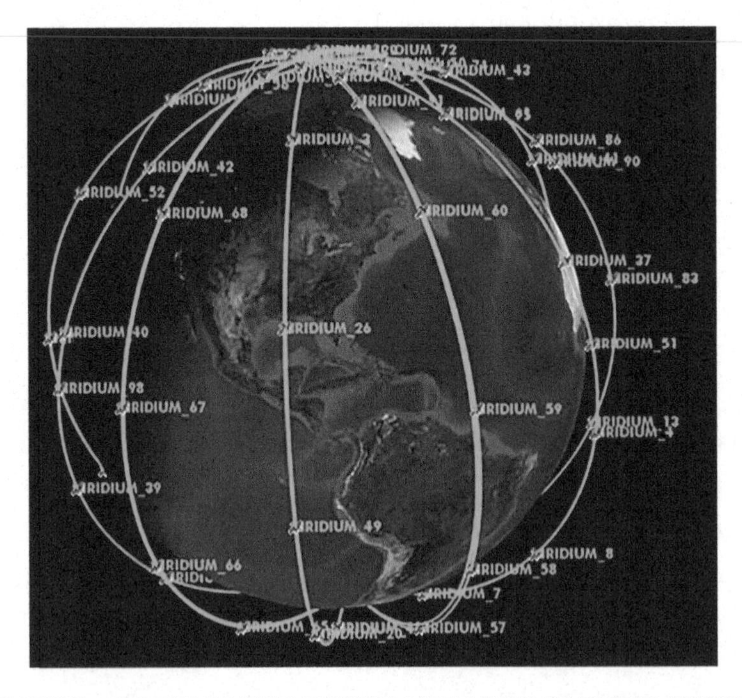

FIGURE 113 LOW EARTH ORBIT SATELLITES IN THEIR ORBITS

Multiple satellites to ensure coverage and a switching system for doing handoffs from one satellite to another as they move are required.

The handoff problem is similar to that of cellular, where the base stations are stationary and the users move around, except that with LEO satellites, the users are mostly stationary and the base stations move around.

9.15.4 Iridium Next

Motorola's Iridium was one of the first LEO projects. Iridium planned to launch 77 satellites (Iridium is element number 77 in the periodic table of the elements), but went live after deploying only 66 satellites.

Gaps in coverage, poor in-building penetration and difficult data communication over the analog radio system led to poor user response and Iridium was a financial failure.

Motorola announced that they would "de-orbit" the satellites and shut down. At the last minute, an entrepreneur who had contracts with mining companies purchased Iridium for 25 million dollars. It cost five billion 1998 dollars to build.

In 2017 and 2018, the SpaceX Falcon 9 rocket was used to launch 75 of 81 Iridium 2 satellites to form Iridium Next. Running on Iridium Next, Iridium OpenPort Internet Service is available at rates of 128 to 512 kb/s. Iridium L-Band broadband service is up to 1.5 Mb/s down, 0.5 Mb/s up.

Iridium Certus is the branding for a new set of features enabled by Iridium Next: mobile satellite communications across maritime, IoT, aviation, land mobile, and government applications.

Among other services hosted by Iridium NEXT is the Aireon aircraft tracking and surveillance system. This system will provide air traffic control organizations and aircraft operators that purchase the service with real-time global visibility of their aircraft. The search for flight MH370 that disappeared in 2014 would have been easier with such satellite tracking.

The original Iridium satellites were de-orbited in 2019.

9.15.5 Orbcomm and Globalstar

Other LEO companies include Orbcomm and Globalstar. Orbcomm was a joint venture between Teleglobe and Orbital Sciences Corporation, intended to provide two-way data communications and the capability to track trucks on highways and tanks on battlefields.

Globalstar is a consortium of telecommunications companies operating a constellation of 48 low-earth-orbit (LEO) satellites. Globalstar phones incorporated both the satellite radio and a cellphone, and could operate on local cellular and/or satellite. Globalstar sold high-capacity inter-city links wholesale to regional and local telecom service providers around the world.

9.15.6 Starlink

The SpaceX Starlink System is being launched 60 satellites per Falcon 9 flight, slated to finish with 12,000 satellites by the mid-2020s.

The satellites are positioned at three orbits: 340 km (210 mi), 550 km (340 mi), and 1100 to 1325 km (690 - 830 mi).

Starlink intends to provide broadband to the masses at prices competitive to terrestrial services. After a launch in 2020, the chief Engineer at Starlink said "we're still a long way from watching cat videos in 4K, but we're on track to get there".

In 2022, it indeed became possible to watch cat videos and guilty dog videos, with download speeds up to 1 Gb/s in bursts, for US$600 to $800 equipment cost and $110 per month service.

FIGURE 114 STARLINK 2ND-GENERATION ANTENNA

Starlink's second generation antenna is a 19 x 12 inch flat-panel antenna that points skyward.

There are actually many antennas under the cover; very sophisticated signal processing adjusts the power and phase of each in real time to perform *beamforming*, focusing the power on the satellite transceivers... as they move.

Similarly, many Starlink satellites work together to focus power on the antenna as they transit overhead. The time a Starlink satellite is visible to an antenna before it fades into the noise horizon is measured in minutes.

Starlink brings Broadband to the Rural and Remote. This will undoubtedly improve quality of life for many, many people, and spur new economic activity worldwide.

10 Fiber Optics

10.1 Fiber Basics

The fundamental idea behind optical transmission is varying some characteristic of a light beam to represent information, transmitting that light beam through a solid tube of glass that guides the light to the far end of the tube, where the light is detected and interpreted.

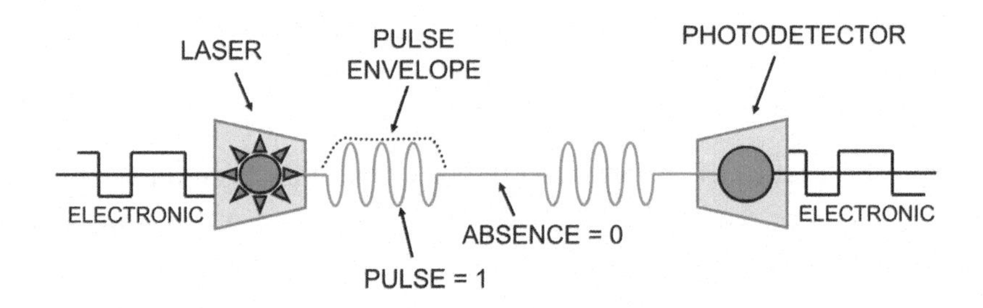

FIGURE 115 COMMUNICATION OF 1S AND 0S USING PULSES OF LIGHT

The information often originates on and is delivered to the end-user on copper LAN cables, involving electronic – optical conversion and a laser at one end, and a photodetector for optical – electronic at the other as illustrated in Figure 115. An *optical transceiver* combines the two devices for two-way communications, usually using two fibers; one for each direction.

Different states of the light emitted by a laser are used to represent 1s and 0s. Most systems use "light on" to represent a one and "light off" for zero. QAM (Section A.6) is used in expensive high-bitrate systems.

On fiber to the neighborhood in many cable TV systems, the amplitude of light within a broad band of frequencies is varied continuously as a direct analog of a continuous electrical signal at the same frequencies on coaxial copper wires. Part of the signal is modems at various carrier frequencies.

10.1.1 Lambdas

Light at a single frequency – or as close to a single frequency as possible, since nothing is perfect – is generated by lasers. Usually, the light is characterized by its *wavelength* rather than its frequency. ℓ is used as an abbreviation for length. In Greek, the letter ℓ is λ (lambda), so wavelengths are called *lambdas*.

There is a direct relationship between wavelength and frequency: $\lambda = c/f$; wavelength = speed of light / frequency.

10.1.2 Pulses and Signals

In an on/off system, the burst of photons vibrating at the specified frequency emitted while the laser is on is called a *pulse,* a type of signal.

The average intensity of the light while on is called the *envelope* of the pulse. This envelope is detected by a photodetector at the far end and used to determine the outgoing electrical pulses of voltage on copper.

The rate at which the laser can be switched on/off determines the bit rate.

10.1.3 Attenuation and Dispersion

Impairments, that is, the factors that reduce the ability to reliably detect the pulses at the far end include *attenuation* and *dispersion,* and usually worsen with distance, so have the effect of limiting the useful range.

Attenuation is the diminishment of signal strength with distance, caused by not-perfectly-transparent glass. This is usually not the limiting factor – unless bad splices or faulty connections severely attenuate the signal.

Dispersion causes the lengthening in time of the pulse envelope while in transit over the fiber. In the extreme, pulses would merge together during transit, making it impossible to reliably detect them.

Before this happens, the signals must be detected and regenerated by a *repeater* in an optical-electrical-optical process, or with very advanced technology, pulses are reshaped (shortened) optically.

There are many ways that dispersion happens: modal dispersion, chromatic dispersion, polarization mode dispersion and others. This is covered in more detail in Section 10.3.

10.2 Fiber Optics and Fiber Cables

Glass fiber is the physical medium of choice for implementing backbone or core networks, for three main reasons: 1) bandwidth, 2) bandwidth and 3) bandwidth. Fibers are capable of supporting the transmission of huge numbers of bits per second.

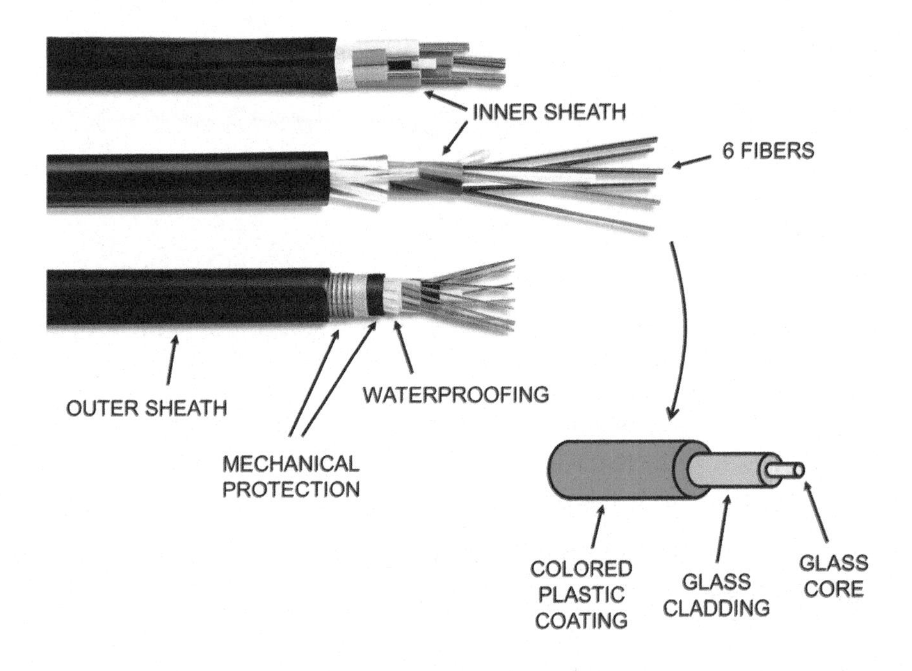

FIGURE 116 FIBER CABLES AND COMPONENTS

A glass fiber is a physical medium for communicating information just as copper wires are. Glass is used because it has good dimensional stability (doesn't kink), strength, cost and transmission properties.

On a glass fiber, binary digits are represented by changing the light from a laser producing energy at a frequency of around 200 x 10^{12} Hz, which corresponds to about 1.5 x 10^{-6} meters = 1.5 micrometer = 1500 nanometers in *wavelength*.

On most systems, light-on represents 1 and light-off represents 0. The number of times per second the laser is turned on and off mostly determines bit rate. Expensive systems use QAM to multiply the bit rate.

The main job of the fiber is to guide the light from one end to the other – without losing any of it.

10.2.1 Core

A fiber consists of two different types of highly pure, solid glass mixed with specific elements called *dopants* to adjust the *refractive index* of the glass, which is one of its transmission characteristics.

The innermost part of the fiber is a solid tube of glass called the *core*. The purpose of the core is to act as a waveguide for the light.

The core diameter is measured in millionths of a meter, called *microns*.

micro (μ) = 10^{-6}; 1 μm = 10^{-6} m.

Fiber core sizes range from about 5 to 50 microns. A human hair is about 100 microns in diameter.

10.2.2 Cladding and Coating

Around the core is the *cladding*, which is also glass, but with a different *refractive index* than that of the core.

The difference in refractive index causes light injected into the core at certain angles to reflect back into the core – thus constraining all of the optical energy which is the pulse to travel in the core and exit the far end.

Around the core and cladding is a colored plastic *coating* to waterproof and identify the fiber.

10.2.3 Cables

A fiber optic cable contains multiple fibers, which are usually organized in bundles in colored soft plastic tubes called the *inner sheath* for identification.

A sticky waterproof compound is placed in the inner sheath to repel water, which can infiltrate glass and change its transmission characteristics.

More layers of hard plastic and metal are added to protect the fibers from water, shovels and backhoes. A *strengthening member* may be present to keep the cable from being bent too sharply during installation, causing micro-cracks or outright breaks.

Mechanical protection - armor - is added to protect against a type of signal degradation on fibers known as *backhoe fading*: being cut with a mechanical shovel.

The outermost layer is called the *outer sheath*. Ripcords – steel and/or nylon – may be incorporated to allow installers to strip away sheaths without damaging the fibers.

A slippery covering can be added to make it easy to pull the cable through long runs of conduit called *ducts*.

A fluorescent orange outer covering may be added to make the cable more visible and lessen the chance of an accidental cut.

10.2.4 Fiber Count

Fiber counts in long-distance transmission cables increased from 6 in the 1990s to 144 and 288 fibers per cable in 2000. Cables are built to spec for big projects, with a fiber count designed for that project. The labor and right-of-way costs far outweigh the cost of the glass.

10.2.5 Redundancy

Cables get cut, particularly by construction crews digging up streets with backhoes for unrelated work.

To maintain *availability* of communications, two fibers on different cables following different geographical routes can be installed.

This is called redundancy or *path diversity*.

In some systems, the same data is transmitted on both cables at the same time, guaranteeing no loss of data.

Other strategies implement *automatic protection switching*, i.e., moving traffic to a different cable after a break, which may involve some loss of data.

10.3 Optical Wavelengths, Bands and Modes

There are five main *windows, bands,* or ranges of frequencies within the light spectrum that are exploited for transmission using fiber optics.

Usually, the frequencies are referred to by their wavelength. In Greek, the letter L is λ (lambda), so wavelengths are called lambdas.

There is a direct relationship between wavelength and frequency: $\lambda = c/f$; wavelength = speed of light / frequency.

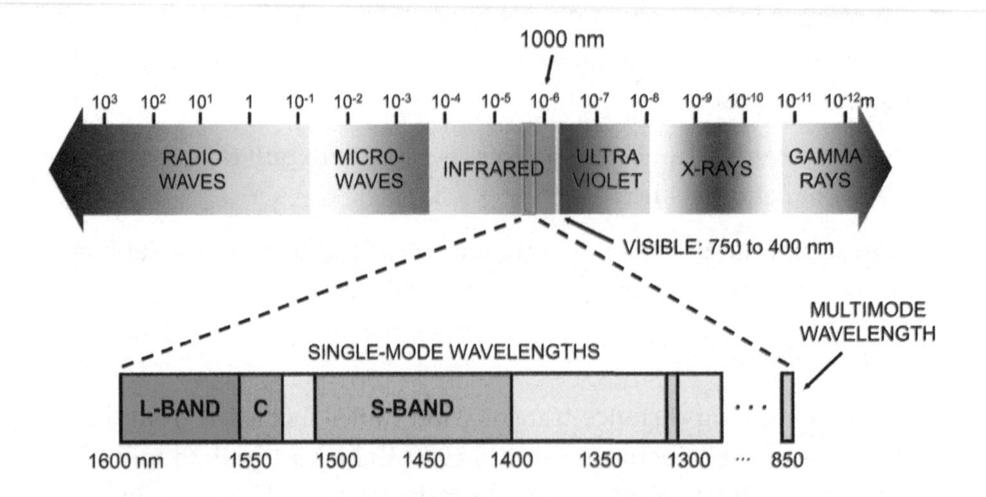

FIGURE 117 WAVELENGTHS AND BANDS

Light wavelengths are measured in nanometers (nm), 10^{-9} m. Wavelengths on the order of 1000 nm are used for communications systems, many at 1550 nm. As the speed of light is 3×10^8 m/s in a vacuum, 1550 nm means a frequency of about 200×10^{12} Hz or 200 Terahertz in vacuum.

10.3.1 Bands

Optical bands are ranges of wavelengths available to be exploited to signal 1s and 0s by pulsing a laser. The first two bands generally used in optical transmission systems were centered around 850 nm and 1310 nm. The 850 nm band was used almost exclusively for short-range *multimode* applications.

So-called *single-mode* fibers were first designed for use in the second window, near 1310 nm. To optimize performance in this window, the fiber was designed so that a type of dispersion called chromatic dispersion would be close to zero near the 1310 nm wavelength.

As the need for greater bandwidth and distance increased, a third window near 1550 nm called the Conventional or C-band has been exploited for transmission. It has much lower attenuation, and is within the frequencies amplified by erbium-doped fiber amplifiers.

More recently, bands above and below the C-band, called the Short or S-band and Long or L-band have been exploited for transmission.

10.3.2 Multimode and Modal Dispersion

Dispersion is the spreading of the duration of the pulse envelope or signal and is caused by numerous factors.

As illustrated in Figure 118, if the pulse envelope lengthens too much, adjacent pulses merge together and cannot be detected at the receiving end. This limits the distance before a repeater is required.

One type of dispersion is *modal dispersion.*

A *mode* is a path that light can follow through the fiber's core. Each mode is essentially a bounce path at a different angle of incidence.

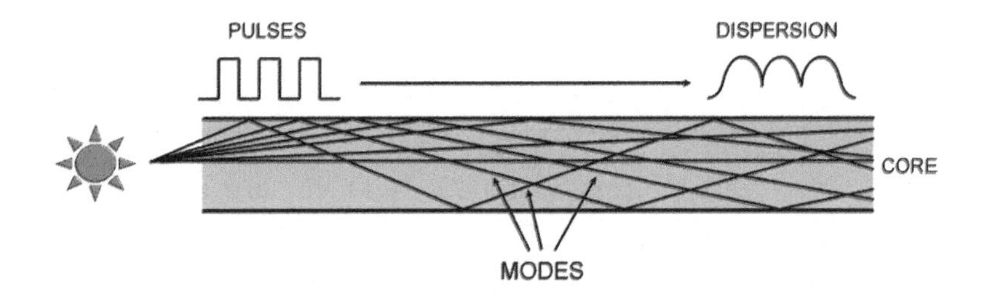

FIGURE 118 MULTIMODE AND MODAL DISPERSION

There are two general categories of optical fiber: *multi-mode* and *single-mode*. Multi-mode, the first type of fiber to be commercialized, gets its name from the fact that numerous modes exist simultaneously in the core.

Light in each of the different modes travels a different physical distance, so arrives at the end of the fiber at differing times, which causes dispersion, the duration of the pulse envelope to lengthen.

Accordingly, multi-mode fiber is not used for long distance applications. However, it is cheap to implement, connector tolerances are generous and low-cost light sources such as Light Emitting Diodes (LEDs) and Vertical Cavity Surface-Emitting Lasers (VCSELs) can be employed.

Multi-mode fiber is usually *graded index*; the refractive index of the core gradually decreases from the center of the core outward. The higher refraction at the center of the core slows the speed of some modes, reducing modal dispersion.

10.3.3 Single-Mode Fiber

The main type of fiber in use, in the beginning called *single-mode* by the marketing department, operates on the same principles, but has a much smaller core, which reduces the number of modes.

A perfect waveguide, i.e., one that would allow only a single mode, in cylindrical form would have a diameter a bit more than half a wavelength.

With wavelengths in the 1500 nm range, that would require a fiber with a core 0.8 microns in diameter; but that does not exist yet.

In practice, single mode meant thinner cores, 5 to 9 microns (compared to multi-mode at 50 microns), resulting in fewer modes. As a result, modal dispersion is greatly reduced and the possible transmission distance or *reach* is greatly lengthened.

All long-distance and high-bandwidth systems use "single-mode" fiber.

Lasers producing wavelengths of 1300 - 1600 nm are used as the light sources for single mode fiber.

Single-mode fiber has a step index; there is a uniform index of refraction throughout the core and a step in the refractive index where the core and cladding meet.

This smaller core and step index both act to reduce dispersion caused by various mechanisms.

10.3.4 Chromatic Dispersion

Lasers are not perfect; they do not produce energy at exactly one wavelength, but rather over a narrow range of wavelengths.

Since the propagation speed of light in glass is affected by its wavelength, this imperfection has the effect of causing some light to take longer to arrive at the far end later than other light, causing dispersion. This type of dispersion is called *chromatic dispersion*.

10.3.5 Polarization-Mode Dispersion

The propagation speed of light in a fiber is also affected by the diameter of the fiber. As light is actually two waves at right angles (horizontal and vertical) propagating forward, slightly oval-shaped fiber causes these two waves to propagate at different speeds, causing *polarization mode dispersion*.

10.4 Wave-Division Multiplexing: CWDM and DWDM

10.4.1 WDM

More capacity can be implemented on a single fiber using multiple wavelengths in parallel, called *Wave-Division Multiplexing (WDM).*

Signals are communicated on multiple single frequencies called *carriers* and referred to by their wavelength.

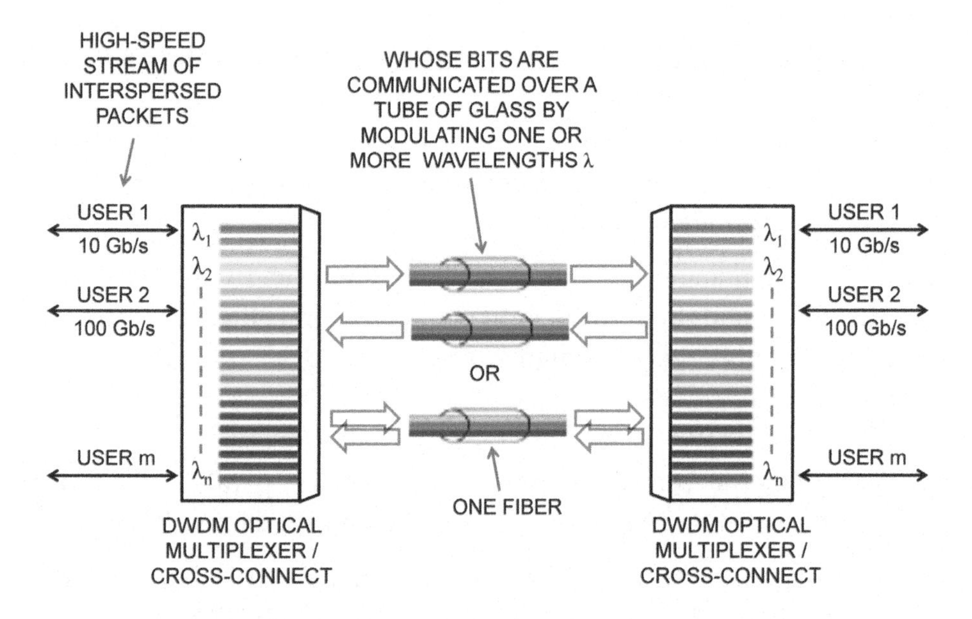

FIGURE 119 WAVE-DIVISION MULTIPLEXING (WDM)

Placing many of these carriers tightly spaced in frequency on a single fiber is referred to as *Dense Wave Division Multiplexing (DWDM).* Placing the wavelengths further apart, with as few as two carriers on a fiber, is called *Coarse Wave Division Multiplexing (CWDM).*

The scientific symbol for wavelength is the Greek letter lambda (λ). When someone refers to "a lambda", they are referring to a particular light carrier signal amongst many on a fiber.

The critical elements are the frequency bandwidth of the carriers – nothing is perfect – and their spacing. Challenges include crosstalk: keeping wavelengths from spreading into neighboring wavelengths during mixing, pulsing and transmission; and channel separation: the ability to distinguish each wavelength.

10.4.2 WDM Multiplexers

WDM multiplexers take multiple optical wavelengths and converge them into one beam of light. At the receiving end, demultiplexers must separate the wavelengths and couple them to individual fibers to be detected.

Multiplexers and demultiplexers can be either passive or active in design. Arrayed waveguide gratings consist of an array of curved-channel waveguides with a fixed difference in path lengths. The waveguides are connected to cavities at the input and output.

When light enters the input cavity, it is diffracted and enters the waveguide array. There, the length difference of each waveguide causes phase delays at the output cavity, where an array of fibers is coupled. Different wavelengths have constructive interference at different locations, which correspond to the output fibers.

A different technology uses thin film filters; the property of each filter is such that it transmits one wavelength while reflecting others. By cascading these devices, multiple wavelengths can be isolated.

10.4.3 Optical Ethernet Paths

An Optical Ethernet transceiver that transmits data in parallel over several wavelengths to achieve high bit rates is one kind of WDM multiplexer.

For example, the 40GBASE-LR4 standard signals 10 Gb/s over four wavelengths in parallel to achieve 40 Gb/s; 100GBASE-SR10 signals 10 Gb/s over ten wavelengths in parallel to achieve 100 Gb/s.

Each of the parallel wavelengths is called a *path.*

10.4.4 Current and Future Capacities

There is a large installed base of expensive DWDM multiplexers implementing 24 or 32 wavelengths around 1550 nm for very high capacity core network connections.

In the future, systems implementing 1,000 wavelengths on each fiber, each signaling at least 10 Gb/s, will result in available capacity of 10,000,000,000,000 bits per second, which is 10,000,000 Mb/s or 10 Terabits per second (Tb/s), per fiber, in the core.

In the not-too-distant future, 10 Gb/s will be a normal speed for business customer services.

10.5 Optical Ethernet

Optical Ethernet is signaling MAC frames (Section 4.4) from one device to another by modulating the light emitted by a laser.

In all but the most expensive systems, the signals that are transmitted are light on representing a 1, and light off representing a 0.

10.5.1 Point-to-Point Connections

Normally, Optical Ethernet is implemented as point-to-point connections: from a port on one LAN switch to a port on another LAN switch, from a port on a LAN switch to a port on a customer edge device, or from a port on a LAN switch to a LAN interface on a computer.

10.5.2 SFP Modules and Connectors

The light flashing on and off is implemented with a laser at the transmitter and a photodetector at the receiver. As illustrated in Figure 120, most systems use two fibers, one for each direction. A device combining the transmitter and detector functions is called an *optical transceiver*.

This device has metal connectors on one side to plug into a slot on a router or switch, and optical connectors on the other side, either factory- or field-installed on the fibers plugged into the transceiver.

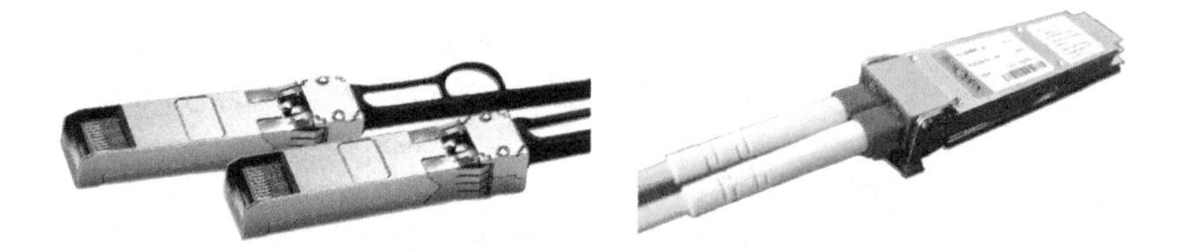

FIGURE 120 SFP OPTICAL TRANSCEIVERS

These transceivers are typically implemented on Small Form-factor Pluggable (SFP) modules, which are hot-swappable in the terminating equipment at each end. 100 Gb/s being communicated through this transceiver is the high end of commercially-deployed technology in 2020.

In some cases, the SFP modules are embedded in the terminating equipment, meaning the fibers are plugged into the terminating equipment. This allows re-use of existing fiber.

In other cases, the SFP modules are attached to fiber cables by the fiber cable manufacturer, meaning the SFP module is plugged into the terminating equipment. This ensures the fiber and transceiver technology are matched and the optical connection is a high-quality "factory" connection.

The SFP module format is not the subject of a standard, but rather described in industry Multiple Sourcing Agreements (MSA).

DESIGNATION	TYPE	REACH
1000BASE-T	COPPER CAT 5e	100 m
1000BASE-SX	2x MULTI-MODE 850 nm	220 m
1000BASE-LX	2x MULTI-MODE 1310 nm	550 m
1000BASE-LX	2x SINGLE-MODE 1310 nm	5 km
1000BASE-LX10	2x SINGLE-MODE 1310 nm	10 km
1000BASE-EX	2x SINGLE-MODE 1310 nm	40 km
1000BASE-ZX	2x SINGLE-MODE 1550 nm	80 km
1000BASE-BX10	2λ CWDM SINGLE-MODE 1310/1490 U/D	10 km
10GBASE-SR	2x MULTI-MODE 850 nm	400 m
10GBASE-LR	2x SINGLE-MODE 1310 nm	10 km
10GBASE-ER	2x SINGLE-MODE 1550 nm	40 km
10GBASE-ZR	2x SINGLE-MODE 1550 nm	80 km
10GBASE-PR (PON)	2λ CWDM SINGLE-MODE 1270/1577 U/D	10/20/30 km
40GBASE-SR4	2x MULTI-MODE 850 nm using 4λ	100 m
40GBASE-FR	2x SINGLE-MODE 1310 nm using 1λ	2 km
40GBASE-LR4	2x SINGLE-MODE 1310 nm using 4λ	10 km
100GBASE-SR10	2x MULTI-MODE 850 nm using 10λ	100 m
100GBASE-LR4	2x SINGLE-MODE 1310 nm using 4λ	10 km
100GBASE-ER4	2x SINGLE-MODE 1550 nm using 4λ	40 km

FIGURE 121 IEEE OPTICAL ETHERNET STANDARDS

10.5.3 IEEE Standards

There are many technologies for transceivers implemented on the SFP module. Some are proprietary, many are standardized by the IEEE. In practice, the same manufacturer's product is used at both ends of the fiber to ensure compatibility.

Most technologies use one fiber for each direction. Some use two wavelengths for two directions on one fiber.

The 40 and 100 Gb/s technologies split the bitstream into subrates and transmit them in parallel on different wavelengths called *paths* or *lanes*.

The table in Figure 121 lists current IEEE standards. More will be published in the future.

10.6 Network Core

Recapping Section 2.5: the network core, colloquially referred to as the backbone, provides high-capacity, high-availability connections between switching centers.

Fiber optics is used as the basis of connections between switching and routing centers since it can support very high numbers of bits per second. Lower-speed (and lower-cost) circuits are used to provide access to this core to users.

A method of organizing the bits for transmission, plus monitoring, alarming, testing and automatic protection switching for fault recovery is required for reliable service.

10.6.1 Optical Ethernet, RPR and MPLS

For new deployments, Optical Ethernet, that is, 802 MAC frames signaled on fiber is used to connect routers point-to-point.

A technology called Resilient Packet Ring (RPR) and/or MPLS is used to implement recovery from broken connections.

10.6.2 SONET and SDH

In the past, the most popular technology for these functions in North America was a standard called *Synchronous Optical Network* (SONET).

In the rest of the world, a very similar technology called Synchronous Digital Hierarchy (SDH) was employed. There is an installed base of legacy SONET and SDH systems (Appendix B).

10.6.3 Fiber Rings

To ensure high *availability*, that is, the possibility of communicating even if a line is cut or equipment fails, it is necessary to provide multiple redundant paths between each point.

The cheapest way to do this is to connect locations in *ring* patterns. This way, there are two connections to every location, but only one extra circuit.

Rings are used to connect COs in a city together. Rings are also installed to connect cities in regions together, and these regional rings are interconnected at multiple places for long-distance communications.

Shortcuts, i.e., connections between non-adjacent points on the ring, are implemented as traffic dictates. The end result is a semi-meshed network, where some locations are directly connected and others are reached via intermediate stations.

10.6.4 Network Core Nodes

At the end of each fiber that makes up the network core is a router. The router is sometimes called a network *node*, after the French word for knot.

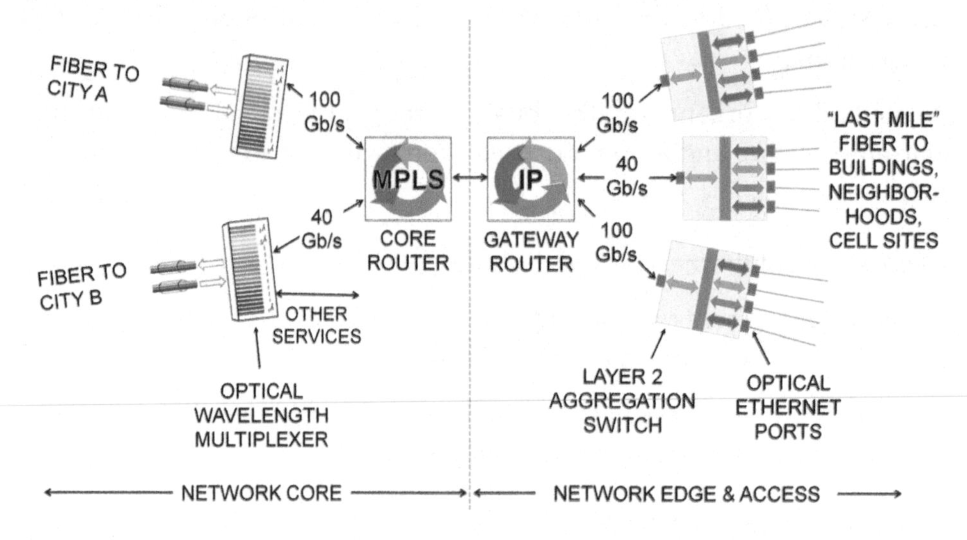

FIGURE 122 NETWORK CORE NODE

A knot, because in addition to connecting core fibers to other cities, the many local access fibers to buildings, neighborhoods, cell sites and everything else are also connected to the core at this node.

Figure 122 provides a closer look at the architecture of a core network node.

DWDM is used on core fibers to increase the capacity connecting routers in different cities to bit rates measured in the Terabits per second. MPLS (Chapter 17) is used to manage network circuit capacity.

Terminating a physical fiber means plugging it into an Optical Ethernet SFP transceiver inserted in a hardware port in a rack-mount device.

Routers are built with a relatively small number of hardware ports between which they can relay packets at line speed. Some of these ports would terminate core fibers, and others terminate aggregation devices.

Layer 2 switches (Section 15.4) with up to hundreds of hardware ports each are used to terminate the access fibers, which can number in the thousands.

Layer 2 switches are also the data concentration or aggregation devices, interspersing traffic from all of the access circuits into high-speed streams to feed to the router. Everything works in both directions at the same time.

Layer 2 switches also implement VLANs (Section 15.5), a critical tool for segregating different users on the same access fibers in cooperation with additional Layer 2 switches connected downstream, as described in the next section.

10.7 Metropolitan Area Network

Metropolitan Area Networks (MANs) are implemented by connecting Layer 2 switches with point-to-point Optical Ethernet connections. For redundancy, locations are connected in ring patterns, implementing two physical connections at each location for the lowest cost.

VLANs are used to separate customers' traffic, so customers cannot communicate between each other directly via the MAN, and customers do not receive each other's traffic.

MANs are a key part of the access network, the "last mile", part of the infrastructure connecting customers to a Central Office, which in turn provides connectivity to the network core.

Many different MANs would originate at a CO, connecting different types of customers. Large business customers, for example an Internet Service Provider or data center could be on their own MAN.

10.7.1 MANs to Office Buildings and Apartment Buildings

In multi-tenant office buildings, carriers implement mini-POPs, a Layer 2 switch in an equipment room to which access circuits going to different tenants in the building are connected. One MAN would connect these mini-POPs in different buildings to the CO.

In apartment buildings, Ethernet to the Suite is implemented in the same way, with a Layer 2 switch in an equipment room and fiber or copper to each suite. Another MAN would connect these buildings to the CO.

10.7.2 MANs to Neighborhoods

In neighborhoods, a MAN would connect Layer 2 switches contained in outside plant enclosures to the CO. An outside plant enclosure is a secured weatherproof cabinet located by the side of the road. From the enclosure, dedicated fiber access circuits are pulled as spokes to business customers.

In residential neighborhoods, a MAN would connect Layer 2 switches contained in outside plant enclosures to the CO. From these switches, fiber would lead to Passive Optical Network splitters on poles or in pedestals, where typically 32 fiber access circuits are pulled as spokes to residences and small businesses, time-sharing the backhaul to the switch in the enclosure.

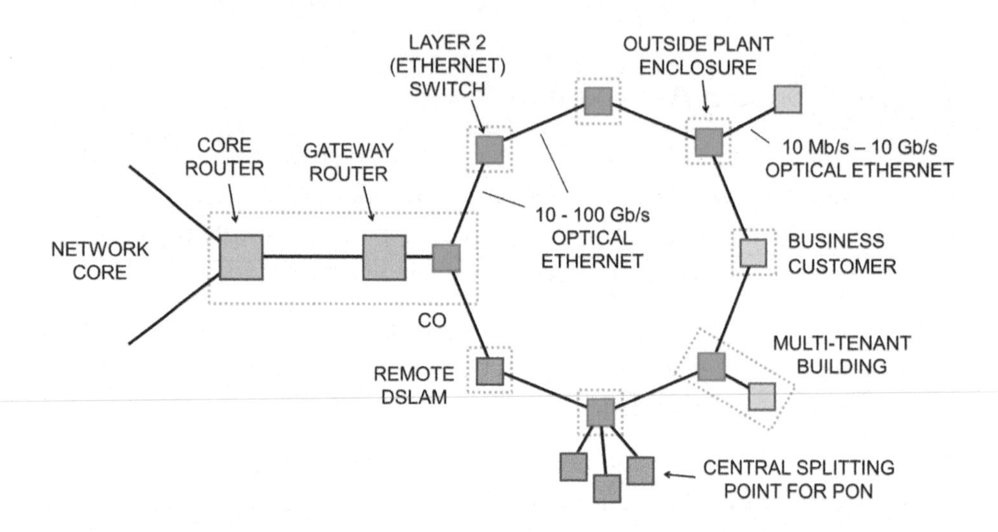

FIGURE 123 METROPOLITAN AREA NETWORK AROUND A CO

A MAN would also connect DSLAMs contained in outside plant enclosures to the CO. The DSLAM contains network-side modems that are hard-wired to subscriber loops to implement communications at up to 200 Mb/s over the last few hundred meters in brownfields, i.e., where copper loops are already deployed.

A business could be a station on a MAN, meaning there is an Ethernet switch owned by the carrier deployed at the customer premise, plus two fiber connections.

Being part of a MAN ring in this configuration means the customer would enjoy high availability - while paying for two optical accesses.

A business could also have a single fiber access as a spoke connecting to a station on the MAN that is located in a CO, POP, in a collocation, in an equipment room or in an outside plant enclosure, as illustrated at the top right of Figure 123.

In this case, the business pays only one optical access charge but would have time to restoration in the case of a cut line measured in hours and days instead of milliseconds.

10.8 Fiber to the Premise (FTTP, FTTH): PONs

Fiber optics are used for the vast majority of transmission systems connecting switching centers.

In the access network, fiber is now used routinely for new installations, both for business services and greenfield residential builds. Deployment of fiber in existing brownfield residential areas is an ongoing project.

This is referred to as Fiber to the Home (FTTH) or more inclusively, Fiber to the Premise (FTTP).

10.8.1 Passive Optical Network (PON)

A popular strategy for residences and small business is a Passive Optical Network (PON), the word *passive* meaning that there are no powered components in the access network, only fibers and light.

Some implementations use proprietary strategies for the optical design, others conform to the Optical Ethernet standards like 10GBASE-PR.

An optical splitter (and combiner) deployed in the access network is also called a Central Splitting Point.

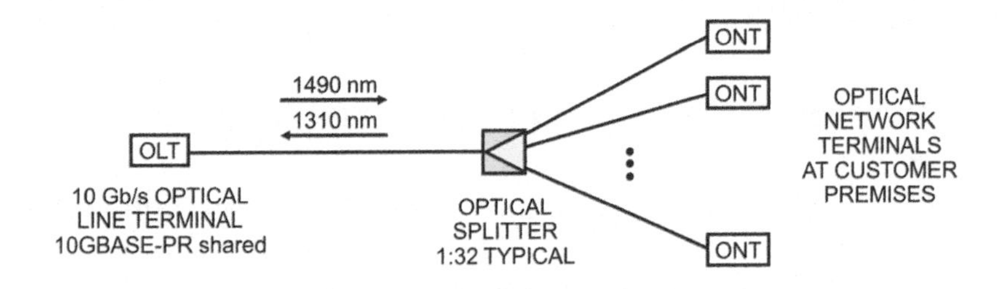

FIGURE 124 PASSIVE OPTICAL NETWORK: SHARED BACKHAUL FROM THE BLOCK

The optical splitter concentrates numerous fibers that lead to customers' Optical Network Terminals (ONT) to one fiber that is connected to the carrier's Optical Line Terminal (OLT). Today, a 1:32 split is common. In the future, that may rise to 1:64 and 1:128.

This is cheaper to implement than a dedicated fiber for each customer, since the 32 customers are sharing a single OLT and a single backhaul from the splitter to the network.

Engineered cables with factory-installed optical connectors at the same spacing as the building lots are often ordered and installed by telephone companies. This lowers the installation labor time and cost and increases the reliability of the connectors.

Typically, two wavelengths are used, one for communications downstream and one for upstream.

This requires encryption of the content downstream, as it is broadcast to all ONTs, and channelized Time-Division Multiplexing on the upstream, since all ONTs are sharing a single upstream path to the OLT.

10.8.2 Active Ethernet

Another strategy for fiber to the premise is called Active Ethernet, or Active Optical Network (AON), where each customer has a dedicated point-to-point fiber terminating on a dedicated OLT at the service provider.

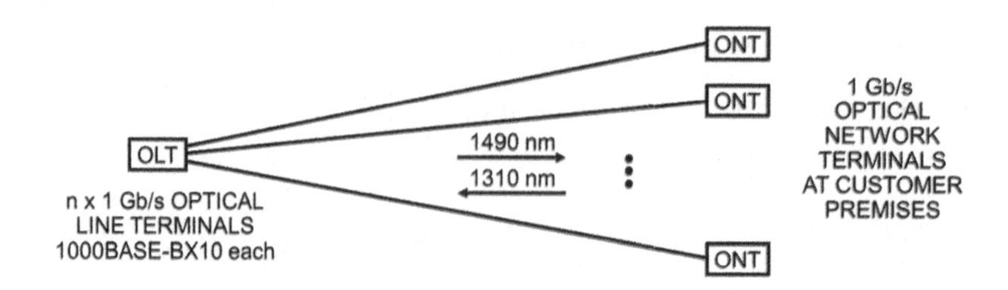

FIGURE 125 "ACTIVE" OPTICAL NETWORK: DEDICATED ACCESS FIBERS

Since point-to-point dedicated connections are the normal configuration in Ethernet, presumably the only reason the word "active" is included in the product name is to differentiate it from Passive Optical Network products.

OLTs are located in the outside plant, providing a fiber to each customer in the area plus a backhaul to the CO or POP. OLTs can also be located in the CO, meaning dedicated fiber from the CO to the customer.

10.8.3 PON Splitter Replaced with Layer 2 Switch

A variation of active Optical Ethernet to the premise is to replace the passive optical splitter of Figure 124 with a Layer 2 switch, turning the PON into something closer to the Metropolitan Area Network (MAN) of Section 10.7.

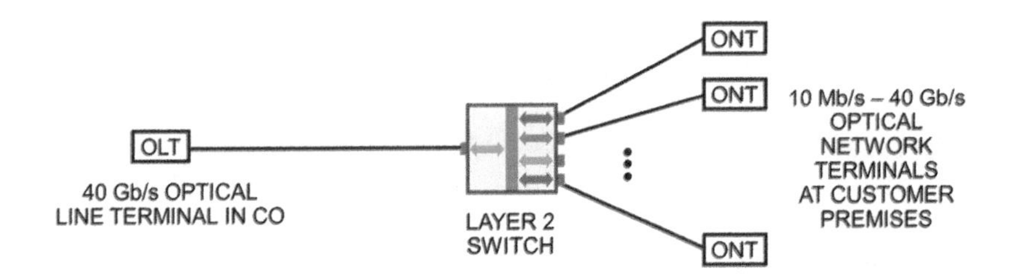

FIGURE 126 UPGRADING PON WITH AN ACTIVE LAYER 2 SWITCH

The switch implements on-demand use of the shared backhaul, improving performance over the channelized shared backhaul in a PON.

This design requires twice as many optical transceivers as a PON, and so is more expensive.

11 Copper and The PSTN

11.1 The Public Switched Telephone Network

Since telecommunications began on copper wires well over a century ago, there is a very large installed base of copper wires.

The locations where copper wires have been pulled past every house by the telephone company are called *brownfields*.

There are countless city centers and suburbs built from 1945 to 2000 that are brownfields, still in full operation.

Even though fiber is a superior choice, and only wireless is available in many parts of the world, this infrastructure is still in wide use, and is still generating hundreds of billions of dollars of revenue for Internet access and telephone service.

In this chapter, we begin with the Public Switched Telephone Network (PSTN) and Plain Ordinary Telephone Service (POTS): copper-wire-based infrastructure.

Understanding POTS and the PSTN is necessary for understanding the newer technology DSL running on the same infrastructure.

DSL is followed by cable modems, T1 and LAN cables... all technologies that operate on copper wires.

11.1.1 Basic Model of the PSTN

We begin with a basic model and will build on it in subsequent discussions.

At the top of Figure 127 is a telephone and a telephone switch. The telephone is located in a building called a *Customer Premise,* and the telephone switch is located in a building called a Central Office or CO. One could refer to the telephone as Customer Premise Equipment or CPE.

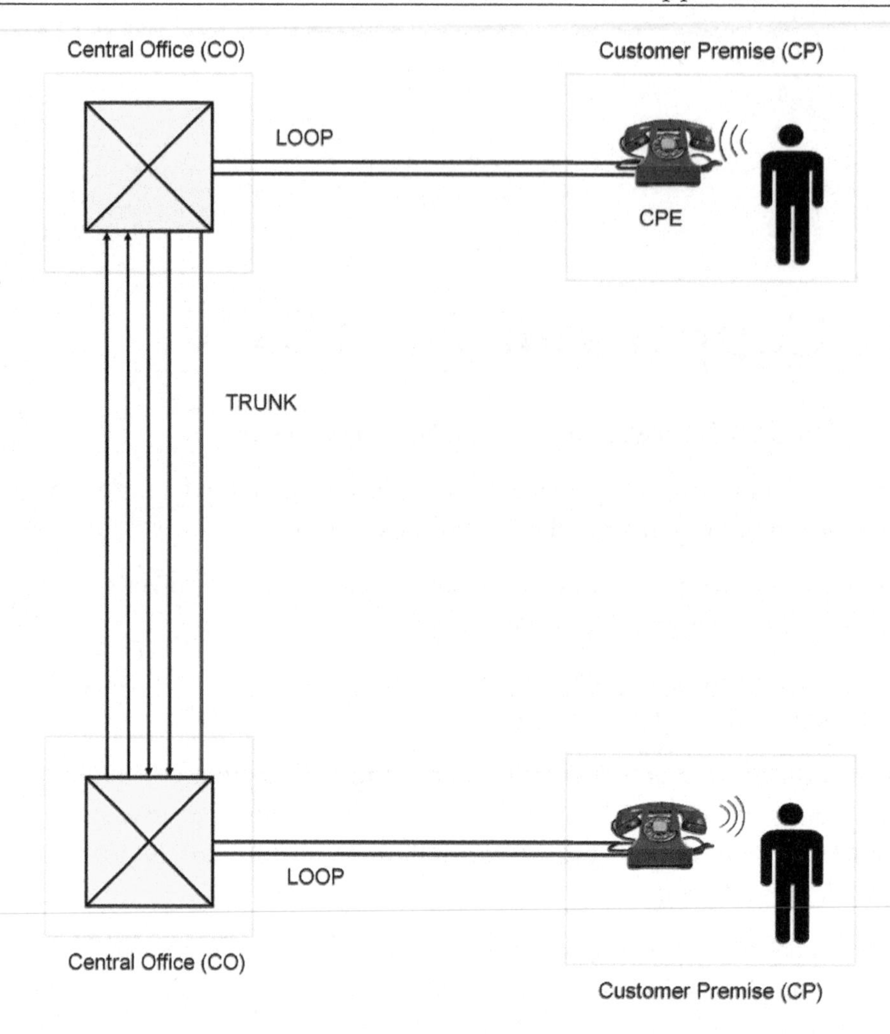

FIGURE 127 BASIC MODEL OF THE PUBLIC SWITCHED TELEPHONE NETWORK (PSTN)

11.1.2 Loops

The telephone is connected to the telephone switch with two copper wires, often called a local loop or a subscriber loop, or simply a *loop*. This is a dedicated access circuit from the customer premise to the network.

There is usually the same arrangement at the other end, with the far-end telephone in a different customer premise and the far-end telephone switch usually in a different central office.

Copper is a good conductor of electricity - but not perfect; it has some resistance to the flow of electricity through it. Because of this, the signals on the loop diminish in intensity or *attenuate* with distance.

The maximum resistance allowed is usually 1300 ohms, which is reached in 18,000 feet (3 miles or 5 kilometers) on standard-thickness 26-gauge cable, but could be as long as 14 miles or 22 kilometers on thicker 19-gauge cable.

This maximum loop length of 3 miles or 5 kilometers defined the traditional serving area around a Central Office, about 27 square miles or 75 km^2.

11.1.3 Trunks and Circuit Switching

Telephone switches are connected with *trunks*. While subscriber loops are dedicated access circuits, trunks are shared connections between COs.

To establish a connection between one customer premise and another, the calling party signals the network address (the telephone number) of the called party over their loop to the network, or more specifically, to their CO switch.

The switch makes a routing decision for the phone call then implements it by *seizing* an unused trunk circuit going in the correct direction and connecting the loop to that trunk.

The called party network address is signaled to the far-end switch, which connects the trunk to the correct far-end loop. When the far-end customer picks up the phone, an end-to-end connection is in place and maintained for the duration of the phone call.

When one end or the other hangs up, the trunk is *released* for someone else to use for connections between those COs. This method for sharing trunks is *circuit-switching*, called *dial-up* when telephones had rotary dials.

11.1.4 Remotes

Figure 127 was a model for the telephone network up to the end of the Second World War. With the subsequent suburban sprawl, it was not cost-effective to build COs every five miles or eight kilometers.

New subdivisions began to be served from remote switches or more simply, *remotes*, which are low-capacity switches in small above-ground buildings or underground controlled environment vaults.

As illustrated in Figure 128, the remote provides telephone service on copper loops in the subdivision and is connected back to the nearest big CO with a fiber *backhaul*.

The electronics and optics in the remote connect the fiber to the copper wires, or perhaps more precisely, take information received over the fiber and transmit it to the residences over copper loops, and vice-versa.

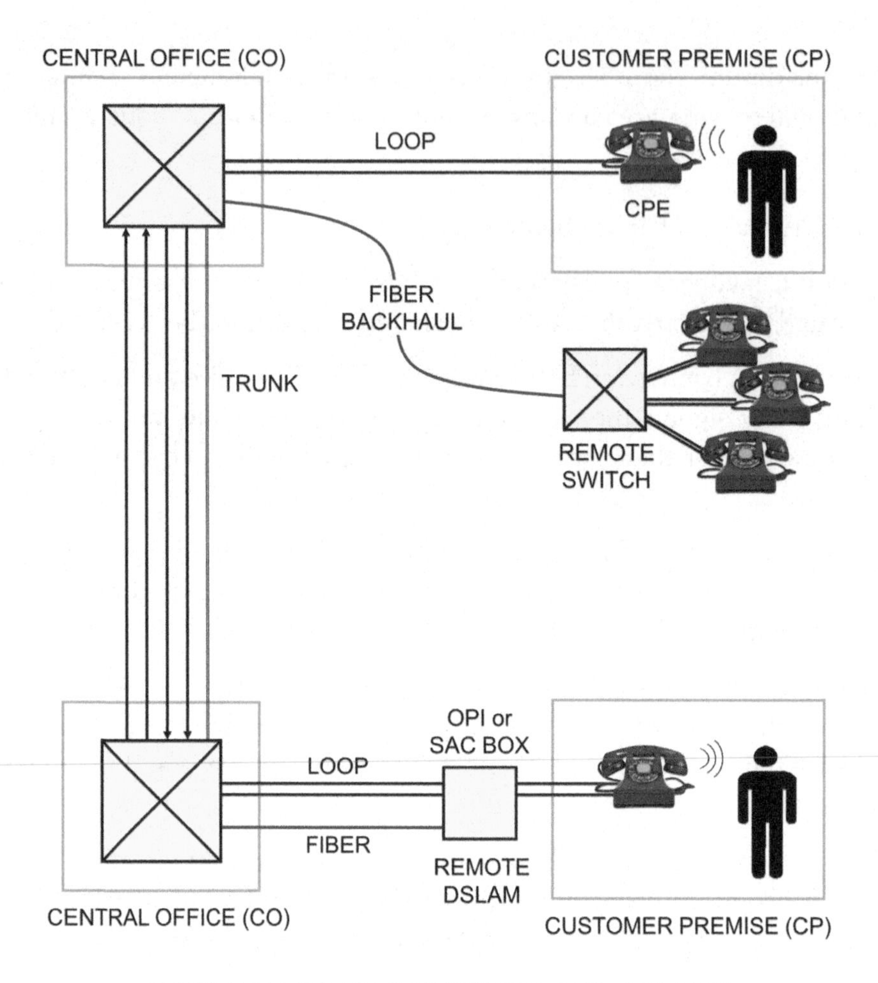

FIGURE 128 REMOTE SWITCHES AND DSLAMS

11.1.5 DSL and DSLAMs in Brownfields

A *brownfield* is a neighborhood where copper wires have been deployed. In the 1990s, modem technology called Digital Subscriber Line (DSL) began to be deployed on existing copper loops to connect a modem at the customer to a modem in the CO, for high-speed Internet access coexisting with telephone service on the loop.

To increase the achievable bit rate, the distance between the modems was shortened by moving the network side modem, contained in a device called a Digital Subscriber Line Access Multiplexer (*DSLAM*) into the neighborhood.

The equipment and wiring in neighborhoods, along with transmission systems carrying trunks is collectively referred to as the *outside plant.*

This remote DSLAM is usually located in a small enclosure bolted on to the side of a larger enclosure called an *Outside Plant Interface* (OPI) or Serving Area Concept (*SAC*) box, illustrated in Figure 129.

FIGURE 129 OPI WITH DSLAM BOLTED ON THE SIDE

The OPI or SAC box is a wiring connection point in the neighborhood where wires in a feeder cable from the CO are connected to wires in distribution cables running down streets. It is a location where a network-side DSL modem can be jumpered on to the existing copper loop.

Fiber-optic and power cables connect the remote DSLAM upstream to the CO and from there to the Internet.

Generically, the DSLAM is a fiber terminal: where the fiber ends. Customers are connected to the fiber terminal with pairs of DSL modems over the existing copper wires for the last few hundred meters.

This is increasingly being seen as a temporary measure while waiting to pull fiber to the home in an older neighborhood. Eventually, most customers will have their own fiber terminal... connecting to copper wires and Wi-Fi inside the house.

11.1.6 Greenfields: PONs on Fiber to the Premise

In *greenfields,* i.e., newly-constructed neighborhoods and multi-tenant buildings, where the cabling is the initial installation, fiber to the premise is routinely installed.

Passive Optical Network (PON) technology is usually employed, where typically 32 customers time-share a fiber connection to the network. One fiber backhaul towards the network is connected in a Central Splitting Point via lenses and mirrors to 32 fibers leading to customer premises.

Only one customer can transmit at a time, so the uplink is shared in a round-robin fashion, and each user is reserved a fixed amount of capacity on the uplink whether they are using it or not. In the Transmission Systems chapter, this is called *channelizing* or *channelized multiplexing*.

11.1.7 Active Ethernet to the Premise

Active Optical Ethernet may also be the physical access circuit. In this case, the customer's fiber terminates on its own port on a Layer 2 switch, located either in the neighborhood or at a wire center. Active Ethernet is routinely implemented for business customers.

Customers have the possibility of transmitting upstream any time they like instead of in time slots. It is more efficient than channelizing and gives users higher upload speeds for the same capacity backhaul... but compared to the PON of Section 10.8.1 requires 31 more network-side fiber transceivers, so is more expensive to install and maintain.

Fiber to the premise is covered in more detail in Chapter 10.

11.2 Analog

The technique for representing information on an ordinary local loop is called *analog*. This term is often thrown about with little regard for its actual meaning, so we will spend a bit of time understanding what is meant by "analog".

11.2.1 Analog Signals

The term analog comes from the use of a microphone in the handset of the telephone. A simple type of microphone, such as those in the handset of a telephone, has a plastic housing, a paper diaphragm and carbon particles between the two.

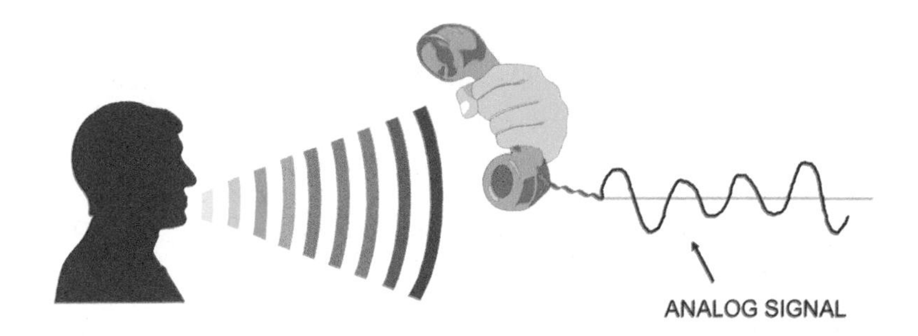

FIGURE 130 THE VOLTAGE ON THE WIRES IS AN ANALOG OF THE STRENGTH OF SOUND PRESSURE WAVES

When someone speaks, sound pressure waves come out of their mouth. The person using the telephone holds the microphone in front of their mouth, so that the sound pressure waves push on the paper diaphragm.

This has the effect of compressing the carbon particles in the microphone, changing its electrical characteristics... the microphone's capacitance, to be precise.

The fact that the electrical characteristics of the microphone change as the sound pressure waves hit it can be used to make a voltage on the telephone wires change.

> This voltage is a direct representation or *analog* of the strength of the sound pressure waves: as a pressure wave pushes on the microphone, the voltage increases; as it stops pushing on the microphone, the voltage reverts to where it was.

This is all that is meant by "analog": representation. The voltage on the wires is an analog of the strength of the sound pressure waves coming out of the speaker's mouth. This voltage could also be called an analog signal.

At the other end, a speaker is used to create sound pressure waves based on the received analog signal. A speaker is an electromagnet glued onto a paper diaphragm.

The voltage that is the analog is applied to the electromagnet, causing the paper diaphragm to move back and forth, creating sound pressure waves, which are hopefully a faithful reproduction of the original sound pressure waves coming from the speaker's mouth.

11.2.2 Analog Circuits

The voltage carried on the loop is an analog signal. People then stretch this terminology and refer to the two copper wires that form the loop as an "analog circuit", which is not very accurate.

The only thing analog in this story is the method for representing speech on copper wires using electricity.

It would be more precise to call the loop "two copper wires that were designed to carry a voltage that is an analog of the strength of the sound pressure waves coming out of the speaker's face".

It is possible to use digital techniques on the same copper wires.

11.3 Capacity Restrictions

Once the method of representing speech on copper wires using an analog is understood, the next question would be: how accurate does this "analog" have to be? What degree of fidelity is required in representing the sound pressure waves coming out of the speaker's throat? How faithful do the sound pressure waves reproduced at the far end have to be?

With a bit of reflection, we realize that the fundamental requirement of the telephone system is to communicate information between people.

The first design decision made was that speech and hearing would be used as the method of communications.

Once this highest-level design decision was made, it is necessary to understand what speech is before the characteristics of the local loop could be designed.

11.3.1 What is Speech?

Speech is a form of sound. A textbook would define "sound" as pressure waves in the air, meaning that at a given point, the air pressure rapidly increases and decreases in a cyclical fashion in time.

Figure 131 is a representation of a speech analog generated by a stereo microphone. The upper part of the diagram shows on the vertical axis the sound pressure, interpreted by your brain as volume; and the lower part of the diagram shows on the vertical axis the frequency of the compression-rarefaction cycle, interpreted by your brain as pitch. The horizontal axis is time.

An increase in air pressure causes the molecules to be pushed closer together or compressed; a decrease in air pressure allows the molecules to spread out or be rarefied. Sound is the *compression* and *rarefaction* of air molecules. Of course, water molecules, plaster, glass and other substances can also support this phenomenon to some extent.

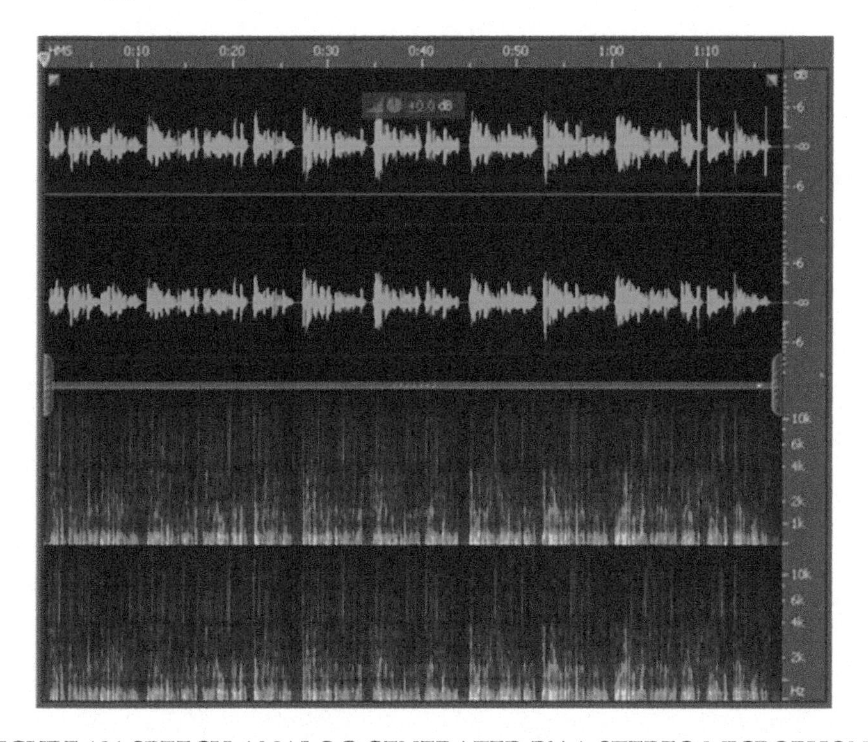

FIGURE 131 SPEECH ANALOG GENERATED BY A STEREO MICROPHONE

Sound pressure waves coming out of the speaker's face vibrate rapidly, that is, go through cycles of compression and rarefaction. If this vibration occurs between 20 and 20,000 cycles per second, the sound pressure waves are said to be audible by the human ear.

11.3.2 Do Trees Falling in the Forest Make a Sound?

Understanding that sound is cycles of varying air pressure, and knowing that if this occurs between 20 and 20,000 times per second it is audible still does not tell us how faithful the voltage analog must be, and how faithfully the sound must be reproduced at the far end.

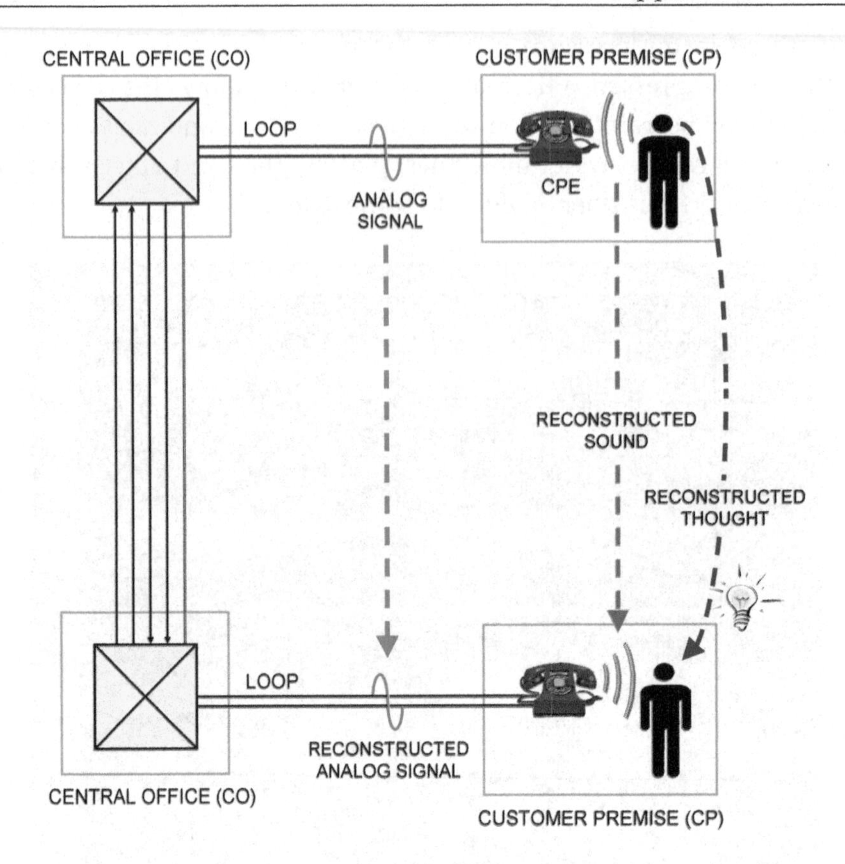

FIGURE 132 RECONSTRUCTING SOUNDS VS. RECONSTRUCTING THOUGHTS

An age-old question is: if a tree falls in the forest, and no one is there to hear it, does it cause a *sound*?

That depends whether you believe sound is pressure waves: air molecules being compressed and rarefied as per the preceding textbook definition; or if you believe sound is the sensation one gets in one's brain when one hears the sound pressure waves.

The two choices in designing the telephone system are then to either:

a) Reproduce sound pressure waves coming out of the speaker in the far-end telephone exactly as they entered the microphone in the near-end telephone; or

b) Reproduce the sensations in the listener's brain the same as they would experience were they speaking directly to the other person.

The difference between these two ideas is that the brain is a hugely complicated processing instrument, and it is possible to play different stimuli at it and get the same response.

Each choice has a dramatically different implication for the cost of implementing the system.

11.3.3 The Voiceband

What answer did Alexander Graham Bell choose? Answer (b).

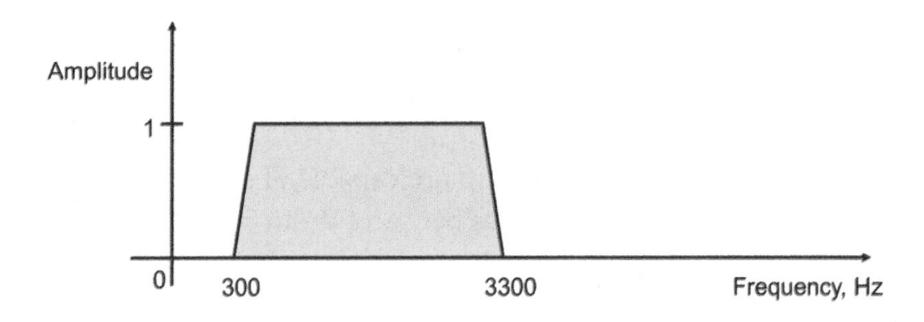

FIGURE 133 THE VOICEBAND

Based on testing human beings' ears, throats and brains, combined with a technical innovation that extended the achievable transmission range, led us to transmit the information in the frequency range between 300 and 3300 Hz.

Hertz (Hz) is the unit for frequency, or changes per second. The range or band of frequencies from 300 to 3300 Hz is called *the voiceband*.

Figure 133 is an idealized representation of the voiceband, with frequency on the horizontal axis and amplitude or intensity on the vertical axis.

It shows that any electricity vibrating at least 300 times per second and less often than 3300 times per second will be passed (indicated by a 1).

Any electricity vibrating less often than 300 times per second will be suppressed (indicated by a 0). Similarly, any electricity vibrating more than 3300 times per second will be suppressed.

Only electricity vibrating within the band 300-3300 Hz will be transmitted.

The suppression of energy outside this frequency band 300-3300 Hz is implemented with simple electrical circuits called filters. There is a filter in the telephone and a filter in the switch in the CO.

11.3.4 Bandwidth

For our purposes, bandwidth means capacity. In the analog world, capacity is measured literally by the width of the frequency band supported on the physical medium by the service you are paying for.

For the voiceband, the bandwidth is the interval between 3300 and 300 Hz, which is 3000 Hz or 3 kHz for short.

This 3 kHz bandwidth is a standard capacity provided for ordinary telephone service by all telephone companies.

11.3.5 Why Does the Voiceband Stop at 3300 Hz?

Why 3300 Hz? It would be technically possible to provide a greater bandwidth for telephone service, resulting in crisper, clearer sound.

The two wires that make up the loop are capable of supporting electricity vibrating more often than 3300 times per second – in fact, DSL technologies require electricity vibrating at frequencies measured in the millions of times per second.

The users' ears and brains are capable of detecting sound pressure waves vibrating more often than 3300 times per second – the human hearing range is traditionally thought to extend up to 20,000 Hz.

So why would the capacity a user is allowed to employ be purposely limited to 3 kHz, even though the wires are capable of more than that, and the users are capable of more than that? The answer is, as usual, money.

This narrow frequency band was chosen based on studying people's ears, throats and brains, to determine the minimum capacity necessary to meet the requirements.

Returning to the question of trees falling in the forest: the sound pressure waves at the far end are not reproduced exactly as they were at the near end; in fact, they are quite muffled and distorted, missing most of the higher frequencies.

The sound is reproduced just well enough so that the listener can recognize the speaker and understand what the speaker is saying, thus meeting the requirement to communicate information using speech and hearing.

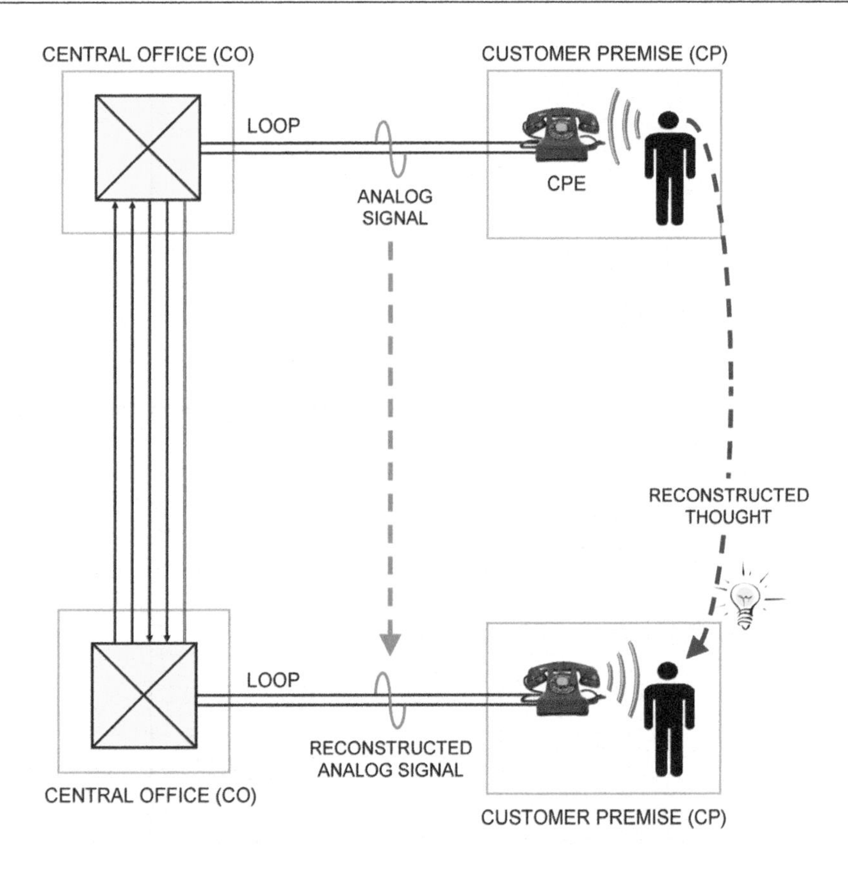

FIGURE 134 COMMUNICATING THOUGHTS

We are interested in transmitting the minimum required to meet that objective since there is a direct relationship between the capacity a user can employ on the access circuit and the cost of transmitting the information long-distance.

11.3.6 Problems With Voiceband Restrictions

It turns out that the voiceband is not quite enough bandwidth to be able to understand everything the speaker is saying!

In particular, it is difficult to tell the difference between "S" and "F" over a telephone. This is because the frequency of sound pressure wave that distinguishes "S" from "F" is above 3300 Hz… which is not transmitted over the phone system.

Phonetic alphabets, such as one used by military forces, use words to communicate each letter, for example, "S as in Sierra" and "F as in Foxtrot".

A	Alpha
B	Bravo
C	Charlie
D	Delta
E	Echo
F	Foxtrot
G	Golf
H	Hotel
I	India
J	Juliet
K	Kilo
L	Lima
M	Mike
N	November
O	Oscar
P	Papa
Q	Quebec
R	Romeo
S	Sierra
T	Tango
U	Uniform
V	Victor
W	Whiskey
X	Xray
Y	Yankee
Z	Zulu

A	Are
B	Bee
C	Cue
D	Djinn (gin)
E	Ewe
F	Fore
G	Gnu
H	Hour
I	I
J	Jee
K	Knot
L	Llama (yama)
M	Mnemonic
N	Not
O	Oh
P	Psalm
Q	Quay (key)
R	Rec
S	Sea
T	Tea
U	Urn
V	Vee
W	Why
X	Xylophone
Y	You
Z	Zee

FIGURE 135 STANDARD AND ALTERNATE PHONETIC ALPHABETS

As illustrated in Figure 135, one could also say "A as in Are", "S as in Sea", "C as in Cue" and "E as in Eye" to liven things up. If that doesn't get the listener confused, there's always "E as in Ewe" and "Y as in You".

11.4 Problems with Analog Transmission

11.4.1 Attenuation and Amplifiers

Aside from bandwidth restrictions, the chief impediments to transmission over analog circuits are noise and attenuation, both of which affect the capacity of the circuit to transport information.

Attenuation on wired circuits is caused by the physical characteristics of the copper wires that make up the circuit. Since copper is not a perfect conductor, and has some resistance, some of the transmitted energy is turned into heat and the signal level decreases or attenuates with distance from the transmitter. On wireless systems, attenuation happens due to the spreading effect of the waves.

In both cases, too far away from the transmitter, the signal will "disappear into the noise", that is, the signal level will become less than the noise level on the line, and it would be impossible to faithfully reconstruct the speech.

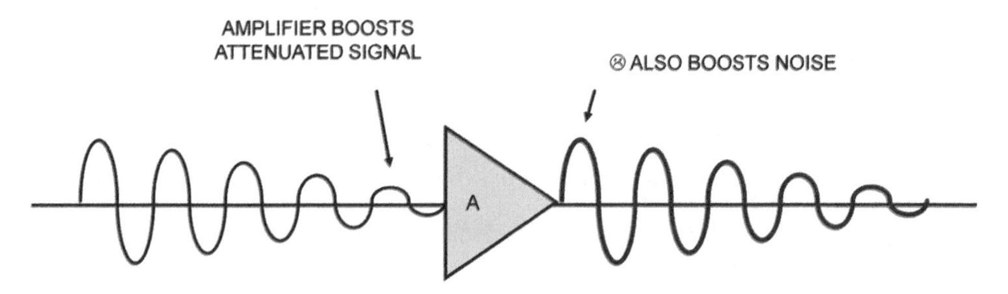

FIGURE 136 ATTENUATION AND AMPLIFIERS

Before this happens, the signal must be amplified at regular distance intervals to boost it back up. The device that performs this function is called an amplifier. It multiplies or boosts the signal on its input by a certain factor.

The problem is noise that is added to the signal during transmission, before the signal reaches the amplifier. The noise and signal are combined; when the signal is boosted up by an amplifier, so is the noise.

This is the fundamental problem with analog transmission: the transmission system both attenuates the signal and adds noise to it; then to boost up the signal, the amplifier also boosts up the noise.

An analog signal becomes noisier and noisier as it passes through each amplifier along a transmission system.

11.4.2 Electro-Magnetic Interference

Noise comes in many forms. On copper-wire access circuits, the most problematic is caused by radio waves, or more precisely, Electro-Magnetic Interference (EMI)

Copper wires act like antennas. When a radio wave impinges on a wire, it induces electricity that adds to the desirable signal being carried on the wire. The source of such interfering additive noise includes television broadcasts, microwave ovens, computer chips, cellular radio base stations, wireless LANs and other sources.

Glass – fiber optics – does not act like an antenna and does not pick up this kind of interference.

11.4.3 Crosstalk

Crosstalk is a specific type of EMI, the transference of energy from one wire to another via electro-magnetic radiation. Usually, this happens when two circuits are in the same cable: a signal placed on one circuit will create a magnetic field that passes through the other circuit and induces current on it.

This is why sometimes you can hear other people talking on a regular wired telephone. The annoyance factor decreases with comprehensibility.

11.4.4 Impulse Noise

Impulse noise appears like spikes of voltage on a circuit. This is caused by lightning striking the wires, by the spark that jumps across the contacts of a switch just before it closes, and when the brushes on an electric motor pass the unpowered portion of its armature.

In days past, this could be seen as white dots on a television screen when a drill or vacuum cleaner is operated in close proximity. Impulse noise is not additive noise – it hard-limits the signal to maximum, and causes a burst of errors to happen.

The most popular way to deal with impulse noise is transmit bits, not analog voltages, to format bits into frames with error detection, and re-transmit a frame if there was a spike on the line that caused a burst of errors to happen.

This is covered in detail in Chapter 4..

11.5 Plain Ordinary Telephone Service (POTS)

Basic telephone service is called POTS: Plain Ordinary Telephone Service in the business, and sometimes referred to as *dial tone* or *individual residence line service.*

As illustrated in Figure 137, this service consists of a rotary dial telephone connected to a line card in a telephone switch by a two-wire copper loop.

11.5.1 Tip and Ring

The two wires that make up the loop are sometimes referred to as *tip* and *ring*. These names come from the first kind of telephone switch, which was a board holding female jacks, to which loops and trunks were connected.

To connect a loop to a trunk, an operator would plug one end of a patch cord into the jack for the loop and the other end into the jack for the trunk.

The connectors on the patch cord were designed so that one of the wires was attached to the metal tip of a male connector and the other wire was attached to a metal ring below the tip.

Plugging the connector into the jack connected the two wires that made up the loop at the same time.

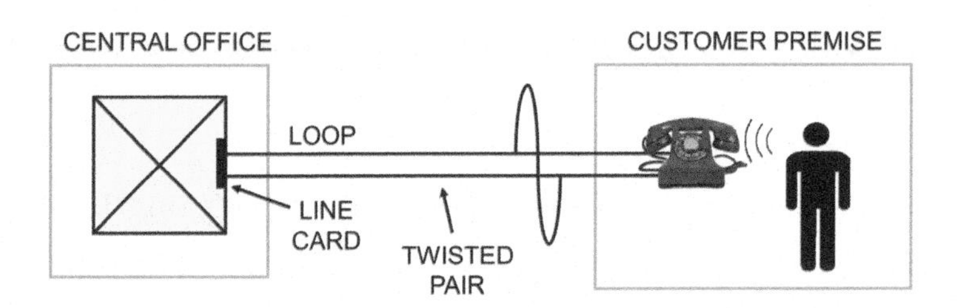

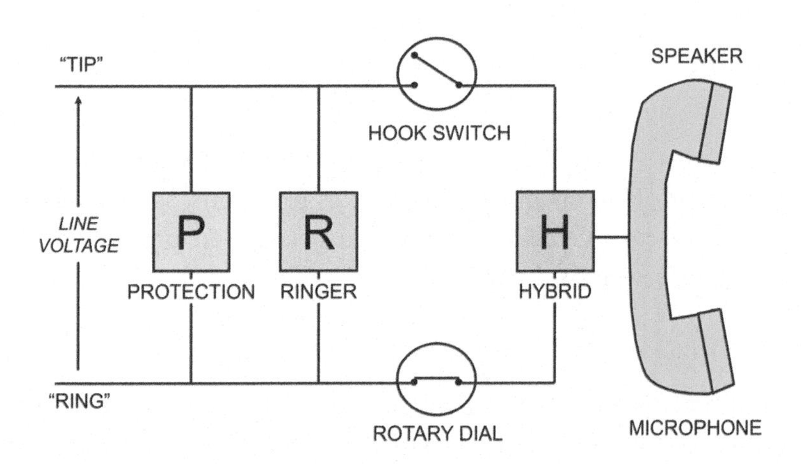

FIGURE 137 PLAIN ORDINARY TELEPHONE SERVICE

The main components of a telephone are the microphone, speaker, hybrid converter, hook switch, dial switch, ringer and protection.

11.5.2 Twisted Pair

A problem with using two-wire circuits is that they act like antennas. Loop antennas. The amount of energy picked up by a loop antenna is proportional to its area... and here, we are connecting a 3 mi. / 5 km diameter loop to the telephone (!).

To minimize the amount of noise picked up on the wires, they are covered in plastic, then twisted together.

Since there is plastic on the wires, they still act electrically like one big current loop, but from an antenna point of view they appear as a series of small loops. The small loops have a smaller area than the big loop, and so this minimizes the antenna effect of the wires.

Since there are two wires twisted together, we call them *twisted pair*. Twisted pair is used for mostly all cabling, including telephone wires on poles, inside wiring and data cabling – LAN cables have four twisted pairs.

11.5.3 Line Card

The twisted-pair loop is terminated on the network side on a *line card*.

A line card is traditionally a small fiberglass board populated with a number of components, integrated circuits and connectors. This line card is plugged into a slot in a drawer, in a shelf, in a rack, that is part of a traditional telephone switch.

In newer applications, the line card might be part of a gateway that converts between POTS and Voice over IP, discussed in detail in Section 12.6.

The line card implements quite a number of functions, sometimes referred to by the acronym BORSCHT: battery, overvoltage protection, ringing, supervision, codec, hybrid and testing.

11.5.4 Microphone and Speaker

The microphone is a kind of transducer, creating a voltage based on sound pressure waves. The value of this voltage is a representation or analog of the strength of the sound pressure waves coming out of the speaker's throat.

The voltage is carried from the telephone over the loop to the line card at the near end, where it is digitized by the codec and transported by the telephone network to be reproduced by the far-end line card and carried by the far-end loop to the far-end telephone.

The speaker, as might be imagined, works in a manner opposite to the microphone: it uses received voltage to create sound pressure waves that are directed into the user's ear.

11.5.5 Balanced Signaling

Voltage is always measured as a difference between the voltage on one object and the voltage on another.

In many cases, one object is the earth and the other is a wire, so the voltage measurement is with respect to the ground.

This is not the case with a telephone loop. On a telephone loop, the voltage is measured between the two wires that are the loop, not between the earth and the wires.

Balanced signaling is used. This means that if the voltage on one wire with respect to ground is some positive value, the voltage on the other wire with respect to ground will be the same value, but negative. Since added noise will be the same on the two wires, when measuring the voltage between the two wires at the receiver, the signal is doubled and the noise is canceled.

11.5.6 Two-Way Simultaneous

The two wires that are the loop are used to transmit information in both directions at the same time.

Both the telephone and the line card cause voltage analogs of sound to be placed across the two wires of the loop. The voltages from the devices at each end are added together.

11.5.7 Hybrid Transformer

The voltage for each direction is separated by a device inside the telephone called the hybrid, which has the two-wire loop on one side and two circuits on the other side, one for the speaker and the other for the microphone.

A similar function is implemented on the line card, connecting the loop to the transmit and receive pins of the codec.

11.5.8 Battery

In addition to the voltage analog of sound, which might be thought of as an AC (or varying) signal, the line card also places a DC (or steady) voltage across the two wires that make up the loop.

This voltage is called battery in the business, and is used to power the telephone. It is nominally -48 volts, measured from ring to tip.

11.5.9 Lightning Protection

Another item on the diagram is the protection circuit across the loop. This is to protect the telephone user from being electrocuted, if lightning hits the loop or a high-voltage electrical transmission wire touches the loop.

There are in fact three levels of protection: a fuse on the line card will blow if too much current passes through it, circuitry on the demarc or demarcation point where the telephone company's wires connect to the customer's wires that will fall to ground if the voltage is too high, and third, inside the telephone a circuit that will short-circuit the loop if the voltage across the loop is too high.

11.5.10 Supervision

Two other components of the telephone, the hook switch and ringer, are used for supervision.

Supervision means regardless of to whom you wish to speak, and regardless of what you are going to say to them, you must indicate to the other end of your loop that you want to start doing all of this.

The hook switch in the telephone is normally open, so the two wires that make up the loop are not connected, and no electricity or current is flowing around the loop.

To initiate communications, the user picks up the handset (goes *off-hook*), which causes the hook switch to close, connecting the two wires together, which then allows the line voltage to push current around in a… loop. This is why they are called loops.

This type of supervision is called *loop start signaling*: the two wires are connected, forming a loop and allows current to flow in a loop.

The line card on the telephone switch detects this current and acknowledges with a dial tone (assuming you have paid your bill).

There are variations on this theme used in other applications such as PBX switches, such as ground start signaling, where one of the wires is plugged into the ground, so the current flows along one wire then back through the ground; reverse battery signaling where the positive and negative line voltage is reversed; and wink start signaling where that is done for a short interval then returned to normal value.

For supervision in the other direction, the switch indicates it wants to initiate communications by having the line card place a ringing signal on the loop.

This is yet another voltage, one that varies 20 times per second. It is applied to the line for two seconds then not for four seconds in a repeating cycle.

When your phone rings, it is on-hook. This means that the hook-switch is open, so the current pushed by this ringing signal flows through the ringer as shown in Figure 137 – originally two brass bells with a clapper between that would move back and forth 20 times per second for two seconds then rest for four seconds. The user acknowledges by going off-hook.

The line voltages are nominally as follows:

- On Hook: -48 Volts DC
- Ringing: -48 Volts DC, plus 100 Volts RMS @ 20 Hz
- Off-Hook: -7 to -12 Volts DC.

11.5.11 Call Progress Tones

Dial tone is a type of call progress tone. There are many others, such as busy, fast busy signals, ringback, congestion, sounder and howler tones. These are generated by the switch to inform the user of different conditions.

Some of the call progress tones, such as dial tone and fast busy are generated by the near-end switch. Busy signals are generated by the far-end switch.

11.6 Network Addresses: Telephone Numbers

Once your request to communicate is acknowledged with a dial tone, it is necessary to inform the network where the call is to be connected.

In general, *network address* is the name given to the piece of information used to identify the final destination of a connection across a network. For POTS, network addresses are of course called telephone numbers.

11.6.1 Dialing Plan

The length of the telephone number, that is, the number of digits that have to be dialed, and how the addresses are assigned to subscribers is called a numbering plan or *dialing plan*.

In days past, the North American Numbering Plan for telephone numbers was composed of digits with specific purposes. Restrictions were placed on the values of various digits so that dumb mechanical and analog switches could distinguish between them.

Addresses were originally of the form NBN-NNX-XXXX, where

- N is any number from 2 – 9
- B is any number from 0 – 1
- X is any number, and
- The first three digits were the area code,
- The next two were the CO code,
- The next one identified the switch in that CO,
- The last four identified the physical pair of wires.

The user had to dial anywhere from five to ten of these digits, sometimes prefaced with a 1 to indicate the desired destination.

All of this has changed with the introduction of computer-based switches, computer control systems for the switching, and the need for more network addresses: the last area code under this plan was assigned in the 1990s!

Today, telephone numbers can be of the form NXX-NXX-XXXX, and the "area code" no longer necessarily corresponds to a unique geographic area nor necessarily means long distance charges will apply.

To provide new network addresses, area codes are split and overlaid, and users are required in these locations to dial ten digits.

The physical destination corresponding to any particular address is now stored in a database in a computer.

11.6.2 Address Signaling

The last main aspect of POTS is address signaling, and in particular, how the network address of the called party is indicated or signaled from the calling party's telephone to the CO switch.

The first kind of CO switch was a person using a switchboard and patch cords to connect loops and trunks. In this case, the mechanism for the caller to signal the network address of the called party was for the caller to use their voice and identify the desired called party by name.

11.6.3 Pulse Dialing

To signal numerical addresses from the telephone to the switch, a rotary dial was added to the telephone. This dial was a metal disc with holes, connected to a dial switch inside the telephone with a spring.

To indicate a digit, the caller placed a finger in the hole in the dial corresponding to that digit, rotated the dial to a stop position, then removed their finger from the hole. As a spring rotated the dial back to its rest position, another spring would cause the dial switch to open and close a number of times corresponding to the desired digit.

FIGURE 138 ROTARY DIAL TELEPHONE

Since the hookswitch is closed at this time, opening the dial switch would momentarily interrupt the flow of electricity on the loop, then closing the dial switch would allow the resumption of the current, then interrupted, then resumed, and so on.

From the line card point of view, this would appear as pulses of electrons coming down the loop; viewed on a voltmeter, it would appear as square pulses of voltage, and so this signaling technique is called *pulse dialing*.

One question that arises is: what is the difference in function between the hook switch and the dial switch, other than the fact that the hook switch is normally open and the dial switch is normally closed?

The answer: nothing. Both switches do the same thing: they either make or break the loop… a Flintstones-era technology called "make-or-break" signaling. Knowing this, it should be possible to signal network addresses using the hook switch on a telephone…

The hook switch must be depressed for 45 milliseconds, then released for 55 ms, which would be one pulse. This is repeated the number of the digit, for example, four times to indicate a "4". Then a pause, the inter-digit interval of 700 ms is required, then the next digit is signaled.

With some practice, it is not difficult to signal "4-1-1" using this method. Be sure to hang up before being charged for directory assistance if you try this and succeed!

There are two problems with pulse dialing: first, it is ridiculously slow – a 0 is not zero pulses, but ten pulses, so it takes 1.7 seconds to signal a 0 including the inter-digit interval. Second, the only device that the signaling goes to is the line card on the switch; the make-or-break-the-loop signaling stops there.

11.6.4 DTMF: "Touch Tone"

The improvement on pulse dialing was called Touch Tone. Pulse dialing is very slow. Touch-tone is faster.

Touch-tone is actually a registered trademark of AT&T. The generic name for this type of signaling is Dual Tone Multiple Frequency or DTMF signaling. This is an address signaling mechanism that uses combinations of tones, i.e., single pure frequencies, to represent buttons being pressed, and the buttons each represent a number.

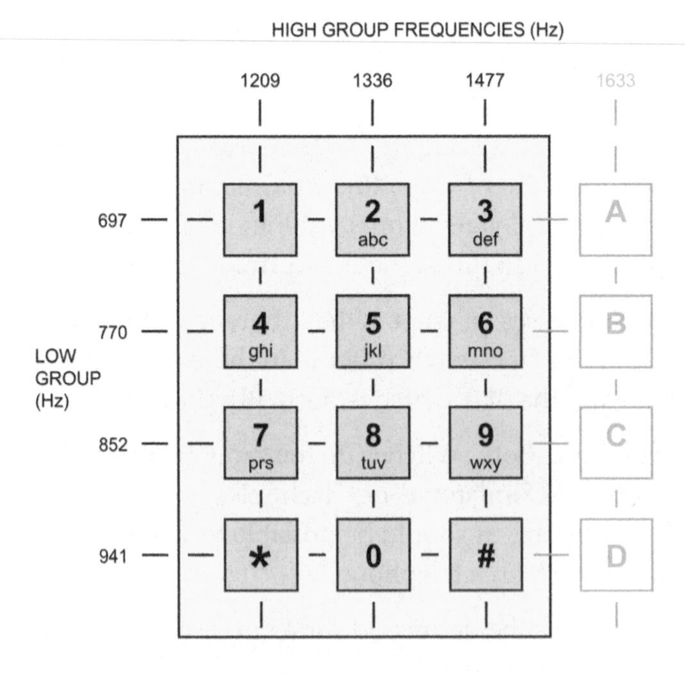

FIGURE 139 DTMF

On a standard telephone keypad, there are 12 buttons: 0 – 9, star (*) and octothorpe (#). Octothorpe is commonly also called the "pound" key. Young whippersnappers call it "hashtag".

The reason this is called a dual tone signaling system is that rather than defining one tone per button, which would require 12 tone generators in the telephone and 12 tone detectors on the line card to represent the 12 buttons on a normal telephone keypad, the tones are instead arranged in a grid pattern, and two tones are generated to represent each button.

For example, to signal the number 4 to the line card, pressing the button marked four causes a tone at 770 Hz and a tone at 1209 Hz to be generated. Two tones per button requires only 7 tones (3 + 4) instead of 12 (3 x 4), and so is cheaper to implement: only 7 tone generators in the telephone and only 7 tone detectors in the line card instead of 12.

DTMF signaling is faster than pulse dialing, as the button must be depressed for a minimum of 50 ms and the inter-digit interval is 50 ms – for all buttons. A zero requires 100 ms (0.1 sec) to signal using this method, compared to 1.7 seconds using dial pulsing.

11.6.5 In-Band Signaling

Another advantage of DTMF is that it is an *in-band* signaling mechanism. All of the tones are within the voiceband: 300 - 3300 Hz. The capability put in place for voice communication is also being used to signal control information, using tones within the frequency band used for voice.

This allows the re-use of DTMF signaling end-to-end between customer premise equipment after the call is completed: for example, from a telephone to a voice mail system.

11.6.6 "Hidden" Buttons

Though a standard telephone keypad has 12 buttons, there are actually 16 buttons defined for DTMF. The "hidden" four buttons are labeled A – D and share the high group frequency 1633 Hz.

These tones are used only for very special signaling situations, like Call Waiting with Caller ID.

11.6.7 Caller ID

Caller ID is another example of in-band signaling. The Caller ID is delivered to the telephone by a 1200 b/s modem in the line card that operates in the voiceband. With standard Caller ID service, the modem transmits ASCII code representing the date, time, calling number and possibly calling name, beginning 0.5 seconds after the first ring and ending before the second ring. During this time, the telephone is on-hook, so the called party does not hear the modem signal being transmitted in the voiceband.

The tones corresponding to the "hidden" four buttons, A – D are used only for very special in-band signaling situations; one example is to support caller ID with call waiting service, also called Call Waiting ID service, where the ID of a second caller is displayed while the line is already in use.

Since the Caller ID is delivered with a modem signal in the voice band, if no special measures were taken, the called party would hear the hissing of the modem signal on the line delivering the ID of the second caller while the first call is in progress. Plus, voice on the line might interfere with the accuracy of detection of the modem signal.

To deal with this problem, a dual-tone CPE alerting signal of 2130 + 2750 Hz is generated by the line card, which instructs the telephone to mute its speaker. The telephone acknowledges with DTMF D. Then the modem signals the call waiting Caller ID and the telephone unmutes the speaker as soon as the modem transmission is completed.

While this allows transmission of a modem signal to communicate the second Caller ID, it also momentarily interrupts the voice communications, which can be annoying to the user.

It is an excellent example of the advantage of an out-of-band signaling system: where the control signals are not carried in the voice band, but are communicated in parallel on a separate control circuit or channel. Having such a capability would make it unnecessary to interrupt the voice conversation to send signals.

11.7 Digital Subscriber Line (DSL)

The need for speed is never-ending. Modulation techniques used by modems that operate within the frequency band defined for POTS – the voiceband, 300 to 3300 Hz – hit practical limits on the number of bits per second that can be reliably communicated.

To achieve more bits per second, a wider frequency band is required.

While it would be theoretically possible to implement a wider frequency band by increasing the size of that defined for POTS, such a move would be unworkable due to the extremely high number of devices with voiceband filters that would become incompatible.

Pulling new fiber cables to every home will be a long process – about half finished as of 2020 in urban and suburban North America – so the existing twisted pair copper entry cable to residences in *brownfields* must be used for some time to come.

However, any technology deployed on existing twisted-pair copper phone lines must be backwards-compatible with POTS.

11.7.1 DSL: Modems Above the Voiceband

The solution for brownfields is the definition of a second, wide frequency band in which modems can operate, above the voiceband as illustrated in Figure 140, on the existing copper twisted pair.

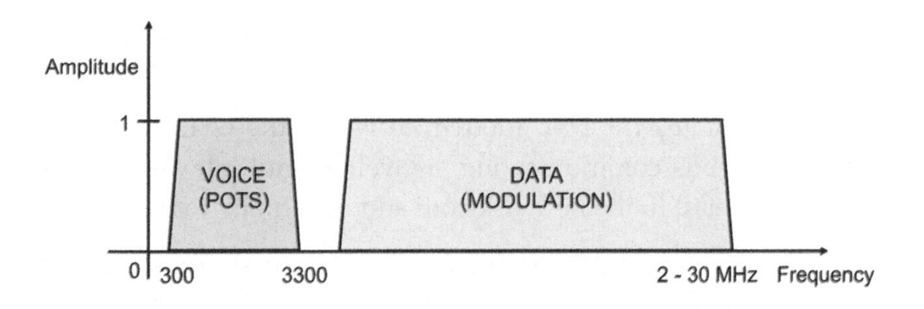

FIGURE 140 DSL: MODEMS OPERATING IN A WIDE FREQUENCY BAND ABOVE THE VOICEBAND ON THE EXISTING TWISTED PAIR LOOP

This is called Digital Subscriber Line (DSL) technology, and allows broadband (high bit-rate) communications of 1s and 0s while still supporting POTS on the same line.

There are a number of DSL modulation techniques, each employing different bandwidths and signaling schemes, with different requirements for loop characteristics and providing different numbers of bits per second… and improving all the time.

Calling this "digital" is inaccurate. DSL does not use pulses, which is the usual definition of digital transmission; instead, DSL employs modulation in frequency channels above the voiceband… a technique more associated with the term "modem" than digital.

"Broadband modems operating in a wide frequency band above the voiceband on the existing twisted pair loop" would be more accurate... but of course, "Digital Subscriber Line" sounds better.

11.7.2 ADSL, SDSL and XDSL

When the downstream capacity (towards the user) is larger than the upstream capacity, it is called Asymmetric DSL (ADSL). This is for residential customers, who are always downloading content like YouTube.

Symmetric DSL (SDSL) has the same capacity in both directions, necessary for a business with servers that are serving up pages. "XDSL" is used to generically refer to the idea of broadband modems on twisted pair, regardless of the variety.

11.8 DSLAMs

Figure 141 illustrates the equipment used for DSL. At the customer premise, the DSL modem is connected to the twisted pair loop, which is connected to a Digital Subscriber Line Access Multiplexer (DSLAM).

The DSLAM contains the DSL modem to which the DSL modem at the customer premise is communicating, as well as multiplexing equipment and a fiber backhaul to the network core and eventually to the Internet.

The DSLAM was originally located in the CO. To shorten the distance between the modems, to increase the bit rate achieved, the DSLAM is now typically deployed in an outside plant enclosure as a type of remote fiber terminal. This is a type of Fiber to the Neighborhood (FTTN).

DSL is markedly different than the old "dial-up" voiceband modem connections, where the dial-up modem makes a phone call and a circuit-switched connection through the telephone switch to a far-end modem, for the duration of the communication session.

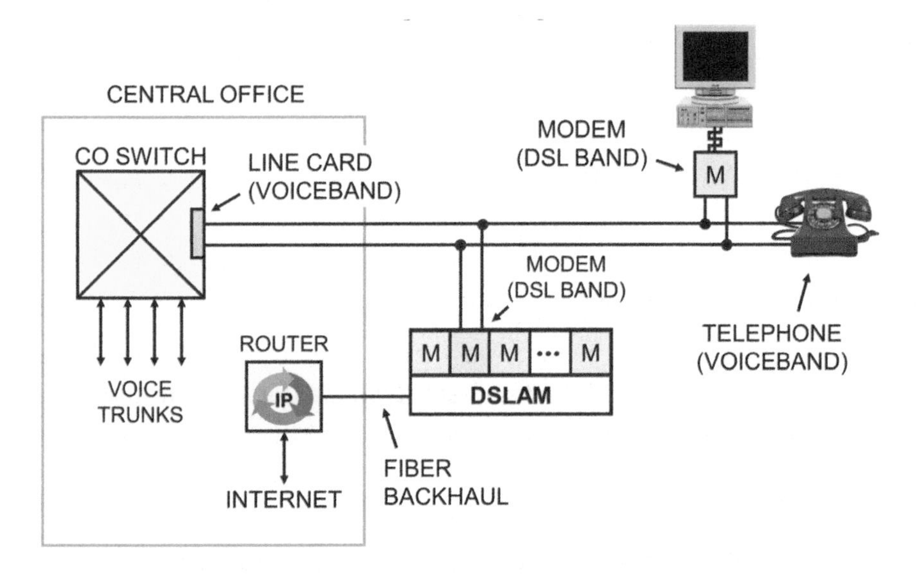

FIGURE 141 FIBER TO THE NEIGHBORHOOD AND DSL MODEMS ON EXISTING TWISTED PAIR TO THE CUSTOMER

With DSL, the customer's modem is communicating with a modem connected to the other end of the customer's loop.

There is no connection through telephone switches; the DSL modems are hardwired together.

This avoids the filters on the CO switch line card and allows the use of a wider frequency band by the modems, and hence more bits per second.

DSL service is referred to as *always on*: the DSL modem at the residence is always connected to the DSL modem in the DSLAM. The connection between the modems is not broken after each communication session like with circuit-switched or dial-up modems in the voiceband.

11.8.1 Coexistence with POTS

The telephone puts energy on the line in the POTS voiceband, and the DSL modem puts energy on the line in bands at higher frequencies. Since these are separated in frequency, the DSL modem does not interfere with telephone service the way a voiceband modem does.

Since the DSL band is much wider than the voiceband, it is possible to communicate more bits per second: tens of Mb/s as a standard service offering today and improving all the time.

11.9 Fiber to the Neighborhood (FTTN), DSL to the Premise

11.9.1 Loop Length

Impairments on twisted pair copper are per foot. As the distance between the customer DSL modem and the network DSL modem in the DSLAM increases, impairments like noise, attenuation and capacitance increase, limiting the achievable bit rate.

To achieve higher bit rates, it is necessary to *shorten the distance between the customer DSL modem and the network-side DSL modem*. The simplest way to do this is to locate the DSLAM in an enclosure in the outside plant.

11.9.2 Remote DSLAMs, OPI and SAC Boxes

Remote DSLAMs are often deployed in cabinets or enclosures bolted onto existing enclosures, which would be the Outside Plant Interface (OPI) or Subscriber Area Concept (SAC) boxes in neighborhoods, spaced hundreds of meters apart and serving perhaps 200 customers each.

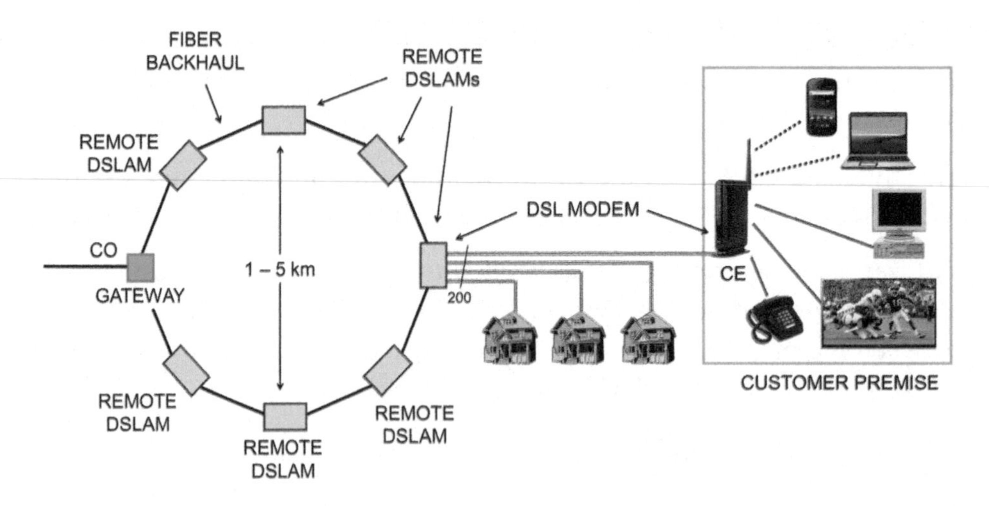

FIGURE 142 FIBER TO THE NEIGHBORHOOD (FTTN)

This provides access to the subscriber loop at a point close to the customer, effectively shortening the distance between the two modems to allow higher bit rates. The remote DSLAM is connected to the CO with a fiber to carry the data and copper wires to carry electricity to power the DSLAM.

11.10 DSL Standards

There are many variations of DSL, achieving ever higher bit rates on twisted pair. Very High Bit Rate Digital Subscriber Line (VDSL) has existed in laboratories for decades (hence the nerdy name). Recently, signal processing technology has increased, and costs have decreased, to the point where VDSL is now routinely deployed.

The different flavors use different modulation and signal processing techniques, different frequency bands and different distance limits.

11.10.1 ADSL2+

ADSL2+, standard G.922.5-2003 from the ITU, is an older high bit rate DSL technology, utilizing a frequency band of up to 2.2 MHz on the loop to implement a maximum of 12 Mb/s downstream and 1 Mb/s upstream with a maximum distance of 4000 feet between the modems.

ADSL2+ employs a legacy technology called ATM to aggregate the traffic on the fiber backhaul. Since the core is now Optical Ethernet and ATM is being discontinued, a gateway or converter is required to interface ADSL2+ DSLAMs to the core.

11.10.2 VDSL2

Newer DSL modems implement VDSL2, standard G.992.3 from the ITU. This standard uses up to 30 MHz of bandwidth on the subscriber loop and can achieve 100 Mb/s symmetric at 500 feet.

Bonding and vectoring increase the bit rate and/or maximum range.

Optical Ethernet and VLANs are used to aggregate the traffic on the fiber backhaul.

D1 U1 D2 U2 D3 D3 U3

.025 3.75 5.2 8.5 MHz 12 MHz 17 MHz 23 MHz 30 MHz

1.3 km 1.2 km 600 m 150 m

4.3 kft 4.0 kft 2.0 kft 500 ft

50/20 68/50 100/50 100/100 MAX Mb/s

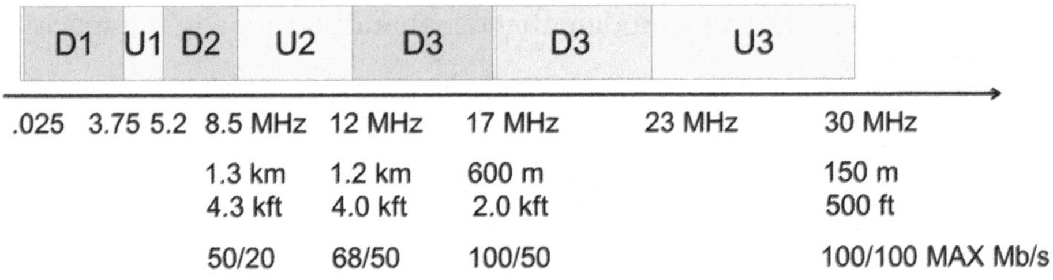

FIGURE 143 VDSL UP AND DOWN FREQUENCY BANDS

11.10.3 VDSL2 Frequency Bands and Profiles

The frequency band for DSL modems is broken into a number of smaller bands, some for uploading and some for downloading.

The higher the bandwidth, the more bits per second can be communicated, but also, the shorter the usable reach.

Figure 144 illustrates the bands and their names and usage.

For comparison, ADSL2+ uses only a bit more than 2 MHz of bandwidth, barely showing up in the D1 band in this chart.

Profile	8 a,b,c,d	12a,b	17a	30a	30a BV	30a BV
Band Name Upstream / Downstream	D2	U2	D3	U3	2 pair bonding + vectoring	2 pair bonding + vectoring
Frequency	8.5 MHz	12 MHz	17.7 MHz	30 MHz	30 MHz	30 MHz
Typical maximum useful range	1300 m 4300 ft	1200 m 4000 ft	600 m 2000 ft	150 m 500 ft	600 m 2000 ft	150 m 500 ft
Downstream MAX	50 Mb/s	68 Mb/s	100 Mb/s	100 Mb/s	100 Mb/s	200 Mb/s
Upstream MAX	20 Mb/s	50 Mb/s	50 Mb/s	100 Mb/s	100 Mb/s	200 Mb/s

FIGURE 144 VDSL PROFILES, SPEEDS AND RANGES

In VDSL2, a *profile* means a modem that uses bandwidth up to the specified number of MHz.

11.10.4 Pair Bonding

Bonding – using two pairs – increases the reach, i.e., the maximum distance between the two modems and/or the total number of bits per second.

11.10.5 Vectoring

Crosstalk – the noise created on a pair of wires by a signal on another pair of wires in the same cable – is a major impediment to achieving very high bit rates with DSL.

Vectoring is a term used to describe very sophisticated signal processing that cancels crosstalk. The crosstalk is determined by examining the signals on all of the pairs in a cable, then the crosstalk on each pair is mathematically subtracted from the modem signal on that pair.

By lowering the noise, vectoring allows the achievement of higher bit rates.

11.11 Broadband Carriers: FTTN & Broadband Coax to the Premise

Cable TV distribution systems were originally known as Community Antenna Television (CATV) systems.

For a city, a television signal would be received by an antenna located at a building called the Head End, then distributed to customers in the city via coaxial cable or *coax*, which is two copper wires, one inside the other.

Today, Head Ends are connected to other Head Ends in other cities with a fiber backbone for digital content distribution.

Coax supports a much broader bandwidth than twisted pair, so these systems can be called *broadband* systems and the operating companies *broadband carriers*.

11.11.1 Hybrid Fiber-Coax Network

This is implemented with a combination of fiber to the neighborhood then coaxial copper cable for the last mile, and so CATV systems are also called Hybrid Fiber-Coax (HFC) systems.

An HFC network consists of a Head End, fiber to the neighborhood terminated on Optical Network Units, coaxial copper feeder cables running down streets, amplifiers, splitters, taps and drop lines into customers' homes where a converter and television are located – along with computers and telephones.

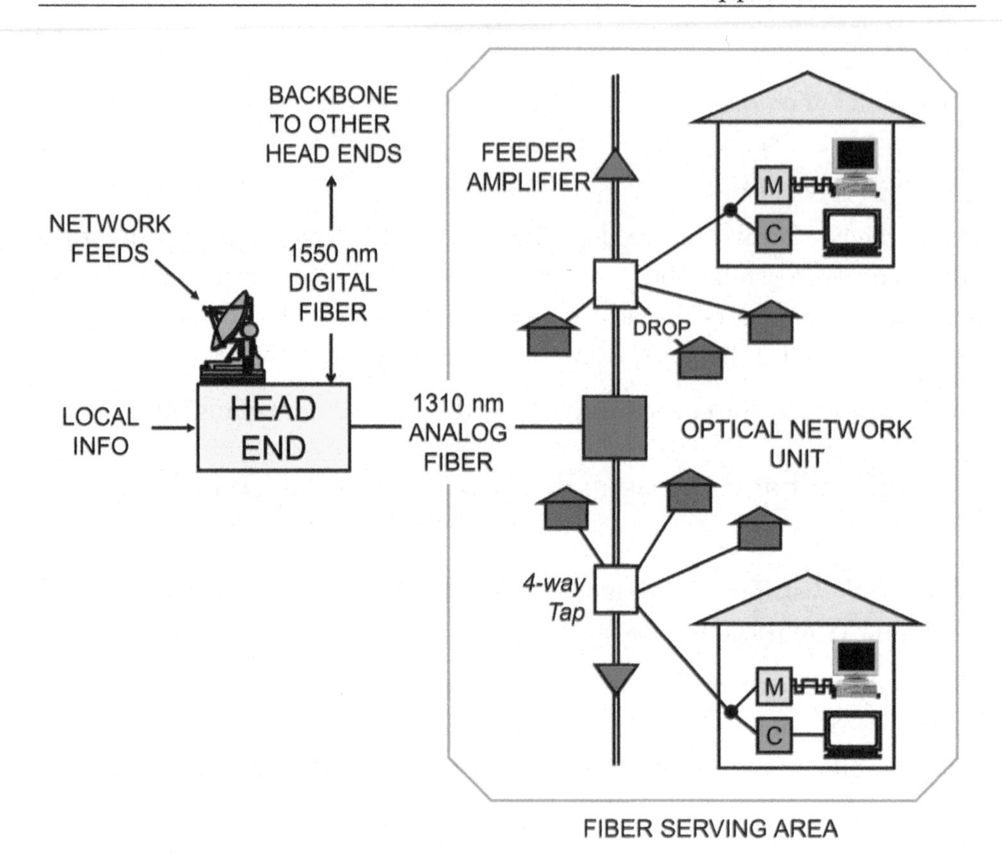

FIGURE 145 CATV HFC: FTTN AND COAX TO THE PREMISE

11.11.2 Frequency Channels

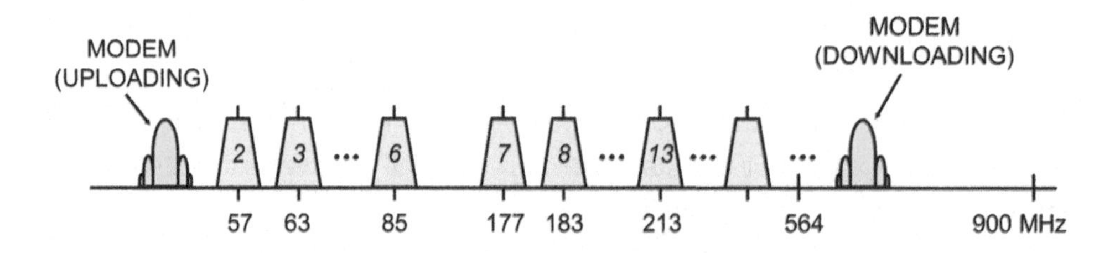

FIGURE 146 6-MHZ CATV CHANNELS

In the previous millennium, Cable TV networks carried multiple analog video signals. The amplitude of the video signal is an analog of the intensity of the light at a point on the screen as it is being scanned along lines left-to-right and top-to-bottom.

The American NTSC standard scans half the screen sixty times per second, resulting in a signal about 4 MHz in bandwidth.

The bandwidth on coax is at least 450 MHz and up to 3 GHz. Many carriers currently use up to 1 GHz.

To make this wide bandwidth usable, it was divided into smaller 6-MHz frequency bands called *channels* by equipment in the Head End. This technique is called *Frequency Division Multiplexing* (FDM).

For standard-definition analog service, the Head End gathered video signals from satellites, terrestrial antennas and local content sources and placed a video signal in each channel using Vestigial Side Band Amplitude Modulation (VSB-AM). The video signal is combined with a single pure sine wave at the frequency of the desired channel, called a *carrier frequency*, to shift the video signal up to the channel frequency.

For those who like details: this actually creates two video signals at frequencies on either side of the carrier. One of the copies and the carrier are suppressed, meaning one copy remains (the *vestigial* copy) on one side of the carrier (the *side band*).

11.11.3 Fiber Serving Area

This entire group of signals, all channels together, is transmitted to the neighborhood using analog techniques on fiber.

The fiber terminates on an Optical Network Unit, located in a cabinet on a pole or on someone's front lawn, where the signals are transferred to copper coaxial feeder cables that run down streets.

The coverage of the coax cables terminating on one ONU is called a Fiber Serving Area (FSA), typically passing 200 – 500 homes.

Taps are installed on the feeder cables at regular intervals. A copper coaxial drop wire is installed from one of the connectors on the tap to the residence.

This has the effect of physically connecting all of the users together and to the ONU, in an architecture similar to the original LAN bus topology. The electrical signal placed on the coax is broadcast to everyone tapped onto the cable.

Since attenuation is more severe on coax than on twisted pair, and more severe at high frequencies, amplifiers are used to boost the signal. Amplifiers are spaced typically every 660 feet, about one amplifier per block.

11.11.4 Television Converters

At a customer, the converter picks the desired frequency channel out of the entire lot, and shifts the signal there back down to the "natural" frequency range 0 - 4 MHz so that it can be displayed on the screen of the television. This "natural" range is called the *baseband*, or sometimes Channel 1.

In days past, a Video Cassette Recorder (VCR) was sometimes used as the converter device, downshifting the desired channel to Channel 3 and the television's tuner would downshift that to Channel 1 and display it.

11.11.5 Modems on CATV Channels

Once a cable TV system is in place, there is no reason why it has to be used only for analog video signals.

Modems may be attached to each end of the system, and one or more channels on the CATV system used for communication of the modem signals.

These modems signal 1s and 0s that can be digitized video, Internet traffic and VoIP telephone service.

"Digital" cable, for example, an HD channel, is video that has been digitized, turned into a stream of 1s and 0s, which are transmitted from the Head End one-way to the set-top box using a modem.

11.11.6 Two-Way Communications Over a Shared Cable

For Internet access, VoIP telephone service or any other two-way communications over this infrastructure, modems are required for each direction.

The main obstacle is the fact that the access circuit is a multi-drop architecture: everyone on the street is connected to and sharing the same cable… methods of rationally sharing the common communication channels are required.

One strategy would be to allocate two 6-MHz channels on the system for each user: one for a modem for uploading and one for downloading, with corresponding modems at the Head End.

That would be a very inefficient way to allocate capacity… there are far fewer available channels than users in a fiber serving area, and 12 MHz of bandwidth is reserved for a subscriber whether they are actually using it or not.

In practice, the users share channels. In many cases, bandwidth is allocated above the television channels for downloading, and below the television channels for uploads from subscribers. This is called a high-low split strategy and makes amplifier deployment easier.

In the downstream direction, a modem at the Head End broadcasts traffic intended for a customer to everyone in their neighborhood. The traffic is encrypted by the Head End and decrypted in the device containing the customer's cable modem. The Head End broadcasts users' data as needed as it arrives, or in a rotation if there are many active users.

Sharing a modem band in the upstream direction is more difficult. The users' modems are all tuned to transmit on the same channel, so they can't all transmit a simple modem signal to the Head End at the same time.

11.12 DOCSIS and Cable Modem Standards

11.12.1 DOCSIS 1: Contention-Based Channel Sharing

The first strategy for sharing, standardized as the Data over Cable System Interface Specification (DOCSIS), defined time slots.

Customers' downloads are encrypted and broadcast to everyone in the Fiber Serving Area.

For uploads, time slots were available for contention by all modems. Collisions can occur and retries are used, similar to the contention-based capacity-sharing strategy for Ethernet.

This worked fine if only one person in a Fiber Serving Area has signed up for cable modem service. The problems started when 50 people signed up, most of them teenagers running bit torrent or other file "sharing" programs, continuously passing on search requests and starving other users of bandwidth. These problems only get worse with the massive increase in traffic from Netflix and YouTube.

11.12.2 DOCSIS 2: Reserved Time Slots on Channels

DOCSIS 2 defined time slots allocated to specific users to guarantee performance, similar to the TDMA strategy for 2G cellular.

11.12.3 DOCSIS 3: CDMA on Channels

To improve efficiency, DOCSIS 3 specifies the use of CDMA allowing variable rates and multiple simultaneous transmissions like 3G CDMA cellular.

11.12.4 DOCSIS 3.1: OFDM

DOCSIS 3.1 specifies OFDM like 4G cellular, 5G, DSL and Wi-Fi.

11.12.5 Wider Channels

Basic cable modems move 30 Mb/s in the 6-MHz channel defined decades ago for standard-definition analog television.

With DOCSIS 3.0 and 3.1 systems, the CDMA and OFDM technologies can use bands wider than 6 MHz to deliver much higher bit rates to customers, heading toward Gb/s.

11.13 T1 and E1

Trunk Carrier System 1 (T1) is a technology that was popular from about 1960 – 2000, designed to carry 24 trunks over 4 copper wires using channelized TDM. Though fiber is now routinely used, there remain thousands of T1 circuits installed and in operation.

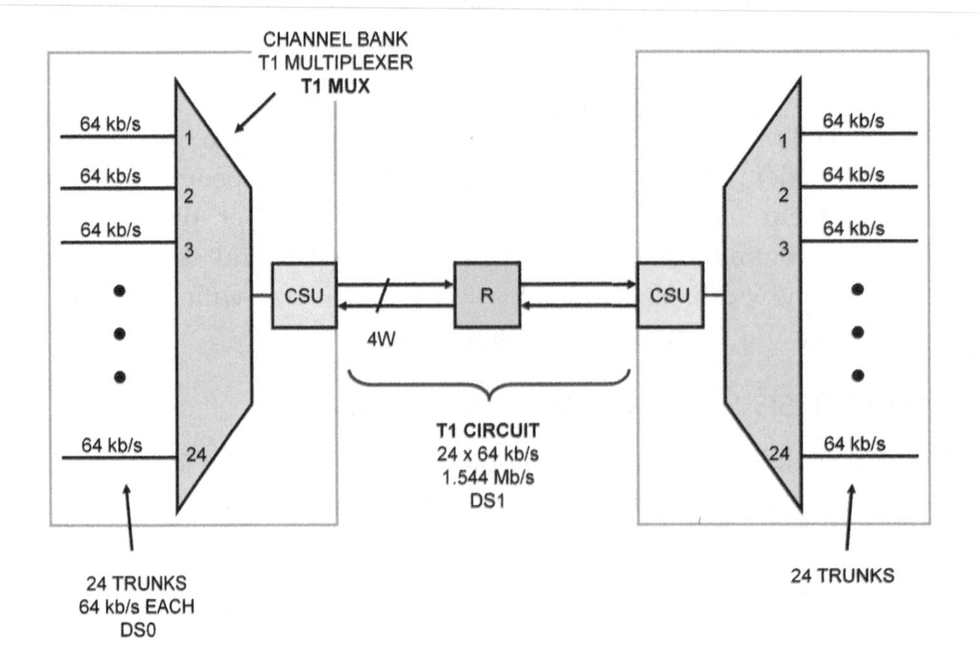

FIGURE 147 T1 MULTIPLEXERS AND CIRCUIT

The Digital Hierarchy: the DS0, DS1 and E1 bit rates, are covered in detail in Appendix B. Full details on T1 are included in Appendix C of this book… material that was in the main body of the earliest editions of this book, now relegated to an appendix, as T1 is now legacy technology.

A brief overview is provided in the following sections.

11.13.1 Time-Division Multiplexers

A basic T1 system consists of multiplexers, Channel Service Units (CSUs) and the T1 circuit: four copper wires with repeaters every mile or so.

To implement channels, a *multiplexer* is attached to each end of a circuit. The multiplexer is variously referred to as a T1 multiplexer, a T1 mux, or a channel bank.

Since T1 was designed to carry 24 trunks, the mux provides 24 low-speed ports, each running at the 64 kb/s DS0 rate; the users' lower-speed access circuits, each on a separate hardware port.

On the other is the high-speed *aggregate port.* The multiplexer intersperses the users' data in a strict order to form a high-speed stream that is transmitted on the aggregate port.

What goes in on a particular hardware port at one end comes out on the corresponding hardware port at the other end.

Each user gets a fixed fraction of the capacity of the high-speed circuit to carry the data on their lower-speed circuit from one building to another. This capacity is their channel.

To implement the channel, each port is allocated a fixed time slot to transmit a single byte across the T1 circuit. This happens 8,000 times per second. The ports do this in a strict rotational order, one after another. The resulting data rate is 24 x 64 kb/s = 1.536 Mb/s.

11.13.2 DS1 Frames

To be able to sort out what goes where at the far end, the transmitting multiplexer sends an extra *framing bit* before the byte from the first port.

The receiving multiplexer uses this framing bit to identify the beginning of the byte for channel one in the incoming bit stream, and direct that byte to low-speed output port number 1 on the far side.

Then the next eight bits are directed to port 2, then the next to port 3 and so one until a byte for each port has been received, then the process repeats.

Framing is covered in detail in Section B.3.

The framing bit brings the bit rate to 1.544 Mb/s, the DS1 rate.

The entire system is two-way simultaneous.

11.13.3 CSUs and Repeaters

The aggregate port on the multiplexer is connected to a CSU. The CSU is the circuit-terminating equipment for the T1 circuit.

This device represents binary digits on the physical wires using pulses of voltage on the copper wires. It performs the same functions as a modem - but is not called a modem since it is a digital device.

In this particular technology, repeaters are required to regenerate the voltage pulses every 6000 feet (6 kft / 1 mile / 1.6 km) along the T1 circuit.

11.13.4 Synchronization

All of the devices have to be synchronized at the bit level to know when a pulse of voltage starts and ends.

In days past, all devices used a clock derived from the US National Bureau of Standards or Canadian National Research Council cesium clock.

Today, clocks are derived from Global Positioning System (GPS) satellites.

11.13.5 Applications for T1

T1 was first used to carry long-distance trunks, then became an access technology for business customers.

T1 was used to carry PBX trunks, used for ISDN PRI services, used to access Frame Relay data services, and to implement private networks made of dedicated lines.

11.13.6 E1 Outside North America

Most places in the world follow "European" standards, not North American standards.

Where a T1 would be used in New York, an E1 is used in London.

E1 is the European Synchronous Digital Hierarchy (SDH) rate in the same category as DS1. E1 is 2.0 Mb/s, which carries 32 channels, some reserved for signaling.

E1 is carried on a copper wiring system called European Conference of Postal and Telecommunications Administrations trunk carrier system 1 (CEPT-1).

E1 can also be carried on fiber and radio.

11.13.7 TDM on Fiber

Synchronous Optical Network (SONET) technology, used for the network core from 1980-2000, operates in the same way as T1.

SONET implements up to 129,024 DS0 channels by transmitting a byte 8,000 times per second for each channel.

The resulting aggregate speed is measured in multiples of 45 Mb/s and is transmitted on fiber using the Optical Carrier (OC) system.

SONET is covered in more detail in Section B.2.

11.14 TIA-568 LAN Cable Categories

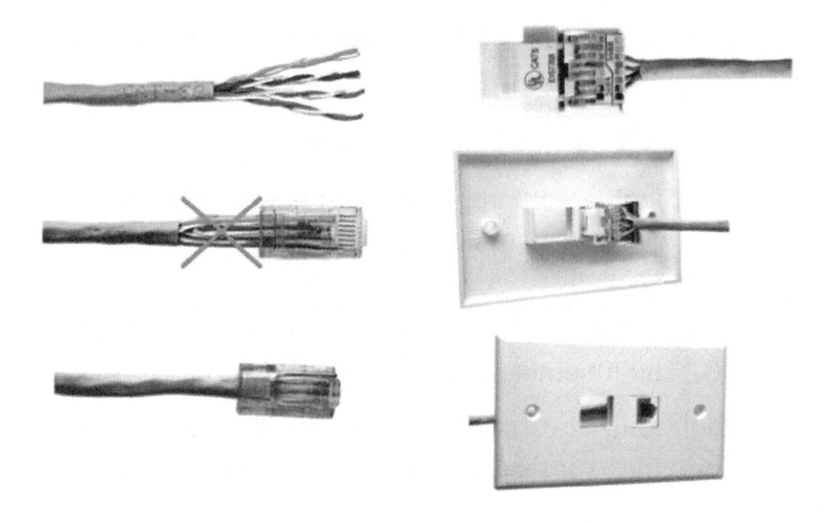

FIGURE 148 BULK CABLE IN WALLS IS TERMINATED ON KEYSTONE CONNECTORS THAT SNAP INTO THE BACK OF A COVER PLATE. PATCH CABLES, LEFT, HAVE RJ-45 CONNECTORS AT EACH END

LANs for the most part run over cables containing eight copper wires inside buildings. Copper is used because it is inexpensive, pliable, corrosion-resistant, and easy to extrude into long, thin wires.

The wires are twisted together in pairs. There are four pairs in LAN cables.

The most widely-followed standard for LAN cables is TIA-568, published by the Electronic Industries Association and its Telecommunications Industry Association sub-group.

Telecommunications Systems Bulletin TSB-67 adds the requirements and methods for field testing installed cable systems. Taken together, these are the authority how to design and install a structured cabling system.

11.14.1 Category 1 through 5

TIA-568 defines *categories* of twisted-pair cabling that support different line speeds.

- TIA-568 Category 1 cable is existing telephone cabling, also called Rusty Twisted Pair (RTP).
- Category 2 cable was 25-pair multiconductor cables for old key telephone systems that had buttons to press to access different lines.
- Category 3 cable was for 10 Mb/s Ethernet on twisted pair, 10BASE-T.
- Category 4 cable was specified for 16 Mb/s token ring.
- Category 5 cabling was for The Future at up to 1000 Mb/s.

Categories 1 through 5 are no longer installed.

Category 5 (*Cat 5*) cable was supposed to support Gigabit Ethernet, but in practice turned out to be missing the specification of some required transmission characteristics.

Enhanced Category 5 (Cat 5e) was subsequently specified to guarantee the operation of Gigabit Ethernet on twisted pair, 1000BASE-T.

Whether a cable can be certified as conforming to a standard is often dependent on the consistency and placement of twists during manufacturing.

Category 6 cable is specified to support 10 Gb/s on twisted pair.

At 1 Gb/s, it becomes necessary to specify the frequency bandwidth supported on the twisted pair, along with all of the other transmission characteristics, to enable signaling at these line speeds.

In theory, Category 7 supports 100 Gb/s on twisted pair. This is in the same league as current mainstream fiber-optic transmission systems, so one could probably expect it will be a while before there is any significant deployment of Cat 7 copper wires.

11.14.2 TIA-568A vs. TIA-568B

There are two specifications for which wires in the cable go to which pins on the connectors: TIA-568A and TIA-568B. There is no difference between the two in terms of performance – but it is necessary to pick one of the two configurations and use it consistently on every jack, every patch panel, every patch cord and every connector.

This is covered in detail in Section 15.3 LAN Cables and Categories.

11.14.3 Maximum Cable Length and Cabling Architecture

All categories specify cables with four pairs (eight wires) and a maximum length of 100 meters.

This means the maximum run length of the cables – including runs through risers, poles, conduits – is 100 m (330 feet).

To be conservative, devices would be connected to a switch located in a wiring closet within a radius of perhaps 200 feet.

These wiring closet switches could be connected to centralized Ethernet switches on each floor and/or connected to a router in the communications room, possibly using fiber.

11.14.4 Difference Between Categories

The difference between the categories rests in guaranteed transmission characteristics of the cable, including specifications for Near-End Crosstalk (NEXT), Attenuation to Crosstalk Ratio (ACR), supported frequency bandwidth, all of which affect the maximum possible information transfer rate, and hence what kind of devices can be successfully attached to each end of the cable.

One of the main factors in getting a cable certified to meet the TIA-568 category is quality control, particularly in the consistency of the twisting and placement of the pairs.

Two pairs will be twisted at a particular number of twists per inch, but offset by half a period to minimize crosstalk between the pairs. The other two pairs will be twisted at a different rate that is not a multiple of the other, and similarly with the twists exactly not lined up.

How well and how consistently this is accomplished during the manufacturing process determines how successful the manufacturer will be in having the cable certified as meeting the standard.

11.14.5 Which Category to Use

When determining which category of cable to use, life cycle and labor cost are determining factors.

For a patch cable connecting a DSL or Cable Modem to a device inside a residence, where we have an expectation that the line speed will not exceed 100 Mb/s in the foreseeable future, then Cat 5 patch cables may be used.

For an extra three cents, a Cat 5e patch cable would allow the continued use of the cable were the line speed to increase above 100 Mb/s up to 1 Gb/s, as it inevitably will.

Since the labor cost is usually far greater than the cable, it is strongly recommended to install cabling inside walls with capacity greater than immediate needs, and twice as many cables as what the conventional wisdom dictates.

Two Category 6 cables to each work area would be the Cadillac solution.

Two Category 5e cables to each work area would be well positioned for the future.

One Category 5e cable to each work area is the minimum.

12 Telecom Equipment

12.1 Broadband Network Equipment: Routers and Ethernet Switches

The core of the broadband network is built by connecting routers in different physical locations with point-to-point fibers. Each location has point-to-point fibers going to many other locations. The router relays IP packets from one fiber to another. Knowing *which* fiber to relay a packet to is the "routing" part of the story.

The point-to-point fiber links follow the Optical Ethernet standards, which specify the kind of fiber, and how the bits are represented by flashing lasers on and off by a transceiver called an SFP.

The physical fiber is plugged into the SFP, which is in turn plugged into a slot on a core router at each end. The router internally moves packets from one SFP to another. The packets are moved between routers by encapsulating them in a MAC frame (Section 4.4) and flashing the laser on and off to represent the bits in the MAC frame.

12.1.1 Carrier-Grade Core Routers

An example of a Cisco carrier grade core router is shown in Figure 149. The chassis and computer running a network configuration system and a very comprehensive set of traffic control functions costs US$200,000. The line cards with the SFP optical transceivers (the lasers and detectors) are extra.

One card to plug into the router with ten 100 Gb/s optical ports costs $1.5 million. One card with thirty 10 Gb/s ports is $500k. For reference, 10 Gb/s is the backhaul capacity required for one cellular base station.

In addition to making route decisions and relaying packets internally fast enough to support Terabits per second passing through it without packet loss, and in addition to the cost of the 100 Gb/s SFP optical transceivers, another reason why the Cisco router is expensive is the traffic management software it comes with, which allows total control over a network of Cisco routers.

Here is the list from the product sheet: Access Control List (ACL) support, Application Visibility and Control (AVC), Bidirectional Forwarding Detection (BFD), Cisco IOS IP Service-Level Agreements (IPSLA), Class-Based Weighted Fair Queuing (CBWFQ), Embedded Event Manager (EEM), Encapsulated Remote SPAN (ERSPAN), Ethernet Virtual Connections (EVC), Hierarchical Quality of Service (HQoS), Location ID Separation Protocol (LISP), Network-Based Application Recognition (NBAR), Source-Specific Multicast (SSM), Static routing, Weighted Random Early Detection (WRED).

FIGURE 149 CARRIER-GRADE CORE ROUTER

12.1.2 Carrier-Grade Ethernet Aggregation Switches

A Layer 2 switch is a less expensive device, typically functioning as a data concentrator or aggregator, connecting many feeder circuits to one or more upstream circuits, which are typically circuits to a router. Layer 2 switches also implement VLANs, an important tool in network security, segregating users' traffic from public view on a shared access network like a MAN.

The Juniper carrier core switch illustrated in Figure 150 is $140,000.

Huawei, a company founded in part by a Chinese intelligence agency, manufactures the same equipment and sells it for less than 1/10 of the cost of Juniper equipment.

Federal Acquisition Regulation 52.204-24 proscribes any entity doing business with the USA government from supplying or using equipment manufactured by Huawei.

FIGURE 150 CARRIER-GRADE CORE ETHERNET AGGREGATION SWITCH

This is due in part to suspicion of hidden methods for Chinese state actors to eavesdrop on communications, and/or remotely reconfigure or disable the equipment.

12.1.3 Enterprise Core Router

The Cisco enterprise core router of Figure 151 is 1/10 the price of a Cisco carrier core router; but it comes with only two 10 Gb/s copper interfaces.

FIGURE 151 ENTERPRISE CORE ROUTER

It includes some, but not all of the traffic management software that is bundled with the $2 million carrier-grade core router.

12.1.4 Enterprise Ethernet Switch

A switch with 48 ports of 1 Gb/s copper each, and no support for VLANs costs $200. VLAN capability costs a bit more.

The type of switch in Figure 152 is deployed in wiring closets to terminate the copper LAN cables that go to work stations.

FIGURE 152 ENTERPRISE ETHERNET SWITCH

It can also be deployed in the "telecom room" to aggregate the traffic from wiring closet switches before the traffic is fed to the enterprise core router of Figure 151 or the enterprise edge router of Figure 153.

12.1.5 Enterprise Small Office Edge Router

The Cisco edge router illustrated in Figure 153 comes with software that makes it straightforward to set up VPNs over the Internet to peer devices in other locations, software not typically included with home devices.

FIGURE 153 ENTERPRISE SMALL OFFICE EDGE ROUTER

Implementing VPNs pairwise between locations over the Internet is called a SD-WAN by the marketing department.

This edge router also includes a built-in Ethernet switch allowing the connection of 15 devices, or 15 aggregation switches, plus the uplink.

It is more than double the price of the consumer edge router of Figure 154, but comes with a better built-in Ethernet switch and VPN setup software.

12.1.6 Home / Small Business Edge Router with Wi-Fi

FIGURE 154 CONSUMER OR SMALL BUSINESS EDGE ROUTER WITH WI-FI

At the lowest end of the scale is a generic edge router with a 5-port Layer 2 switch and Wi-Fi built in as illustrated in Figure 154. This device is plugged into, or built into, customer-premise network equipment like a DSL modem, Cable modem or PON fiber terminal.

12.2 Broadband Customer Premise Equipment

This section covers broadband equipment at the customer end of the circuit.

12.2.1 Fiber Terminal

Figure 155 depicts a fiber terminal, supporting the highest bandwidth, beside a twisted-pair loop terminal, supporting the lowest.

The fiber, which is the network-side interface to the customer-premise fiber terminal, is in the bright green cable on the top left of the fiber terminal.

The LAN jack, which is the user-side interface for Internet and/or video programming from the ISP, is on the right of the blue box. The user-side interface for POTS are the other jacks that have phone cords plugged into them, as well as the red and green screw terminals below.

12.2.2 POTS Terminal

Figure 155 also depicts a POTS terminal. The lowly POTS terminal serves as a metallic connection point and as a legal point of demarcation or *demarc* between the carrier's and the customer's wiring.

FIGURE 155 FIBER TERMINAL AND POTS TERMINAL

POTS runs on two-wire loops; this terminal has space to terminate two loops. Only the left half is being used. DSL runs on the same loop as POTS.

To metallically connect an access circuit loop to the inside wiring, both are terminated on a screw post, separated with washers.

Typically, red is used for one wire, and the screw post the red wires are connected to may be labeled R for Ring. Green is traditionally used for the other wire in the pair, and its screw post might be labeled T for Tip, old terms from the Flintstones era of patch cords.

12.2.3 Cable and DSL Modems

DSL modems are for twisted pairs of copper wires, and use frequencies up to 30 MHz. Cable modems operate on coaxial pairs of copper wires, and use frequencies of 60 MHz - 1 GHz and more.

Both devices pictured in Figure 156 are often called "modems". They are more precisely called *Customer Edge (CE) devices* that include the modem, plus an Ethernet switch, DHCP client and server, Network Address Translator, Wi-Fi Access Point and more.

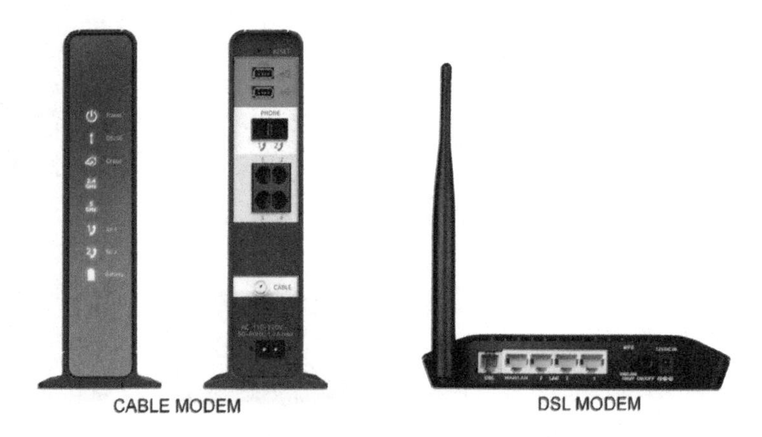

FIGURE 156 CABLE AND DSL MODEMS

The cable modem is identified by the coaxial screw connector in the white area; the DSL modem has an RJ-45 jack to connect a twisted pair loop.

12.2.4 Wireless Terminals

Wireless Internet service is available virtually worldwide via the Iridium Next LEO service, Globalstar, Orbcomm, military satellites and StarLink.

Satellite-based service is very expensive compared to service from land-based operators of 3G, 4G and 5G cellular, fixed wireless and Low Power Wide Area networks.

Starlink's objective is prices similar to landline.

Figure 114 depicts the 19 x 12 inch flat-panel antenna for broadband Internet service from Starlink. There are actually many antennas under the cover; very sophisticated signal processing adjusts the power of each in real time to perform beamforming, focusing the power on the satellite transceivers.

Download speeds approach 1 Gb/s peak for $110 / month.

Figure 157 illustrates customer-end broadband wireless equipment for terrestrial fixed and mobile networks.

In the 2020s, mobile network operator networks are not designed so that everyone can watch guilty dog videos in HD at the same time at their homes on Sunday evenings.

The 3.5 GHz fixed wireless broadband Internet is designed for this purpose and is deployed both by mobile operators and their competitors.

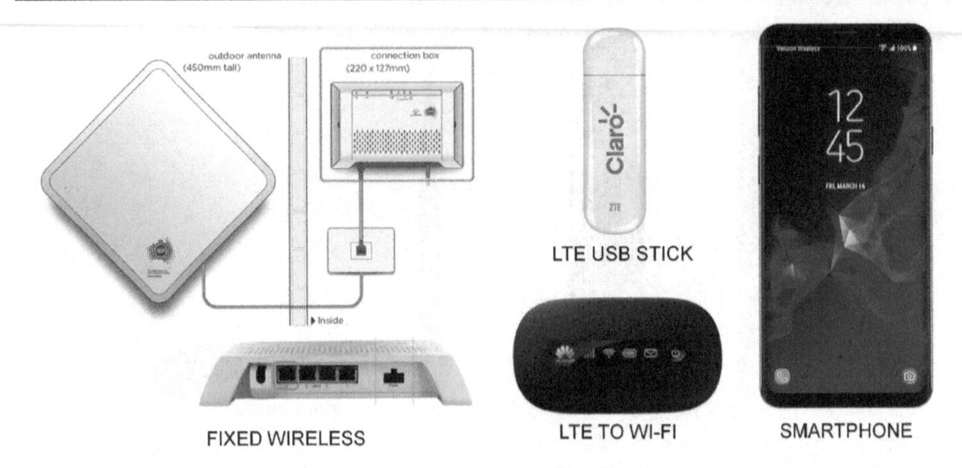

FIGURE 157 BROADBAND TERRESTRIAL WIRELESS USER EQUIPMENT

At the left is the diamond-shaped antenna and indoor unit for 3.5 GHz fixed wireless that supports 25 Mb/s service to 100 homes and businesses per sector.

The other devices in Figure 157 are equipment for service from a mobile network operator.

The LTE USB stick is used with a "data only" Internet plan. It appears like a LAN connection in Windows network connections.

The LTE to Wi-Fi device is also used with a data-only plan for Internet access over the mobile network, and makes the connection available to anyone nearby via the Wi-Fi hotspot it creates.

The smartphone can also act as an LTE or 5G to Wi-Fi device, with the advantage of mobility, by activating the mobile Wi-Fi hotspot feature.

While many people will connect a computer, a VoIP gateway and/or a television set-top box to a wireless terminal, others will use their smartphone as an all-in-one device.

12.3 Call Managers, SIP, Soft Switches, Hosted PBX and IP Centrex

12.3.1 Hard Switches

In its simplest form, a switch is a device that enables communications from one point to one other specific point, usually when there are multiple points to choose from.

A traditional Central Office telephone switch or PBX might be called a "hard" switch. The vast majority of the floor space taken up by this kind of switch is line cards, the physical termination for customer access circuits.

The telephone is plugged into the telephone switch, or more accurately, the loop is connected to a line card in a drawer, in a shelf, in a rack that is part of the physically large rack-mount computer system called a telephone switch.

The telephone switch physically moves the speech inside the switch from one line card to another to implement the connection for the duration of a phone call.

12.3.2 Soft Switches

In the Voice over IP (VoIP) world, access circuits are not terminated on a switch, and the voice communications do not flow through a switch.

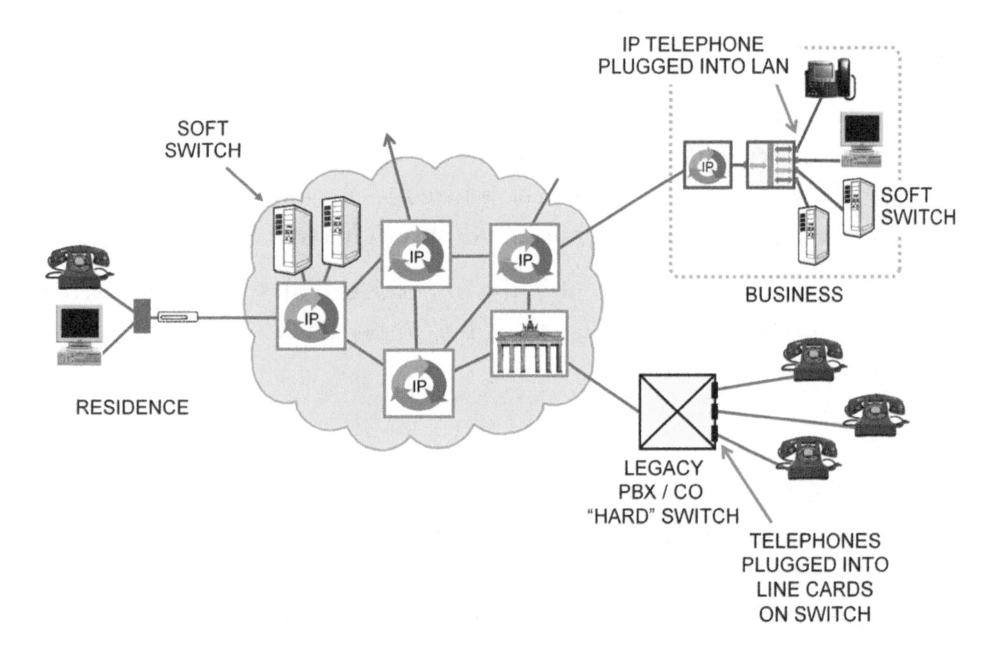

FIGURE 158 SOFT SWITCH VS. "HARD" SWITCH

VoIP telephones are connected to a LAN in the same way as desktop computers, and the digitized voice is carried in IP packets directly from one telephone to another over the LAN. Therefore, no line cards are required on a VoIP switch.

If all of the line cards are removed from a "hard" switch, what is left over is the "soft" part of the switch or *softswitch*, which is the software that performs call setup functions.

Soft switches are deployed by carriers to provide network-based call setup, replacing CO switches and toll switches – and by end-user business customers to perform in-building call setup, replacing PBXs.

A business customer can physically implement the software that is the softswitch on a computer at the customer premise, or can outsource this to a third party who implements the software that is the softswitch on a computer at a remote location.

There are many terms used for soft switches. *Call manager* is more accurate. Most VoIP systems today support the Session Initiation Protocol (SIP), uses the terms *proxy* and back-to-back user agent. Terms used by product manufacturers for their products that might implement these and other functions include call manager, call server, VoIP switch, communication server and hosted PBX.

Regardless of what it is called, the main function of a softswitch is call setup, and the essential function is to inform the telephones at each end of the call of the other telephone's IP address, since the voice goes in IP packets directly between the phones, not through a telephone switch.

For both privacy and flexibility, the IP address of the called party's telephone is usually not published. It has to be determined before voice communications can begin.

12.3.3 SIP

During a VoIP telephone call, the telephones send IP packets containing digitized speech directly to each other. To be able to do this, the telephones must know each other's IP address. In a standards-based system, the SIP protocol is used to inform the telephones of each other's IP address.

Each phone is associated with a SIP server, which acts on behalf of or is a *proxy* for the telephone to set up the call. Instead of a telephone number, each person has an Address of Record, which in the SIP standard has the same format as an email address, for example, someone@theirserver.com.

For interoperability with traditional systems, the Address of Record might be translated to a format that looks like a traditional telephone number... but this has to be resolved behind the scenes to a SIP standard format.

Everyone's Address of Record is made visible to the public. This would be printed on business cards and included in email signature blocks.

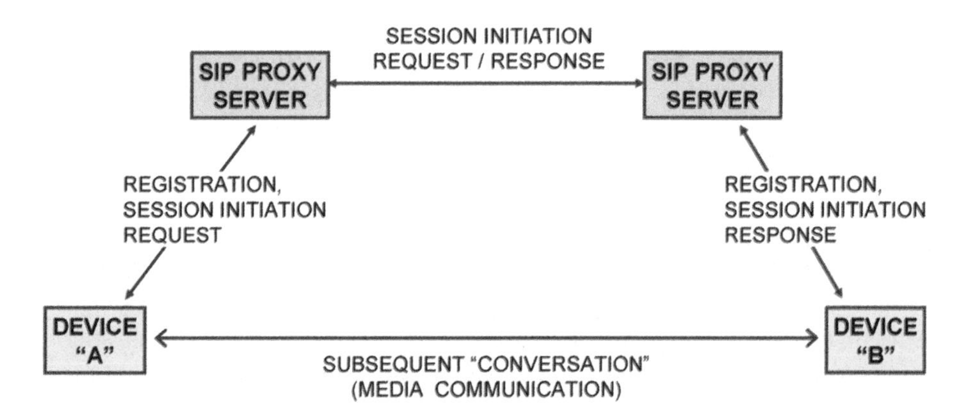

FIGURE 159 THE SIP TRAPEZOID FOR CALL SETUP

To make a VoIP phone call to someone, it is necessary to find out what their telephone's IP address is.

It is possible to find out the IP address of their *SIP server* by looking it up in the Domain Name System (DNS) just like a web server... but it is not possible to determine their telephone's IP address. Only their SIP server knows their phone's IP address.

When a phone is plugged in or restarted, it is assigned an IP address like any other computer. Then the phone *registers* with its SIP server, that is, informs its SIP server of its current IP address.

To establish a call from telephone A to telephone B as illustrated in Figure 159, caller A asks their SIP proxy server to initiate the phone call to B's Address of Record.

A's SIP server looks up the IP address of B's SIP server in the DNS, then sends a session initiation request to B's SIP server.

Since telephone B previously registered with B's SIP server, B's SIP server knows the IP address of telephone B. B's SIP server then passes on the incoming call request to B's telephone.

If B indicates to B's SIP server that they will take the call, B's SIP server transmits the IP address of telephone B to A's SIP server, which in turn relays it to A's telephone.

At that point, the two telephones can send IP packets containing digitized voice (called *media communication* in SIP) directly from one to another, and the SIP servers are no longer involved.

This is necessarily a simplified explanation of the SIP call setup protocol, but hopefully conveys the essential idea. VoIP is covered in Chapter 8. Voice digitization is Chapter 7. IP addresses and packets, DNS and other protocols are covered in Chapter 16.

12.3.4 Additional Functions

In addition to running a SIP server for call setup, the softswitch may also perform authentication, authorization and accounting functions such as generation of Call Detail Records, and potentially hundreds of other call setup and processing functions, such as voice mail, integrated messaging, call pickup groups, Interactive Voice Response (IVR) functions and Automated Call Distributor (ACD) functions.

12.3.5 Location Independence

The SIP messages between the telephone and its SIP server are short and simple, requiring very low bandwidth compared to the subsequent exchange of digitized speech between the telephones.

This means that the softswitch can be located anywhere on the planet; as long as IP packets containing the SIP call setup messages can be communicated from a telephone to its SIP server with suitable maximum delay and packet loss, it is irrelevant where the SIP server is physically located.

The SIP server could be located at the customer premise, at a telephone company building, or at some third-party data center.

12.3.6 Customer Premise Softswitch

When the hardware and software implementing the softswitch is located at the customer premise, typically purchased by the customer, it is usually called a softswitch, call manager, unified communications system or IP phone system.

12.3.7 IP Centrex

When the hardware and software implementing the softswitch is located at a telephone company, its functions are provided as a service by the telephone company, and might be called IP Centrex or Hosted VoIP by the telephone company's marketing department.

Alternatively, the phone company might continue to the call the service Centrex to avoid confusing anyone. The fact that it is implemented IP packets and SIP for new customers is just a detail.

12.3.8 Hosted PBX

When the hardware and software implementing the softswitch is located at a third party, the software and the hardware it runs on is provided as a service by the third party and is usually called a Hosted PBX, similar in concept to web hosting and virtual web servers.

Of course, telephone companies may be in the business of providing Hosted PBX services in addition to IP Centrex.

12.3.9 Cloud-Based Softswitch as a Service

The evolution of the marketing term "Hosted PBX" is to "Softswitch as a Service" and "cloud service".

In reality, the "Hosted PBX" of the previous section is a software program, running on top of virtualization software, running on servers in a data center... which is the definition of "cloud" computing.

The differentiator between different services is reliability and support. This spans the range from open-source software with no reliability and no support with the customer supplying the phones and LAN; to the telephone company taking care of everything, including the phones, and a telephone company employee support tech embedded at the customer.

12.4 Telephone Circuit Switches

Telephone switches are hardware that implements connections between circuits that are plugged into it. Two main types of circuits are loops: twisted pair to the customer; and trunks: circuits to other neighborhoods and other cities.

Connections are established for a phone call by *seizing* trunks to make a physical path all the way from near-end switch to far-end switch, reserving them for this phone call, for the duration of the phone call.

When the phone call is ended, the trunks are released for someone else to use. This is called circuit-switching.

Three types of circuit switches are Central Office switches, where loops connect to trunks; PBXs, where loops inside a commercial building connect to trunks going to a CO; and Toll switches, which make connections between trunks for long-distance toll calls.

As carriers' core networks are now VoIP, the toll switches internal to their networks have been replaced with SIP, and the trunks are 100 Gb/s Optical Ethernet connections between cities.

Due to the inertia of an immense number of regulations and the very large amounts of money and number of interested parties involved, toll switches are still used to interconnect LECs and other carriers. In the future, they will be replaced with VoIP.

All new business telephone systems are VoIP.

There is an installed base of millions of legacy CO switches and PBXs still in operation, waiting to be replaced by broadband IP service and SIP.

This section covers the installed base of legacy telephone circuit switches and the associated services. This was Chapter 3 of the early editions of this book; now demoted to Chapter 12 as broadband IP takes over.

12.4.1 Circuit Switching

In the case of a CO switch, the connection is full-time for the duration of the call, between a loop and a trunk, or between two loops for a call local to that switch. In a toll center, this would be trunk to trunk connections.

To understand what a telephone switch does, it is useful to think of railroad switching yards. Switches in a railroad yard connects tracks so that a particular spur line (analogous to a loop) is connected to a particular trunk line for the duration of a train, then the switches are changed so that a different spur line is connected to the same trunk for the next train.

The same principle applies to telephone switches: a route decision is made, loops are connected to trunks to implement a connection for the duration of a phone call, then a different loop is connected for the next call.

This type of switching: a full-time connection for the duration of a call, is called *circuit-switching*. We sometimes use the term *traditional* to refer to this kind of switching, as it is being replaced with Voice over IP, soft switches and packet switching.

TRUNKS

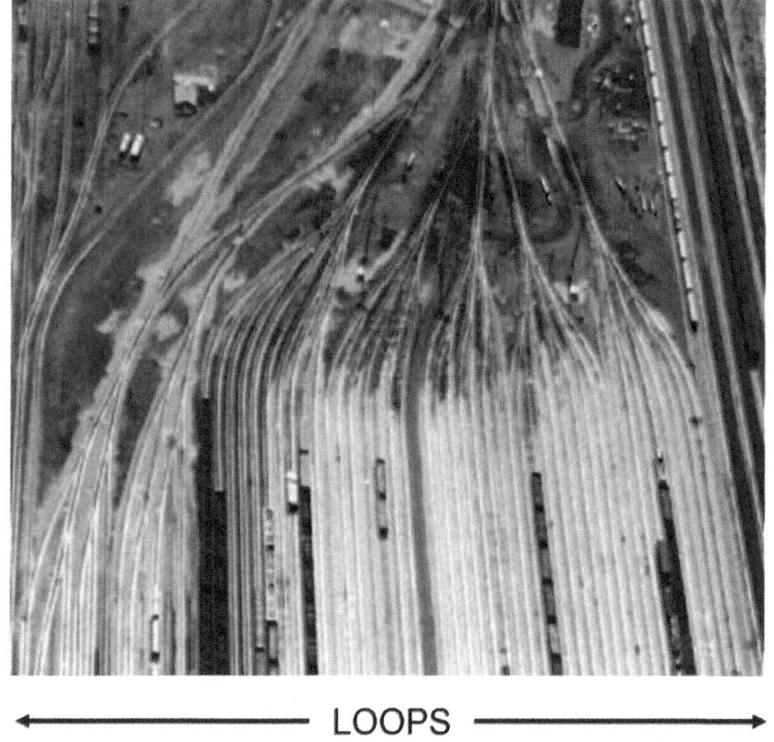

FIGURE 160 RAILROAD SWITCHING YARD: CIRCUIT-SWITCHING

12.4.2 CO Switches

Telephone switches are computers. They are often constructed as rack-mount systems enclosed in cabinets. Nortel switching equipment was traditionally painted brown, though is now a light gray color. Lucent equipment was traditionally painted off-white. Others are beige.

Other than that, all telecom switching equipment looks similar: like large upright filing cabinets with many wires connected on the back and indicator lights on the front.

Alcatel/Lucent/Nokia's CO switch product was the Class 5 Electronic Switching System (5ESS). Nortel's main product was the Digital Multiplex Switch model 100 (DMS-100). The DMS product line and the servicing of its installed base was acquired by Avaya following Nortel's bankruptcy.

These switches are capable of handling up to 100,000 loops, but are usually built up to a maximum of 60,000 loops per switch.

FIGURE 161 FRONT BAY OF A DMS-100 CO SWITCH

There are many other switch manufacturers and products.

12.4.3 Line Cards

The twisted pair loops are carried into the switch on a Main Distribution Frame. The component of a switch that terminates a loop is called a *line card*.

Just as a PC can have an adapter card that allows a telephone line to be plugged into the PC, a telephone switch has line cards to allow the connection of loops to the switch.

Individual line cards are implemented as small Printed Circuit Board (PCB) line card modules, plugged onto a larger PCB, mounted in a drawer in a shelf in a rack as illustrated in Figure 162.

There are 48 line card modules in the drawer illustrated, meaning that there can be over 1,000 line card drawers as part of this CO switch.

Line card drawers make up most of the footprint of a CO switch. Figure 161 illustrates only the first row of racks of a DMS-100 switch. There are at least ten full-length rows of racks packed with line card drawers behind it.

FIGURE 162 LINE CARD DRAWER

12.4.4 Digital Switching

All of the communication of voice information inside the switch is digital. As covered in Chapter 7, the analog voice signal on a loop is digitized at 8,000 bytes per second, or 64 kb/s on the line card.

The fundamental task of a traditional telephone circuit switch is to transfer a byte through the switch from one input to one output, and vice versa, eight thousand times per second, for the duration of a call.

12.5 Traditional PBX and Centrex

Traditional or *legacy* Centrex and PBX means systems developed and installed before VoIP was a reality. This section includes discussion of legacy PBXs, as well as functions that are more important than ever like IVR, ACD and Web Call Center – Customer Care integration.

Both are based on a large rack-mount computer system called a telephone switch. The telephone is connected to the switch, which establishes connections on a call-by-call basis.

There are three main differences between Centrex and PBX: location, location and location. It depends whether the telephone switch is at the customer premise (PBX), or in the Central Office (Centrex).

12.5.1 PBX

If the switch is in the customer premise, this is called "having a PBX", a Private Branch Exchange. The customer can buy, rent or lease one, and can provide their own dial tone, in-building dialing plan and in-building switching.

If the dialing plan involves assigning every telephone in the building a four-digit "extension" number, a user enters just the four-digit number to make a call to another phone connected to the PBX in-building.

Many features such as no-answer transfer, call pickup groups, Interactive Voice Response menuing and call center functions like Automated Call Distributor (ACD) can be provided by the PBX.

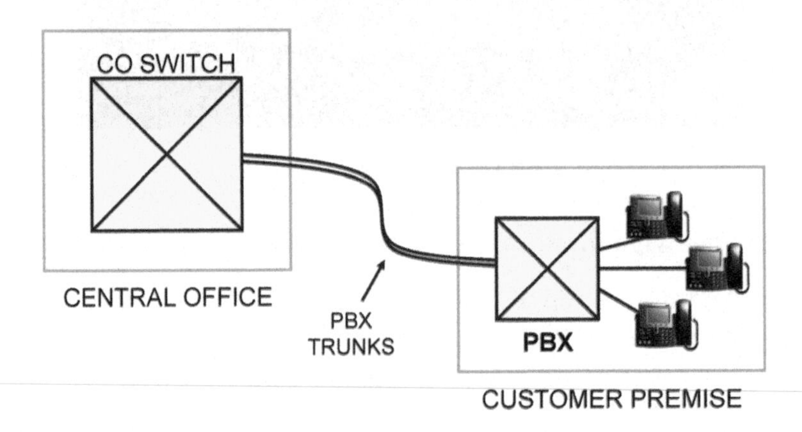

FIGURE 163 PBX AND PBX TRUNKS

12.5.2 PBX Trunks

A PBX is connected to the telephone network with PBX trunks between the PBX and CO switch. Usually, one trunk is provided for approximately every ten telephones.

When a user goes off-hook, they hear a dial tone from the line card in the PBX. If they dial 9 for an "outside line", they are assigned a PBX trunk, a circuit-switched connection is made to the CO switch, and the user hears a second dial tone generated by the CO switch.

PBX trunks can be ordered as one-way incoming, one-way outgoing, or both-way. This does not refer to the voice communications capability, which of course happens in both directions in all cases; it refers to whether these trunks are used to receive phone calls, initiate outgoing calls or both.

12.5.3 Digital Telephones: Electronic Business Sets

Both Centrex and PBXs support analog and digital telephones. Digital telephones, which are often called Electronic Business Sets, are far more popular as they support a much richer user interface and feature set.

"Digital telephone" means that the voice is digitized in the telephone and communicated as 1s and 0s along with call control messages from the phone to the switch, represented by pulses of voltage on copper wires.

For the traditional Centrex and PBX described in this section, the coding and formatting of the digitized voice and the call control messages are not standards-based, meaning that only telephones supplied with the Centrex service or PBX device will work. This is on purpose, to lock the switch customer into buying all phones and upgrades from the PBX or Centrex vendor. This is a source of profit for the vendor.

New-generation Voice over IP systems are much more likely - but not guaranteed - to use standard methods of coding and call control messaging, which would allow the use of third-party telephones.

12.5.4 PBX and PABX

The term *Exchange* is an older term for a circuit switch. *Private* means that the customer has the switch, not the telephone company. *Branch* refers to the topology of PBX trunks looking like branches off the main telephone company tree trunks.

In the beginning, a PBX was implemented with a board with jacks terminating loops and PBX trunks, and an operator connecting loops to PBX trunks manually with a patch cord, like the first CO switches.

Like CO switches, PBXs came to be implemented with mechanical systems then computers. For a while, computer-based PBXs were called Private Automated Branch Exchanges (PABXs). This term is not used much today.

12.5.5 Attendant

Even though the switch is implemented with a computer that connects PBX trunks and telephones, an operator or *attendant* is required to route inbound calls, to connect an incoming call to the correct telephone in-building.

Typically, all of the inbound trunks will be associated with a single telephone number valid on the public telephone network. When a caller dials that number, the CO switch will connect the caller to an available incoming trunk, terminating on the PBX.

In the simplest implementation, the PBX by default connects all incoming calls to an *attendant console*, where the attendant answers the call and asks the caller to whom they would like to speak.

The caller identifies the desired called party by name to the attendant using their voice, the attendant looks up the corresponding extension number on a piece of paper or computer screen, enters the extension number in the console and presses the "transfer" button.

This instructs the PBX to now connect the incoming trunk to the line card corresponding to that extension number and start it ringing.

12.5.6 Automated Attendant

An attendant is expensive, and can route only one call at a time. A computer program running on the PBX performing the attendant function, called an *automated attendant*, is much less expensive than an employee, and can handle more than one incoming call at a time.

Incoming calls are first terminated on automated attendant software on the PBX, which plays a recorded message to the caller requesting that if the caller knows the extension number they would like to be connected to, that the caller use in-band DTMF signaling to indicate the extension number.

Once received, the automated attendant software conveys the extension number to the PBX switching software, which connects the incoming trunk to the appropriate line card and starts it ringing.

If the caller does not supply the extension number, possibly because they have a rotary-dial phone, or they don't know it, or selected a preconfigured option like 0, the automated attendant software will route the call to a human attendant.

In a low-budget implementation with no backup human attendant, the caller might be transferred to a voice mailbox and asked to leave a message.

12.5.7 IVR

The automated attendant function is usually implemented in practice as part of an *Interactive Voice Response* (IVR) system running on the PBX.

An IVR provides more ways for the caller to have their call routed through the PBX to a particular telephone without knowing the extension number.

The most common implementation involves a recorded message asking the caller to signal a number corresponding to one of a number of menu choices. The result will either be transfer to a particular extension, or to a second menu where the process is repeated.

Speaker-independent word recognition has become reliable enough that the caller may have the option to speak words to navigate the menu instead of using in-band DTMF signaling.

In addition to determining the extension to which to transfer the call, an IVR may be used to have the caller enter information, for example, their account number at the called organization.

A sophisticated IVR might be integrated into the called organization's customer care system, allowing the caller to retrieve information without speaking to a person.

An example would be calling an airline and the caller entering their frequent-flyer number, then having the IVR do a query on the airline's customer care system to determine that account's mileage balance and communicating it to the caller using recordings of someone speaking numbers.

12.5.8 Direct Inward Dialing (DID)

The telephone company controls the telephone numbers and charges per number, per month to assign numbers to users.

The lowest-cost configuration with a PBX is to pay for only one telephone number for the PBX. All inbound PBX trunks are associated with that one telephone number in a *hunt group*. When a call is placed to that number, the CO switch hunts through the group of trunks to find the next one available and connects to call to the PBX on that trunk. The caller must as a second step indicate to an attendant or automated attendant to whom they wish to speak.

If an organization desires to have a PBX but eliminate the two-step process of first dialing a number then dealing an attendant or automated attendant to have calls connected, the organization can pay the phone company for *Direct Inward Dial* (DID) service.

With DID, the hardware configuration of PBX connected to CO with PBX trunks of Figure 163 remains unchanged. The telephone company assigns a telephone number called a DID number that is valid on the public telephone network, for each of the extensions on the PBX.

When a caller dials one of those DID numbers, the call is connected from the CO to the PBX over a PBX trunk as usual, plus the CO switch indicates to the PBX the DID number that was called. The PBX can then look up in a table to determine the extension number associated with that DID number and switch the call to the correct line card without any further interaction by the caller.

This service is billed per DID number, per month by the phone company.

12.5.9 Automated Call Distribution (ACD)

The destination of a call could be an extension that identifies not a particular line card, telephone and person, but instead identifies an *Automated Call Distributor*.

When running on the same hardware as the PBX switching function, an ACD is a computer program that deals with situations when there are more callers than there are people or *agents* to answer the calls.

The ACD is configured to have queues associated with extension numbers. The queues are associated with specific activities, such as a particular type of caller wanting to perform a particular activity.

Upon being transferred to the ACD, the caller is placed in a queue and recordings are typically played to the caller to keep them interested. Which queue the caller is placed in might be determined by the number they dialed, or choices they made in an IVR before being transferred to the ACD.

Agents are associated with queues. An agent can be dedicated to answering one call queue, or able to answer multiple call queues. When the ACD determines an agent is available to answer the next call in a queue, the caller is switched from the ACD's recording to that agent.

12.5.10 Call Centers

Inbound call centers are places where customer service agents receive calls from customers and access the customer's account information via a terminal connected to a customer care system, a data processing system.

Traditionally, this has been a customer-premise-based solution. The end-user company buys, integrates and maintains a PBX to handle incoming calls, an IVR to get information about the caller, and an ACD to route the call to an agent, and a customer care system to store and manipulate information about customers and orders.

The agents sit in a large "call center" room with supervisors and may have to raise a paddle to request to go to the bathroom... and might not be allowed to if the call volume is heavy.

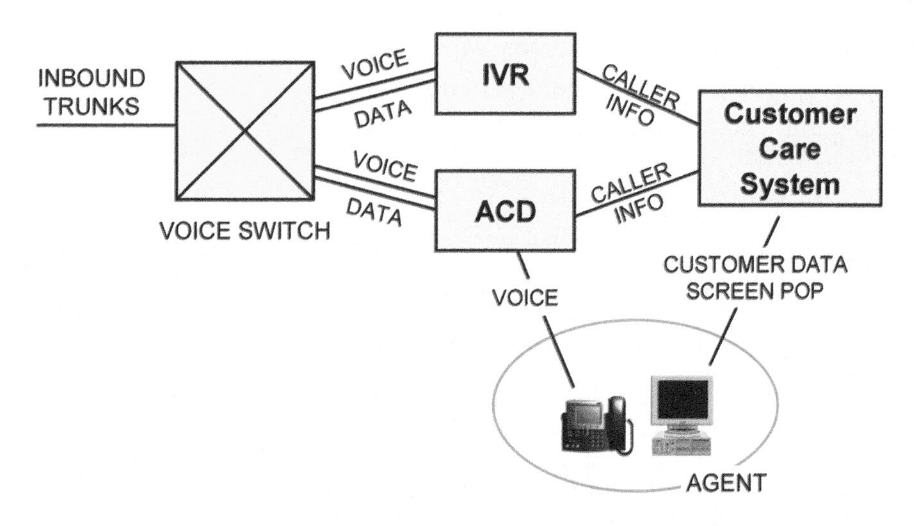

FIGURE 164 INTEGRATED CALL CENTER

A sophisticated integrated system would first pass the caller through an IVR to determine their account number and desired activity, then to the appropriate queue on an ACD, then when the call is finally switched to an agent, send a message to the customer care system. The customer care system would then cause the caller's account information to appear on the agent's screen at the same time the agent answers the call.

One of the main purposes of this integration would be to reduce the amount of time the agent has to spend with the caller to determine who they are and what they want to do, which reduces the number of agents required and thus saves money.

The infrastructure can of course be outsourced, using network-based call center services, where a third party has the IVR and ACD and handles the call queuing and call setup for agents at a different location.

Using Voice over IP and SIP call setup, the agents can be anywhere... even working at home. This would provide significant benefits in flexibility of staffing to handle peak call volumes to meet call-answer-delay requirements.

A recent trend was to locate the agents in countries where there were good IP telecom services, salaries low and employment standards less stringent than elsewhere. This was in some cases abandoned after companies decided that low-budget customer service provided by people on the other side of the world was a revenue-negative idea.

The next step is multimedia contact centers, where there are a number of different ways that a customer can contact the agents in the center, including speaking to the agent, e-mail, web chat, web collaboration, click-to-talk, and click-to-see.

12.5.11 Advantages of PBX

The main advantages of a PBX system are the service pricing model and the ability to control the hardware and features.

With Centrex service, described in following sections, the switch is at the CO, and the telephone company provides telephones connected with individual lines to the CO switch.

With a PBX, the connections are trunks, not loops, with something like one trunk for every ten telephones. PBX trunks cost more than individual lines, but not ten times as much. This means that the monthly service cost is less.

Moreover, the cost of value-added features like call forwarding and voice mail are notionally per PBX, not per line, which is a definite cost savings.

Plus, the customer determines which features are available, based on selection and configuration of the PBX by them, not by the phone company.

Another advantage of having a PBX is not having to pay the phone company for moves and changes. If a person moves to a new cubicle, and wants to keep their extension number, a technician has to reconfigure the switch so that the extension number is associated with the line card or wires going to the new cubicle, and not the old cubicle.

With a PBX, the organization can perform moves and changes with in-house staff instead of paying the phone company to do it.

12.5.12 Disadvantages of PBX

The main disadvantages of having a PBX are capital cost, scalability, support and maintenance. When an organization gets a PBX, they are going into the local telephone business inside their building, and must perform all the functions of a local telephone company.

This means the organization must decide which manufacturer and which model of PBX to get, how many of them to get – one for each location is the starting point – plan for future growth and future features and technologies, and finance the hardware.

Planning for the future is especially important considering that the connection from the PBX to the telephone is historically not standards-based, meaning that only particular telephones made by the same manufacturer will work with the PBX.

The implication is that once a PBX is purchased, the customer is obliged to purchase all future telephones from that manufacturer, which may turn out to be costly.

Having a PBX means the organization must have a help desk, trouble ticket system and skilled staff to operate, maintain and repair it, and to deal with the carriers providing local and long-distance telephone service.

The organization must also decide how long they want their telephones to keep working after the big ice storm, hurricane or earthquake knocks out all of the main power distribution for thirty miles around the building. A minute? An hour? A week?

The latter requires a contract signed before the disaster happens for guaranteed delivery of fuel for generators... when the city is blacked-out and everyone wants fuel, this organization is the one that will get it.

Finally, unless the organization is willing to pay per extension per month for DID service, callers have to go through a two-step process to connect a call: first dialing a phone number, then dealing with an attendant or IVR before the call is connected.

12.5.13 Centrex

Service with the exact same look and feel to the user as having a PBX can be provided by the telephone company. This is generically referred to as *Centrex*. Every telephone company has their own brand name for this service or bundle of services.

Centrex means that the telephone service is provided by a CO switch, rather than a PBX. Typically, a part of the CO switch will be partitioned in software and dedicated to a particular customer, making it appear to the customer as though they have their own switch, with the same features as a PBX such as four-digit dialing and having to dial "9" for an outside line.

With Centrex service, the connections between the telephone company and the customer are loops - one for each telephone. If a particular Centrex customer has many telephones, the telephone company will carry the loops not on many pairs of wires, but instead on a single fiber optic *loop carrier system* between the buildings.

The loop carrier system does not add any value to the service; it is simply a mechanism to carry the information for the individual loops together on one fiber instead of on many sets of copper wires.

In this case, as illustrated in Figure 165, the fiber terminating equipment at the customer premise for Centrex has line cards connected to telephones by copper wires.

PBX trunks are carried the same way, on fiber, so the fiber terminating equipment at the customer premise for PBX also has line cards connected to telephones by copper wires.

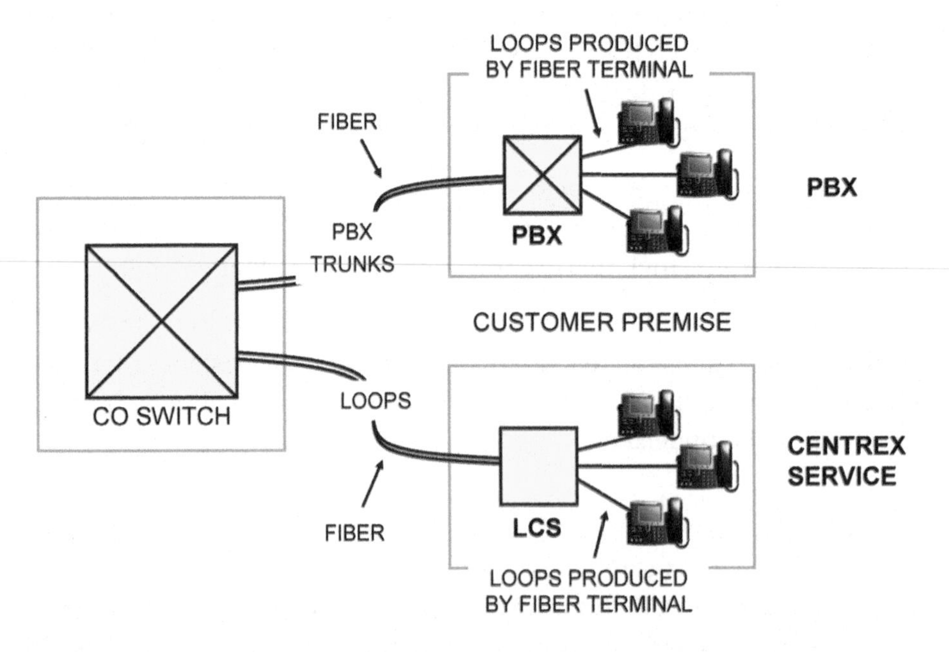

FIGURE 165 TRADITIONAL PBX AND CENTREX SOLUTIONS HAVE THE SAME PHYSICAL LAYOUT. THE LOCATION OF THE SWITCHING IS THE DIFFERENCE

In fact, comparing the two architectures in Figure 165, the choices are identical except for location, location and location: whether the switching is at the customer premise (PBX), or at the CO (Centrex).

12.5.14 Advantages of Centrex

The main advantage of Centrex is that the phone company will take care of planning, purchasing, installation and maintenance of the telephone switch and telephones, and provide a service agreement specifying the availability of service and time to repair. The customer does not need to have experts on staff to configure and maintain the telephone system.

With Centrex, there is no capital cost for the switch, though there may be for the phones. Monthly payments with a fixed-length contract are typical.

In addition, the phone company deploys many switches in different geographical areas, facilitating the implementation of seamless regional and national service, and ensuring that there is enough switching capacity for each of the customer locations.

12.5.15 Disadvantages of Centrex

The downside of Centrex is cost. Centrex is not a money-losing business at the phone company; it's part of their bread-and-butter.

The pricing model for Centrex service is per line. The monthly service charge for dial tone is per line. Cost for voice mail and features like call forwarding is per line. This ends up being more expensive than service implemented with a PBX, where the pricing model is more per-PBX.

In addition to monthly charges, another cost with Centrex is moves and changes.

When someone changes cubicles and wants to keep their phone number, the switch has to be reconfigured to associate the phone number with a different line card, terminating the wires going to the new cubicle.

This costs in the neighborhood of $100 per change... every time someone changes cubicles.

PBX vs. Centrex

The question of PBX vs. Centrex often boils down to this question: what business is the customer in? Do they want to devote part of their energy to providing local phone service, and save some money by doing it themselves; or do they want to use their energy for their printing business and pay the phone company to do phone service, knowing that the phone company makes a profit doing it.

12.5.16 Key Systems

A *key system* is a low-capacity, low-budget combination of Centrex and PBX functions. A key system terminates lines from the phone company, not PBX trunks, but, like a PBX, allows the connection of more phones than there are lines. A 3x8 key system would support up to 8 telephones in-building connected to one of 3 phone lines.

In the old days, mechanical key systems used telephones with a row of transparent buttons across the bottom to select which line the phone was connected to.

More recent electronic key systems use Electronic Business Sets with programmable buttons and displays, more or less identical to those used for Centrex and PBX.

In the future, all call setup will be done with SIP.

12.6 Gateways

In the previous millennium, the telecom network core was based on SONET, a technology implementing channels on fiber optics.

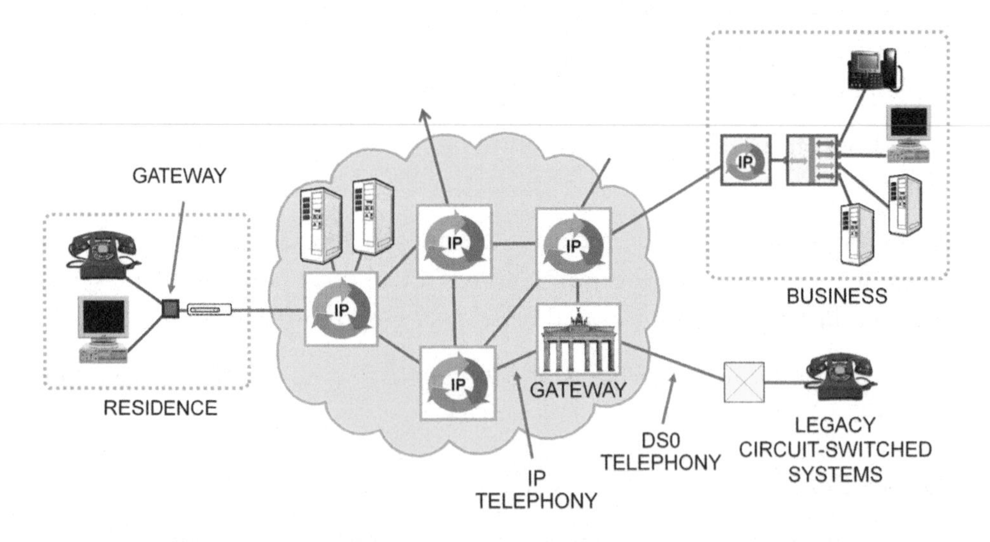

FIGURE 166 GATEWAYS CONVERT VOIP AND VOICE OVER DS0S

Telephone calls were converted from analog on loops to 64 kb/s DS0 streams by the line cards in the CO switch, then switched to trunks carried long distance in DS0 channels on the SONET backbone.

PBX trunks connected a business telephone system to the CO switch, one trunk per DS0 channel on a T1. ISDN PRI service could be implemented on the T1 for enhanced call control functions.

Going forward, the telecom network core is everything in IP packets carried on Optical Ethernet and managed with MPLS.

To transition the huge installed base of analog loops converted to DS0 channels by CO switches, and PBX trunks on PRI service, to the new world of VoIP, *gateways* connect old systems to the IP network, and in some cases connect new systems to old.

Converting between legacy channelized DS0 circuit-switched systems and the modern IP network carrying VoIP, consists of two main tasks: media conversion and signaling conversion.

12.6.1 Media Conversion

Media conversion means converting between the format in which voice, video, fax and other types of content ("media") are represented on one system and the format used on another system.

For example, voice might be converted between the coding method, bit rate and IP packet format of a VoIP system on one side, and streaming 64 kb/s DS0 channel format on the other. Media conversion could equally involve transcoding voice from one standard to a different standard, even though both are carried in IP packets.

Media gateways are processor-intensive, and need to be able to convert in real time without introducing any delay into voice conversations.

12.6.2 Signaling Conversion

Signaling conversion means translating between connection setup protocols used on a VoIP system and those used on other networks. For example, the signaling gateway might support SIP call setup messages on the VoIP side, and convert these to ISDN User Part (ISUP) messages for the legacy side.

In the longer term, the gateway function will be moved to the customer premise and the analog loops, CO switches and DS0 trunks will completely disappear.

The network will be all-IP; all traffic to and from the customer premise will be voice, video, data, Internet traffic and anything else, interspersed in IP packets.

13 Carriers and Interconnect

13.1 IX: Internet Exchange - Interconnect for Internet Traffic

The Internet is a huge, amorphous collection of Autonomous Systems (ASs), connected at buildings called Internet Exchanges (IXs).

An AS is a collection of IP routers managed by the same organization.

FIGURE 167 INTERNET EXCHANGE - EXTERIOR VIEW

An ISP is an Autonomous System. Google is an AS, as is UCLA. Backbone and bulk carriers are ASs. Every Autonomous System gets a number: its Autonomous System Number (ASN). The list of ASs is public and easy to find.

13.1.1 Fiber to an AS

The Internet Exchange is a building run by a neutral third party where connections between ASs are made by IP routers.

The connections are implemented by defining a route, in software, internal to the router, between two optical ports. One port has an Optical Ethernet fiber leading to the first AS, and the other port has a fiber to the second AS.

Figure 168 is a photograph of cables containing actual such fibers inside an IX building.

These connections are implemented for business reasons, between carriers.

FIGURE 168 FIBER CONNECTIONS INSIDE AN INTERNET EXCHANGE

There are two types of business arrangements: transit and peering.

13.1.2 Transit

Transit is a service that smaller players buy from bigger players, assured packet delivery to anywhere in the world, specified in a Service Level Agreement (SLA), priced per Gb/s per month.

13.1.3 Peering

Peering is settlement-free in the terminology of the wholesale business, meaning no money is exchanged. There are no service level guarantees with peering. It is implemented between ASs that have similar volumes of traffic in both directions, and between Content Delivery Networks and ISPs. The Internet was built with peering.

Transmitting a packet over the Internet means giving the packet to another AS, and **hoping** they will give it to another AS or deliver it. There is no source to destination guarantee available on the Internet.

Services including an end-to-end guarantee are Virtual Private Network services, not Internet connections.

13.2 Telephone Network Architecture

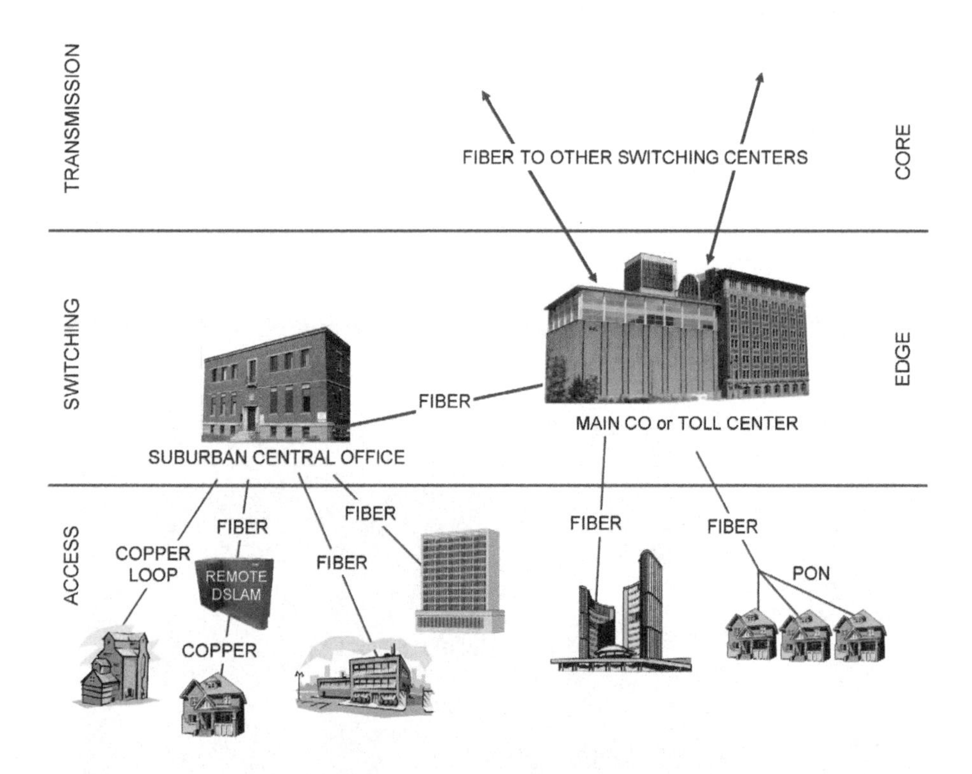

FIGURE 169 TELEPHONE NETWORK ARCHITECTURE MODEL

The telephone network is traditionally described as being made of three parts: access, switching and transmission, sometimes called the access network, switching network and transmission network.

13.2.1 Access Network

The access network, also called the *outside plant*, is the equipment and cabling used to connect the customer to the switching network, typically to a Central Office.

This is also referred to as the "last mile" - though, of course, the people who work in this part of the business prefer to call it the "first mile". It may in fact be very much shorter or longer than a mile.

Historically, this was implemented with twisted-pair copper wires in feeder cables with a thousand or more pairs leading from the CO to wiring connection points in neighborhoods called Outside Plant Interfaces (OPIs) or Serving Area Concept (SAC) boxes.

From there, distribution cables with a hundred or more pairs run down streets, with terminals where a drop wire to the customer premise is connected. Physical connections are made in the terminal and OPI/SAC box to implement a metallic connection of two wires between the CO and the customer premise. Since electrical current flows in a loop on these two wires, the pair is also called a loop.

Neighborhoods with this infrastructure installed are now referred to as *brownfields.*

Fiber to the Neighborhood (FTTN) then DSL to the subscriber is used to implement high-speed Internet access in the very large installed base.

FIGURE 170 FIBER TO THE NEIGHBORHOOD OPI/SAC BOX

A fiber is pulled from the CO to each OPI/SAC box, which may be generically referred to as an *outside plant enclosure.*

Inside the enclosure, the fiber is connected to a DSLAM, which houses banks of DSL modems. A short pair of wires is used to connect one of the DSLAM's modems to one subscriber for high-speed Internet.

The customer's network access is fiber to the enclosure, then a short run of copper to the customer premise.

The shorter the run of copper at the end, the more bits per second can be communicated. VDSL2 technology achieves 200 Mb/s with a maximum run length of 150 meters (150 yards).

In new neighborhoods, called *greenfields*, fiber to the premise is deployed. For residences and small business, a Passive Optical Network (PON) strategy may be employed, where typically 32 customers share a fiber backhaul using time sharing. Medium and large businesses might be connected with a dedicated fiber.

13.2.2 Switching Network

The switching part of the network was traditionally organized into a five-level hierarchy, with the Central Office at the lowest level in a hierarchy of switching centers.

A Central Office is the *wire center*, where all of the access wires converge and are connected to switching equipment. This equipment is usually owned by the telephone company, but might also be equipment owned by a competitor collocated in the CO. In the past, this switching equipment was a circuit switch, establishing connectivity to an outgoing circuit for the duration of a phone call. Going forward, this switching equipment is a packet switch or router, forwarding one packet at a time.

This equipment is called *edge equipment* by network engineers, as it is notionally the edge of the telephone company's core network. This equipment provides a data concentration function and converts between the physical media of the access circuit and the physical media of the connections between switching centers and the transmission network.

The COs in a city are connected to its *toll center*, a building at the second level in the switching hierarchy and the interconnection point with transmission networks owned by the same telephone company or by a competitor.

13.2.3 Transmission Network

The transmission network connects switching centers, providing high-capacity and high-availability connectivity between COs and between cities. This part of the network is called the *network core* by transmission engineers. In the past, the capacity was organized into fixed 64 kb/s channels, with switches or routers directing traffic onto the channels. Going forward, traffic on the core is all packets, transmitted on demand.

13.3 PSTN Switching Centers, COs and Toll Centers

Figure 171 depicts the Bell System's five-level switching center hierarchy.

The bottom two levels, Central Office (Class 5) and Toll Center (Class 4) are of most interest.

13.3.1 Class 5: Central Office

At the bottom level is the End Office, End Serving Office, Serving Office, Class 5 Office, Number 5 Office or *Central Office* (CO).

In all cases, these words refer to the building that contains the switch to which a POTS telephone set is connected with a loop, or with a loop carrier system if the phone is connected to a remote switch. It is owned by the Incumbent Local Exchange Carrier (ILEC).

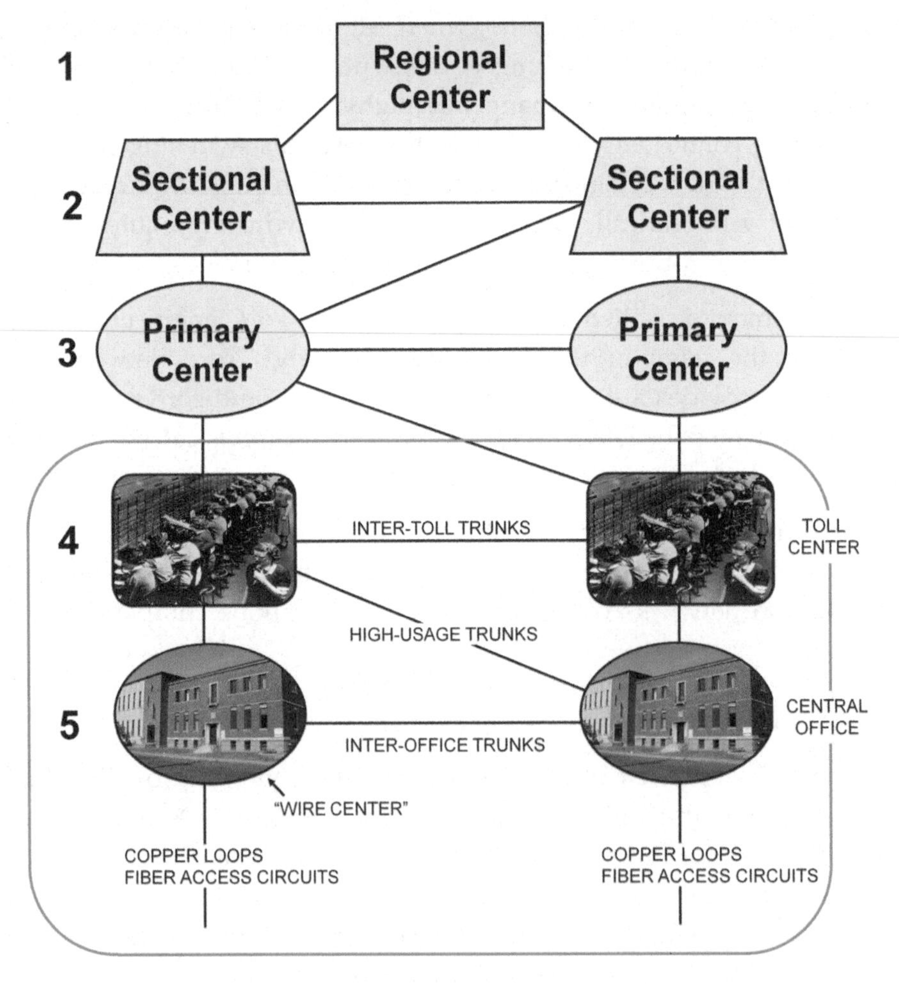

FIGURE 171 FIVE CLASSES OF SWITCHING CENTERS

13.3.2 Wire Center

The CO is also called a *wire center* because it is the physical point where access wires converge for connection to the switching network and network core. This is where CLECs get access to the wires.

Many cables carrying many wires and fibers arrive in a cable vault in the basement of the CO, illustrated on the left of Figure 172. These are carried on *risers* up to a room in the center of the building where individual copper pairs and fibers from the street are connected via the Main Distribution Frame (right of Figure 172) to cables leading to hardware ports on the switches, routers or muxes in the building.

FIGURE 172 CABLE VAULT AND MAIN DISTRIBUTION FRAME IN A CO

Most of the cables leaving the building lead towards customers. These cables contain the copper subscriber loops and fiber access circuits.

Some of these cables contain fibers connecting to other COs, and fibers to the Toll Center for interconnect with Inter-Exchange Carriers as described in the following sections.

13.3.3 Local Calls

When you place a call to your neighbor, their loop is usually terminated in the same CO as yours, and so the call is handled within the Central Office. If you place a call across town, your call will be routed across trunk circuits to another CO switch and then on to the far-end loop. The Central Office is level 5 in this hierarchy, so CO switches are called *Class 5 switches*.

13.3.4 Class 4: Toll Center

Directly connecting the thousands of Central Offices together would not be possible due to the enormous numbers of connections that would be required. A hierarchy was needed.

Generally, each metropolitan area has a building called a *toll center*, containing switches to which all of the CO switches in that city are attached.

To make a phone call to another city, your call is routed from your CO to your city's toll center, and on the far-end city's toll center, to the far-end CO and then the far-end loop.

The toll center is level 4 in the hierarchy, so switches in toll centers are called toll switches or Class 4 switches. With competition, the ILEC's toll switches are connected in tandem to those of the IXCs and other LECs, so they are also called *tandem switches*.

13.3.5 Class 1, 2 and 3 Switching Centers

Because the Bell System was so large, there are more levels in the hierarchy. Each state had a *primary center* (class 3) to which all of the toll centers in that state were homed. The USA was always divided in seven sections, and the primary centers in each section were homed to a *sectional center* (class 2). Sectional centers were connected via *regional centers* (class 1).

13.3.6 High Usage Trunks

In practice, connections are installed between switching centers where traffic warrants. If there is high traffic between two COs not homing on the same Toll Center, then a High Usage Trunk might be installed directly between those COs.

This practice moves the actual implementation of the network from the strict hierarchical model shown to more of a meshed network, with many different paths between switching centers.

Traditionally, these trunk circuits were carried as reserved channels on SONET fiber-optic transmission systems organized in ring patterns around town, around the region and around the country.

In modern systems, phone calls are carried when needed in IP packets in Ethernet frames on fiber optics.

13.4 Implementing Competition: LECs, POPs and IXCs

13.4.1 LECs, ILECs and CLECs

The official parlance for a company that provides the "last mile" to the customer is a Local Exchange Carrier (LEC).

The Incumbent Local Exchange Carrier (ILEC) is the local phone company, the owner of the cabling between the CO and the customer premise.

CATV companies providing phone service are also LECs, owning the Head-End and the Hybrid-Fiber Coax network.

Cellular companies are also LECs, owning or leasing the spectrum, base stations and towers for the last mile to the customer.

Competitive Local Exchange Carriers (CLECs) are companies that can co-locate equipment in the CO and provide services over the ILEC's local access network to customers.

13.4.2 Inter-Exchange Carriers: IXCs

Long-distance networks connect LECs different cities. Also called *long lines*, they are owned and operated by Inter-Exchange Carriers (IXCs).

A *facilities-based* IXC is one that for the most part owns their own physical transmission facilities, typically fiber-optic cables and equipment.

A *reseller* leases capacity from a facilities-based carrier to form a network. Both of these companies are Inter-Exchange Carriers.

Due to the pricing structure at the time, many reseller-type IXCs sprung into business once competition was introduced, leasing high-capacity services from IXCs, and signing up customers who route their calls over the high-capacity leased service… buying wholesale and selling retail.

With a subsequent drop in retail prices, the profit margins for resellers were dramatically reduced, to the point where many went out of business, or transitioned into more facilities-based operations.

13.4.3 POP: Point of Presence

The place where the LECs and IXCs connect is the toll center building. Inside the building, the place where an IXC's equipment that terminates its fibers is physically located is their *Point of Presence* (POP).

This term originated when law and regulation forced the Bell System to provide physical space in the toll centers for IXCs like MCI to house equipment to terminate their fibers and connections to the LECs.

Everyone who worked in the building would refer to it as "the MCI room". This room was their physical point of presence in the building.

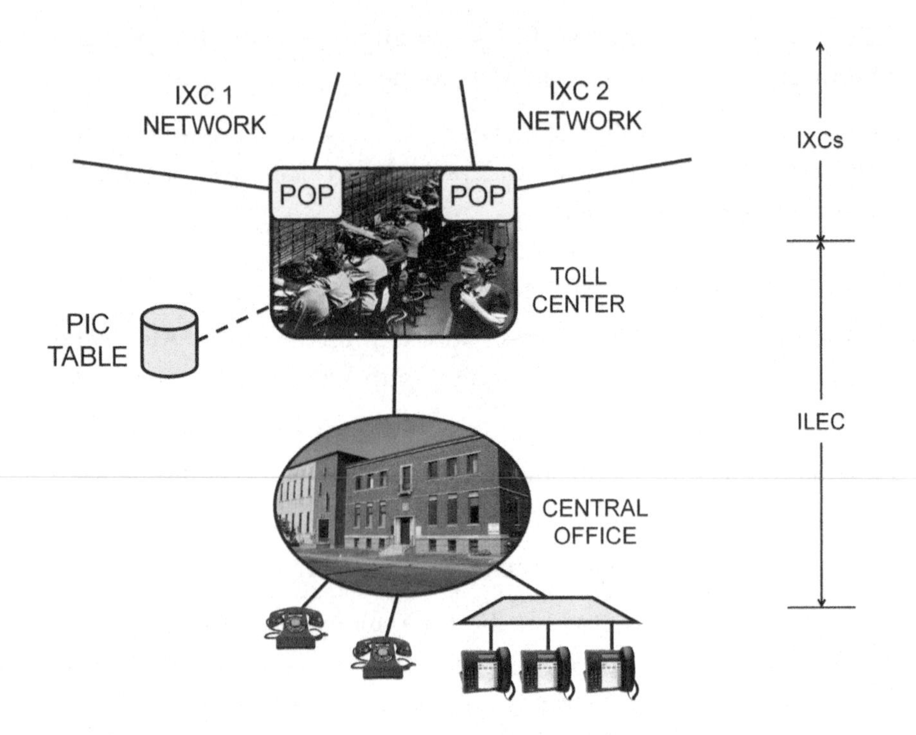

FIGURE 173 SWITCHED ACCESS ILEC CUSTOMERS TO IXC POPS

From the ILEC's point of view, it is the beginning of the IXC's network. From the IXC's point of view, it is the end of their physical circuits.

The term *POP* has now moved into general usage, to mean a building where a competitive carrier terminates at least two fiber-optic cables: a station on a regional ring, to connect to other carriers.

This building is today often not the toll center, but a different building sometimes called a *POP hotel*, across the street or across town, connected to the toll center with fiber.

13.4.4 Switched Access

ILECs are required to provide their customers equal access to competitive IXCs. CATV and Cellular LECs are not.

The customer of the ILEC can select any IXC, and the ILEC will connect the customer to that IXC on a call-by-call basis. This is called *switched access.*

The LEC at each end bills the IXC a per-minute *switched access charge* for that last mile connection.

13.4.5 Equal Access and PIC Codes

Equal access means that a customer of an ILEC can select in advance the Inter-Exchange Carrier that will handle that customer's long distance, and the routing through the toll center and the POP on to the IXC's network is transparent to the customer.

This is implemented with an entry in a database maintained by the ILEC called the customer's *Preferred Inter-Exchange Carrier* (PIC) code.

Each IXC has a Carrier Identification Code. AT&T's three-digit carrier code is 288 (ATT on a telephone keypad). MCI is 222, Sprint is 333, Global Crossing is 444. There are many others.

When a customer of the ILEC changes long-distance companies, the customer's PIC code is changed to the carrier code of the new IXC.

When making a call from someone else's telephone, such as a payphone at an airport, it is possible to manually route the call through a particular IXC on a call-by-call basis by dialing 1010 then a three-digit carrier code, or 101 then a four-digit carrier code.

Dialing 1010288 connects to AT&T's POP; 1010222 connects to MCI, 1010333 to Sprint and so forth. Many of these companies also have 1-800 numbers that accomplish the same thing.

In 1984 in the USA, the ILEC and IXCs became strictly separate companies. Following changes in law and regulation, they were able to rejoin. Today typically the holding company that owns the ILEC is also an IXC.

This was always the case in Canada, where the phone companies were not split into separate local and long-distance companies. Equal access to competitive IXCs began in Canada in 1992.

13.5 Wireless and CATV Local Exchange Carriers

Companies using a coaxial entry cable, primarily to residences, were historically called Community Antenna Television (CATV) companies and more recently cable companies, and more technically, Hybrid Fiber-Coax (HFC) companies.

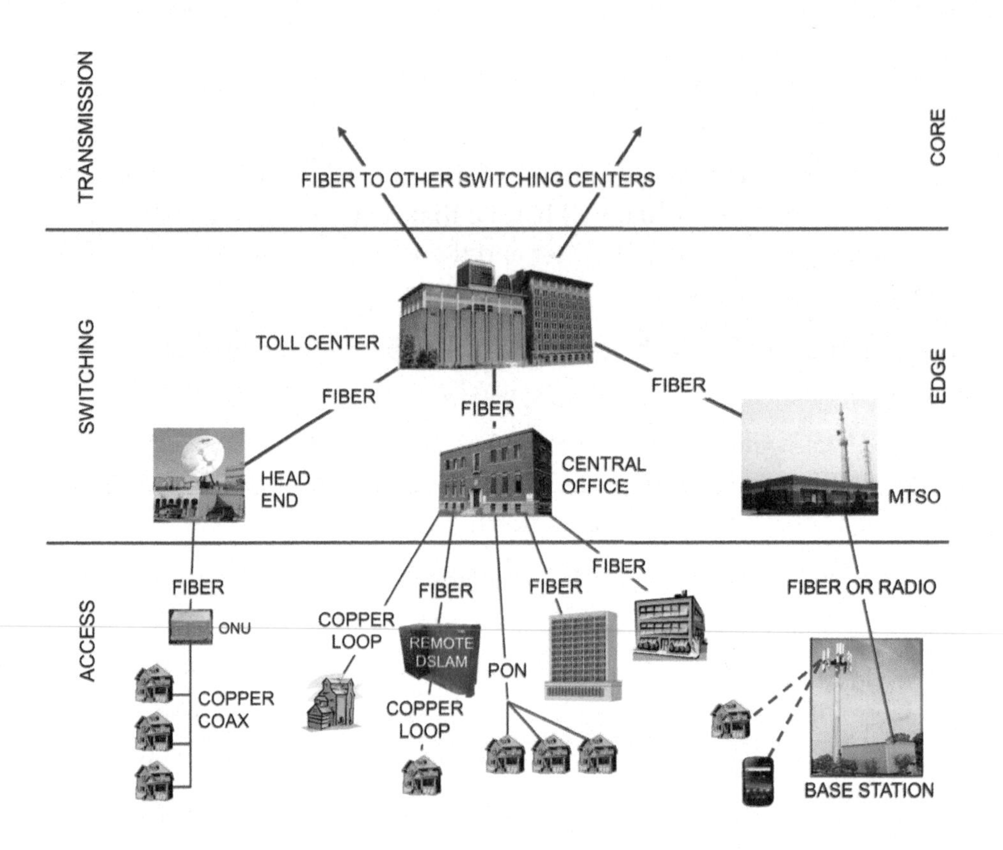

FIGURE 174 WIRELESS AND CATV LOCAL EXCHANGE CARRIERS

With the delivery of telephone service and Internet via broadband modems, and backbone fiber networks, cable companies are now full telecom service providers and carriers, for both business and residence, and are properly referred to as *broadband carriers*.

Wireless and CATV operators are a new type of LEC. These operators connect at the toll center for phone calls between customers of connected LECs, for outbound long distance, and for inbound phone calls from IXCs. Wireless and CATV LECs are not required to provide switched access to competing IXCs for outbound long-distance calls.

13.6 CLEC: Collocations and Dark Fiber

13.6.1 Unbundling

Paying the ILEC for tariffed switched access or dedicated line service to connect the last mile from a competitive carrier's POP to the competitive carrier's customer started in 1984.

Subsequent legislation and regulatory decisions *unbundled* the ILEC's physical cables from the ILEC's services provided on those cables.

This enables competitive carriers to lease just the ILEC's physical cabling to the customer instead of paying for a tariffed service from the ILEC on the same cable.

The regulatory rationale is that the ILEC built the wire center and access network when they were a monopoly. The community was *obliged* to pay the ILEC to build the access wiring. Therefore, in a way of thinking, the community has a degree of ownership of the resulting physical access wiring... and therefore, the community has the right to use it without being obliged to also have the ILEC provide them services over it.

13.6.2 Dark Fiber and Dry Copper

If the ILEC is providing copper wires without electricity on them, i.e., not attached to a CO switch line card, this is called a *dry* circuit. A fiber not attached to anything is called *dark fiber*.

13.6.3 Competitive Local Exchange Carrier (CLEC)

The next more sophisticated solution to connect a competitive carrier's customer to their POP in the same city is to rent two fibers from the ILEC: one from the CO to the customer, and one from the CO to the POP.

The competitive carrier must also rent space inside the ILEC's CO - at the wire center closest to the customer, where they can locate equipment to operate the fibers.

The competitive carrier installs fiber-terminating equipment at the customer, in the ILEC's CO and at their POP, then connects the POP to the CO with one ILEC fiber and the CO to the customer with another.

An organization doing this is said to be a *Competitive Local Exchange Carrier (CLEC)*.

It should be noted that large carriers use this technique for the last mile in areas where they are not the ILEC.

In this case, it is not really appropriate to call the large carrier "a CLEC", as this is a very small part of their business. It is more correct to say that the large carrier "has collocations".

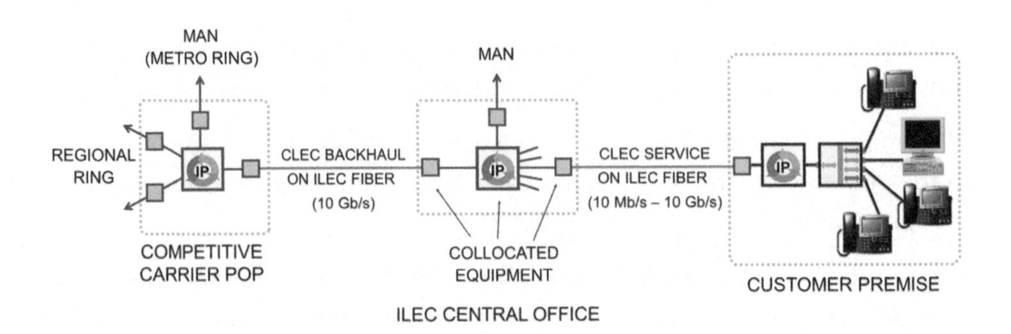

FIGURE 175 COLLOCATION OF EQUIPMENT AT THE WIRE CENTER

13.6.4 Collocations

In addition to being required to lease dark fiber to their competitor, the ILEC is required to build *collocation facilities* in its COs.

Collocation facilities are rooms in the CO, often with separate entrance doors, where the competitive carrier can locate their own equipment.

The competitive carrier thus gains access to the wire center: the termination point for the fibers and copper wires in cables leading out to the street and ultimately to the customer premise.

In the collocation facility, the competitive carrier places or *collocates* network equipment like Optical Ethernet switches and routers.

The ILEC's dark fiber is connected to optical transceivers in the collocated equipment in the CO at one end and at the customer premise at the other end, implementing a "last mile" connection over the ILEC's fiber.

13.6.5 Advantages

There are two key benefits of leasing dark fiber and collocations compared to tariffed services from the ILEC for the last mile:

1) Cost: leasing a dark fiber is a tiny (!) fraction of the cost of paying for a 10 Gb/s dedicated service from the phone company.

2) Performance: The competitive carrier is now in control of the optical technology: the optical transceiver technology, WDM strategy, the manufacturer, the bit rates, transmission characteristics and so forth.

13.6.6 Disadvantages

Collocation requires the competitive carrier to have Engineers to design the system, select the equipment and determine how it should be installed and configured. This is a never-ending process, as newer, better products and technologies are constantly becoming available.

Another disadvantage is that there may not be any dark fiber available from the ILEC, and no time frame for new cable construction in that area.

The CLEC does not have any control over the physical connection and how it is provisioned, maintained and especially repaired.

The CLEC is relying on their competitor to provide and maintain the physical fiber, and to repair the fiber and restore the service in a timely fashion after the fiber is cut.

13.6.7 Application

Collocations would be implemented when there is enough business around a particular CO to justify it.

Collocations and dark fiber can be used to provide 10 Mb/s to 10 Gb/s Optical Ethernet services over fiber. In theory, they can also be used to provide DSL and even POTS and T1/PRI services on twisted pair copper wires to a customer.

For the POP to CO portion of the physical connection, sometimes called the *backhaul*, the same technique of leasing a dark fiber from the ILEC between the POP and collocation may be used.

13.7 SS7

SS7 is a global standard defined by the International Telecommunication Union (ITU) Telecommunication Standardization Sector (ITU-T).

It defines the protocols by which network elements exchange information for call setup, routing and control, both wireline and wireless.

The ITU definition of SS7 allows for variants including the American National Standards Institute (ANSI) and Bellcore standards used in North America, and the European Telecommunications Standards Institute

(ETSI) standard used in the rest of the world, called "Europe" in the business.

Signaling System 7 allows the communication of messages used to set up phone calls between telephone switches. The set of messages, what they mean, what order they are used in and many other things are specified in standard protocols that telephone companies use.

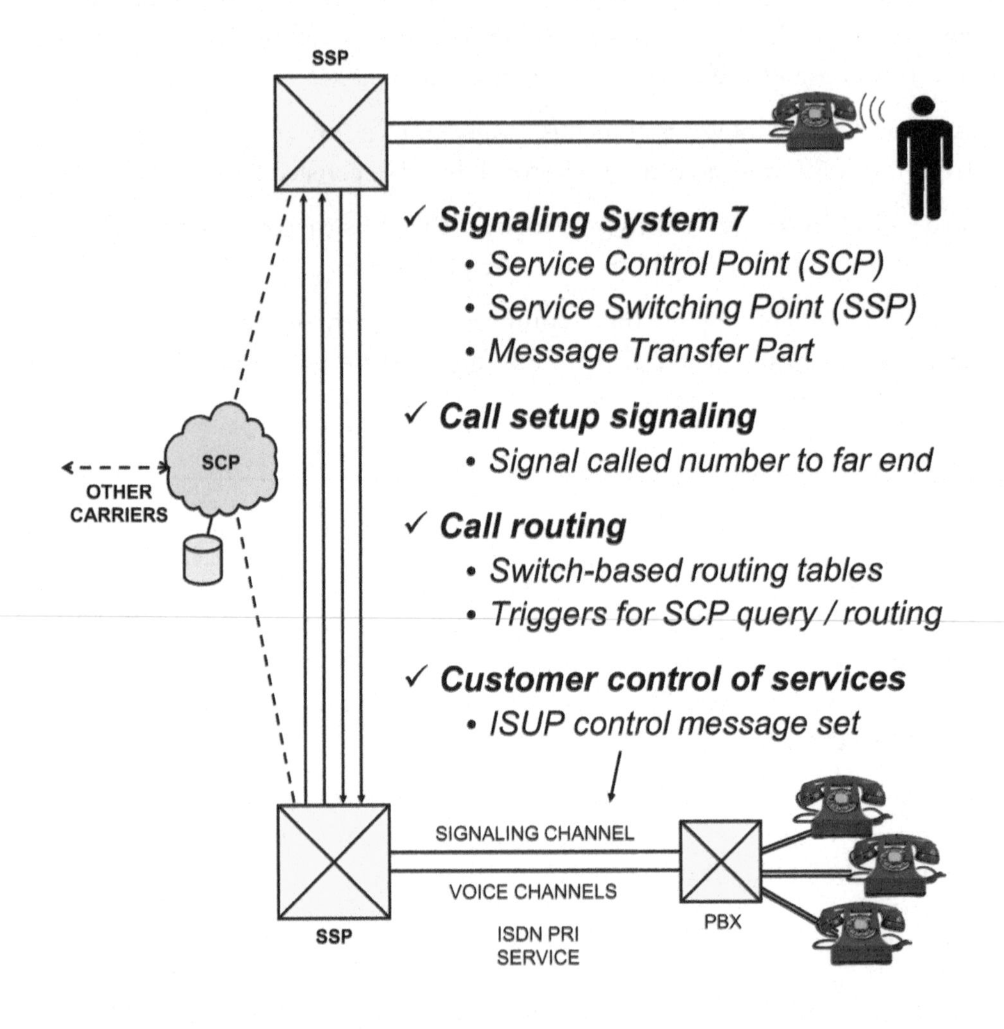

FIGURE 176 SS7

There are two uses of this function by a carrier:

- Primarily, communication of call setup messages with other carriers.
- Formerly, communication of call setup messages between their switches and call routing software for calls on their network.

13.7.1 Carrier Interconnect

SS7 is the protocol that LECs and IXCs use to exchange call setup messages for calls between customers of different LECs.

As of 2022, call setup messages associated with circuit-switched Tandem Access Trunks connecting carriers at toll centers is the only tariffed protocol for phone call interconnect in North America. Connections between carriers via POPs at Toll Centers is described in Section 13.3.

Connections to non-cooperating competitors can only be implemented with a tariffed service, where the provider is forced to provide service to any person that applies under the terms of the tariff.

The tariff will eventually be changed to include SIP and IP in addition to SS7 and Tandem Access Trunks, but there is a lot of work to do renegotiating many agreements.

This tariff is the legal and regulatory basis for competition, and the basis for countless millions of regulations on monthly charges, per-call charges and monthly settlements between thousands of carriers.

There are billions of dollars involved, so it will take a few years.

The remainder of this chapter is a primer on how a POTS phone call is set up with SS7.

13.7.2 Call Setup

SS7 was used for call setup within a carrier's network. This is being replaced with SIP. When using SS7 for call setup, once the caller has signaled the desired called party's address from the telephone to the near-end switch, the next two functions are routing the phone call and signaling the called number to the far-end switch.

To connect a phone call over the PSTN, it is necessary to have a physical connection established all the way, end-to-end, between the two LECs.

Amongst other things, that requires signaling the value of the *called* number to the far-end telephone switch. The called number has to be forwarded to the far-end LEC so that it is able to connect the incoming trunk to the correct far-end loop, CATV modem or cellphone.

In the old days, this was done using Multifrequency (MF) tones similar to DTMF on the trunk circuits.

Beeping numbers with tones on the trunks is a slow way to communicate, especially considering that there are multiple switches between the near-end switch and the far-end switch, and the whole number would have to be signaled using tones from the first switch to the second, then once that was completed, from the second to the third, then once that was completed from the third to the fourth and so on as the call is routed to the far end.

Today, Signaling System 7 (SS7), also known as Common Channel Signaling System Number 7 (CCS7 or C7) is used to do this address signaling function.

13.7.3 Out-Of-Band Signaling

SS7 signaling is *out of band*, that is, using digital messages on separate data channels, not using tones on the voice communication channels.

13.7.4 Service Control Points and Service Switching Points

In practice, SS7 is centralized computers and databases (Service Control Points, SCPs) connected via the Message Transfer Part (MTP), which is data circuits and packet switches called Signal Transfer Points (STPs), to telephone switches (Service Switching Points, SSPs).

SS7 implements an infrastructure and standard protocols for the exchange of control messages or signaling between control computers and switches. The set of call control messages is called the ISDN User Part (ISUP).

A company's SS7 system will exchange ISUP messages with their switches, with other companies' SS7 systems, and with customers' control systems. Messages to and from customer systems are usually communicated over an ISDN Primary Rate Interface (PRI) signaling channel.

13.7.5 Advanced Intelligent Network (AIN)

In a perfect world, called the Advanced Intelligent Network (AIN), all telephone call routing decisions would be made by the centralized computers, the Service Control Points, and not the switches.

This has large advantages for the network service provider, since it allows the rollout of features on the one or two sets of centralized computers, rather than on the hundreds of CO switches.

However, having the SCPs perform all call routing introduces a single point of failure into the telephone system… proved during a nine-hour complete failure of the telephone system on the East Coast of the United States some years ago.

13.7.6 Switch-Based Call Routing

Due to this failure mechanism, in practice, most telephone companies use a call routing computer program from a supplier like Nokia (Lucent and Nortel) to update CO switch-based routing tables every ten seconds or so.

The switch uses this table to determine the call routing, rather than a table in the SCP. This allows the continued functioning of the network if the call routing computer crashes.

13.7.7 SS7 In Practice

SS7 is used for call setup between different carriers, for example, communicating the calling number and called number from the local phone company's system to a wireless carrier when a call is placed from a home phone to a cell phone.

SS7 was used by big telephone companies for call setup signaling, to support database inquiries, and for high-end call routing features. This is being replaced with SIP and the converged IP telecom network.

Call setup signaling is indicating the *called number* to the far-end switch, and possibly the calling number for caller ID purposes.

An example of a database inquiry message is credit authorization for billing phone calls, such as when you use your telephone company calling card from a payphone, or roam with your cellphone.

13.7.8 Residential Service Application Example

High-end value-added call routing features are sometimes called AIN services. An example for residential service is call forwarding. When you press *72 on your phone and hear four beeps, this indicates that you are now communicating with the SCP, perhaps indirectly.

When you enter the number you want your phone forwarded to, an entry is made in a database, and a trigger is placed on your line card. The trigger is a bit set in a status register associated with your line card in the computer called the telephone switch.

When a call is to be routed to that number using the basic switch-based routing, the fact that the *trigger* on the line card is set causes the far-end CO switch to not terminate the call on that line card, but instead to do a query on the SCP to get the routing information – which will be to the number you forwarded your phone to.

This sort of service control will eventually disappear to be replaced with SIP, which brings advantages like being able to set up call-forwarding of your phone from anywhere. It is necessary to be physically present in front of the phone to call-forward it with legacy technology.

13.7.9 Business Service Application Example

For businesses, examples include both basic 800 service and sophisticated call routing services that change where an 800 number is terminated based on time of day, geographic location of the caller or the call volume.

An example of the latter is an airline that has two call centers in different parts of the country, for example, one in Utah and one in Georgia. There is a single 800 number 1-800-AIRLINE for that airline that is valid everywhere in North America.

By default, calls are routed based on geographic location of the caller; callers in the West are routed to the call center in Utah, and callers in the East go to Georgia. However, the airline pays their Inter-Exchange Carrier for a service that allows them to do load balancing: if for example the call center in Utah becomes busy and the call center in Georgia is not, the airline can signal the network to route phone calls to Georgia, regardless of where the caller is geographically located... and then signal the network to change the routing back to normal a minute later.

This idea is sometimes referred to as "customer control of the network", perhaps more accurately "real-time customer control of their call routing". It is a sophisticated service enabled by SS7.

Going forward, it will be implemented with SIP.

14 OSI Layers and Protocol Stacks

14.1 Protocols and Standards

We use the term *protocol* in the data communications business the same way it is used in the diplomacy business: it is a plan for how two different systems will interact.

FIGURE 177 A PROTOCOL IS A PLAN

In diplomacy, protocol officers get together in advance and hammer out the plan: it says who is going to greet whom at the bottom of the steps of the aircraft, what color the carpet is going to be, what music the band will be playing, are you allowed to be sitting down while the president of the United States is in the room… the plan on how two countries will interact.

To communicate, it is necessary to have a set of conventions that specifies how the systems are going to communicate. This is the definition of a protocol. Mutual adherence to an agreed protocol or set of protocols makes communication possible.

14.1.1 Functions To Be Performed

Quite a number of areas and functions must be specified in a communication protocol.

Taking e-mail as an example, first, it is necessary to agree what the format of the message will be.

How will the message be coded into 1s and 0s? Will it then be encrypted? There had better be an agreed plan for that, or not much communications will be happening.

Most communications today is client-server… and e-mail is an easy example. When checking Outlook-type email, it is necessary to log on to the mail server with a username and password and be authenticated… so part of the protocol has to be how to transmit usernames and passwords to the server.

One could imagine the mind-numbing complexities created if it is desired that the password not be transmitted as clear text, but encrypted as a measure against eavesdropping… how to transmit the decryption key for the password without encrypting *it*?

Once authenticated, then it is necessary to transport the message from the server to the client, and there are a number of things that have to be figured out.

Segmentation and reassembly are usually required, breaking up the message into manageable pieces for transmission and putting it back together at the receiver… in the correct order.

The segment of the message has to be *encapsulated* in control information. An example of control information for a segment of data is a network address.

Once a packet with a network address is created and transmitted to a router, how are routers going to make routing decisions based on those network addresses? And how is the route decision-making kept up to date as new links are added, others are removed or become busy?

Probably the most important aspect is error control: sending data with errors and not knowing about it is probably worse than not sending any data at all. Sometimes error control is performed on each link. Sometimes not. It ends up being necessary to check errors end-to-end between the sender and recipient.

How is *flow control* implemented: when one system can't process information as fast as the other, and has to have a way of temporarily interrupting the flow of data.

How is *access control* implemented – when there is more than one station on the link, which gets to transmit next?

At the bit level there are things that have to be specified: what physical medium to use, and how to represent the bits on the physical medium.

How will conversions between different media and different bit rates be implemented? All this and more has to be part of the plan.

14.1.2 Monolithic vs. Structured Protocols

There are two basic choices for the plan: *monolithic* or *structured* protocols.

A monolithic protocol would embody all of the required functions in a single plan. The problem with this approach is that it becomes unwieldy when all possible variations are included in the single package, and makes maintenance impossible.

A structured approach, where the totality of functions is divided into easy pieces, then individual protocols covering each of the pieces are developed is more workable.

This allows mix and match of protocols for different functions on the systems: for example, everything could be the same on all systems, except the access circuit at one end will be wireless, the network is implemented with fiber, and the access circuit at the other end is copper wires.

Other than that, everything from frame and packet format and addressing to the message format and coding is the same across the entire system.

14.1.3 Open Systems and Standards

In an open system, the protocols are published: all information necessary to implement communications is available to the public.

It would be possible for an individual to develop a set of open, structured protocols for communications. This would only be useful if everyone, or at least a critical mass of users, agreed to use that particular set of protocols.

We are always interested in implementing **standard** protocols.

A protocol is a plan.

A *standard* is when everyone agrees on a particular plan.

14.2 ISO OSI Reference Model

One approach to implementing structured data communications protocols is the *OSI Reference Model.* In 1983, the International Organization for Standardization (ISO) adopted a "Basic Reference Model for Open Systems Interconnection".

The purpose of this model was to "provide a common basis for the coordination of standards development for the purpose of systems interconnection, while allowing existing standards to be placed into perspective within the overall reference model".

A key point is that this is a **model** for discussing protocols and standards. It does not specify **how** to actually perform a function, but instead describes **what** functions must be performed, and organizes these functions into manageable groups or *layers*.

14.2.1 Layers

A layer is a subset of the totality of functions that must be implemented to interwork diverse systems. Protocols are established for each layer.

FIGURE 178 LAYERS IN A STACK

The physical connection between the systems is specified by one layer, and implemented in hardware and signaling using electricity, radio waves or light to communicate 1s and 0s between the systems.

All of other functions, all of the other layers, are implemented in software. A particular software package may implement one or more layers.

14.2.2 Separability of the Layers

The choice of functions included in each layer was made so that the layers were separable: the functions performed by one layer are independent of the functions performed by another layer.

This allows systems to choose a protocol for a particular layer without having to take into consideration the choices made for other layers.

For example, the choice of email message format is independent of the choice made for network packet format.

Dividing functions into separate layers is also useful for understanding the different functions that must be performed and how they are implemented, being able to discuss separate issues separately and not get things confused.

14.2.3 Protocol Stacks

In the OSI Model, a piece of software implementing a layer following some protocol will perform its task, then call on a lower-layer piece of software to perform some utility function for it.

The lower-layer piece of software takes the data it is handed, performs its function following another protocol, then asks a yet lower layer to perform some utility function for it in turn.

For example, the top layer might generate an email message, then ask the next layer down to encrypt the message.

The lower layer takes the message and encrypts it, then might ask the next layer down to transport the encrypted output reliably to the far-end destination.

This process repeats until the bottom layer is reached, which transmits bits one at a time on an outgoing circuit.

Since the layers and the protocols implementing them work in this chain-like fashion, they are often depicted as sitting on top of each other in a stack like a layer cake, and the collection is called a *protocol stack*.

14.3 The OSI 7-Layer Model

The OSI Reference Model is referred to as a *7-layer model* because the total set of functions required to interwork diverse systems was defined and then broken up into seven groups or layers, arranged in a hierarchy. Each layer has a name and a number. The numbering starts at the bottom.

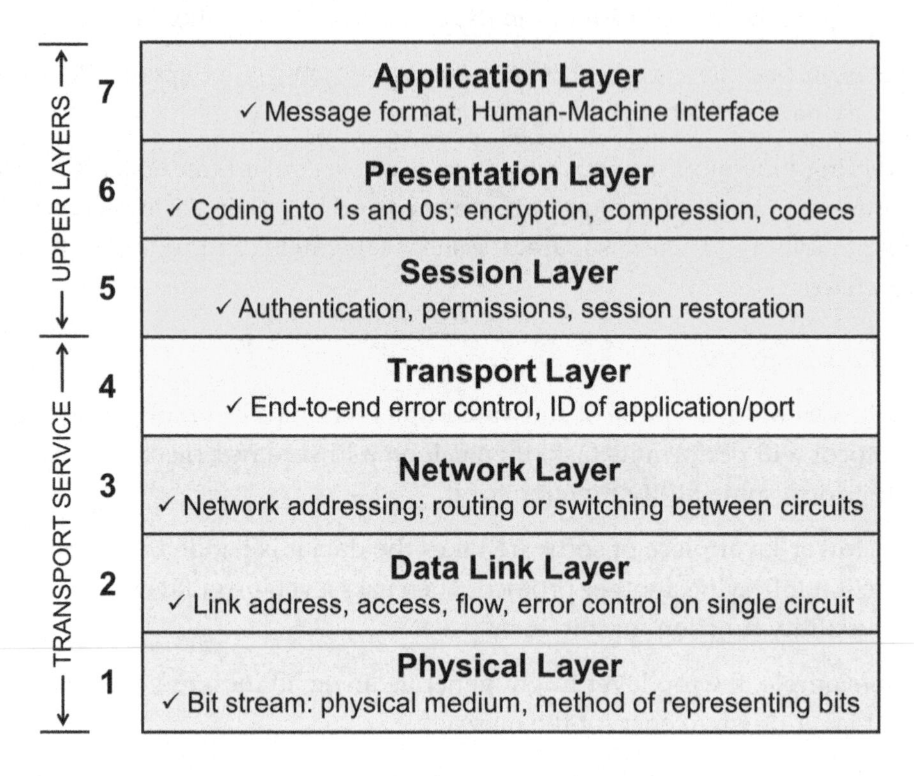

FIGURE 179 THE OSI 7-LAYER MODEL

1: Physical Layer The physical layer provides a raw bit stream service. It moves 1s and 0s between the systems. This is all it does, but it has to do this completely. The physical layer includes the mechanical, electrical, functional and procedural specifications for moving binary digits over a physical medium.

2: Data Link Layer The data link layer manages communications on a single circuit, a single *link*. There may be several stations connected to the circuit as is the case with a wireless LAN, or there may be just two stations on the link, as is the case with a LAN patch cable. The data link layer performs access control, flow control and error detection on the link, transmitting frames on the physical medium. This allows communications of blocks of data from one device to another that are on the **same circuit**.

3: Network Layer The definition of a *network* is multiple data links connected by network equipment. Instead of broadcasting data to all stations on all of the links, data is relayed from one link to the next to eventually be delivered to the correct link to which the desired station is attached. A *router* moves packets from one link to another, essentially a forwarding function. Knowing **which** link to forward the data on is the *routing* part of the story. All of these functions are the network layer.

The first three layers working together form a communication network, giving the user the ability to send data to a destination on a **different circuit**.

4. Transport Layer The transport layer implements two major functions. One is reliability. The other is network connection sharing.

Some network protocols, IP for example, do not provide guaranteed delivery of packets. The transport layer communicates between the source and destination across the network to verify that each segment of a message is successfully received, and in the case of file transfers, retransmits lost segments.

The second function performed by the transport layer is to identify the software application the data is intended for at the far end. There may be many apps running on the far-end computer. The *port number* in the transport layer header indicates **which app** the segment of data is for.

This allows multiple applications to use the same network connection, for example, an email program and a browser can both receive packets over a single shared network connection. The port number indicates whether an incoming packet is for the email application or the browser application.

These four layers working together provide a *transport service,* moving data reliably from an application on one system to an app on another system. This is also called a *socket* in the IP / UNIX worlds.

The remaining three layers are called the *upper layers:*

5. Session Layer The session layer manages *sessions* between applications, including initiation, maintenance and termination of information transfer sessions. Usually this is visible to the user by having to log on with a password in the case of client-server sessions.

6. Presentation Layer The presentation layer is very important: this is the coding step, representing the message to be communicated in 1s and 0s. ASCII is an example of a presentation layer protocol. Compression and encryption also fit into the presentation layer – they are methods of coding messages into 1s and 0s, as are codecs for voice and video digitization.

7. Application Layer Sitting on top of all of this is the application layer. The application layer defines the format of the messages that will be exchanged, and usually implements a Human-Machine Interface.

Using the application layer is a person.

The person interacts with the system via the Human-Machine Interface implemented by the application layer, that lets the person create a message.

In turn, the application layer would ask the presentation layer to code it, and then that would ask the session layer to open a session with the far-end piece of software, and in turn ask the transport layer to move it reliably to a particular application on the far-end system.

The transport layer would ask the network layer to move it to the far-end computer, perhaps on a best-efforts basis, then the network layer will move a packet to the next hop, the next router, by putting the packet in a frame and transmitting the frame one bit at a time on a physical connection like a LAN cable or wireless frequency.

At the far end, the network-layer packet is received in a link-layer frame over a physical-layer connection. The content of the packet is extracted and passed to the transport layer, which would perform error recovery if necessary, then pass it to the correct computer program.

That computer program would pass the data to its presentation layer to decode it, then to the top layer, the application layer to display it as a message to a human.

14.4 Physical Layer: 802.3, DSL, DOCSIS, Wireless

Layer 1, the physical layer, provides a *raw bit stream* service to higher layers.

The physical layer includes the *mechanical, electrical, functional* and *procedural specifications* for moving binary digits over a physical medium.

The mechanical specification includes which type of physical medium will be used. This could include copper wires - shielded cables, twisted pair, or coaxial cable; it could include space (radio), or optical fiber. The connectors or antennas are also specified.

The electrical specification dictates how binary digits will be represented on the physical medium - the modulation technique or digital line code.

The functional specification indicates how many individual wires or circuits will be used to make up a single communication channel, and the function of each circuit.

The procedural specification specifies the relationship between the circuits: are some for data, some for control; is there a circuit that has to operate first, one second and so forth.

There are many different physical layer protocols.

Any kind of modem implements a physical layer protocol by specifying the physical medium and how bits are represented on it. This includes DSL modems that operate over twisted pairs of copper wires, DOCSIS cable modems operating in channels on a Hybrid Fiber-Coax system, all kinds of wireless systems where modems communicate 1s and 0s over radio channels, and of course, old-fashioned dial-up modems.

LANs include a physical layer protocol – the LAN interfaces that provide the familiar LAN jack implement signaling using pulses of voltage on twisted pairs in Category 5, 5e (illustrated in Figure 180) and Category 6 cables following the 802.3 standard.

FIGURE 180 A PHYSICAL LAYER PROTOCOL SPECIFIES THE PHYSICAL MEDIUM AND HOW BITS ARE TO BE REPRESENTED ON IT

Optical Ethernet employs optical transceivers to signal using pulses of light on fiber to communicate bits between devices.

The older SONET includes a physical layer protocol, specifying how the laser is turned on and off to signal anywhere from 500 Mb/s to 10 Gb/s.

Repeaters, amplifiers, pulse shapers, DWDM frequency spacing and anything related are part of a physical layer protocol.

ISDN Basic Rate Interface (BRI) - digital telephone lines - included a physical layer, specifying pulses on the loop.

The old T1 technology implemented a physical layer protocol, moving 1.5 Mb/s over four copper wires using the AMI line code, CSUs and DSUs.

The old serial port standard RS-232 is a physical layer protocol.

The list goes on and on.

14.5 Data Link Layer: 802 MAC

Layer 2, the data link layer, is concerned with communications between devices on the same physical circuit, or more accurately, between devices in the same *broadcast domain*.

There may be just two stations, as in a point-to-point link; or there may be many stations, as in a wireless LAN, but in both cases, the stations are directly connected on the same physical circuit, and are in the same broadcast domain.

Most implementations of a data link protocol do only error detection, and discard data that has errors in it. Some implementations do both error detection and error correction.

14.5.1 LANs, Frames and Layer 2 Switches

The data link protocol encapsulates segments of data to be transferred into *frames*: adds a link address and control information in a header, a frame check sequence in a trailer, and framing around the whole lot.

The frame is then broadcast on the physical circuit, and any stations in the same broadcast domain might receive it.

At each receiver, errors are detected by the data link software performing a Cyclic Redundancy Check (CRC). If it passes, the contents or payload from the frame is extracted and passed up to the next higher layer.

If it fails the CRC, the frame is discarded and will have to be re-transmitted somehow. This is usually implemented by a higher layer protocol like TCP, but can also be done by the data link protocol, as in fax machines.

The dominant method of implementing the Data Link Layer is with LANs. A LAN includes Layer 1 (physical connections and LAN interfaces) and Layer 2 (frames and link addresses).

LANs move frames between computers on the same physical circuit, or connected to the same LAN switch. The most accurate way of saying this is LANs move frames between devices in the same broadcast domain.

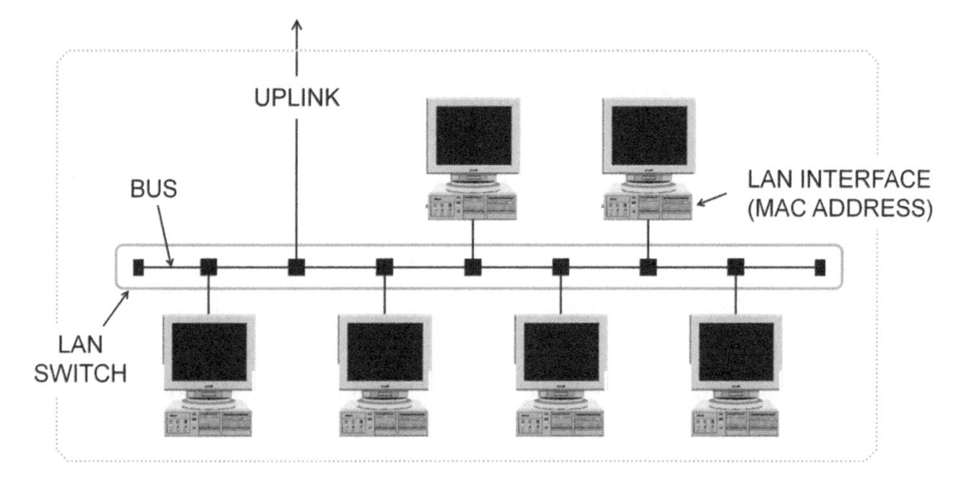

FIGURE 181 LANS IMPLEMENT LAYER 1 AND LAYER 2

Since LANs are implementing Layer 2, LAN switches, particularly the very high capacity ones in carrier networks, are also referred to as *Layer 2 switches*. This is covered in detail in Section 15.4.

14.5.2 MAC Frames and MAC Addresses

The frame format for LANs is IEEE standard 802 Media Access Control (MAC) service plus 802.2 Logical Link Control (LLC), hence the use of the term *MAC frame* and *MAC address* in conjunction with LANs.

IEEE standard 802.3 defines particular implementations of the MAC service on twisted pair, coaxial cable, radio and fiber, specifying the framing, timing, method of representing bits and other Layer 1 functions. 802.3 is commonly referred to as *Ethernet*.

14.5.3 Other Data Link Protocols

The ISO High-Level Data Link Control Protocol (HDLC) is the mother of all data link protocols. The ITU Link Access Procedures (LAP-) for public communication networks, the ANSI Advanced Data Communication Control Procedures (ADCCP), IBM's Synchronous Data Link Control Protocol (SDLC) are all data link protocols derived from or similar to HDLC. The legacy Frame Relay network service from carriers was usually discussed as a Layer 2 protocol.

14.6 Network Layer: IP and MPLS

The data link layer handles communications between devices on the same physical circuit. What if there is not a single physical circuit, but 86 of them, and it is not desired that data be broadcast to all stations on all 86 circuits, but rather routed or switched and delivered to a particular destination? This is the definition of a network, and Layer 3 of the OSI model.

A network is made up of many network devices like switches or routers connected with high-speed data links. Access circuits are provided to the network equipment to allow users to send data into the network. **The network equipment moves data from one circuit to another, essentially a relay function.**

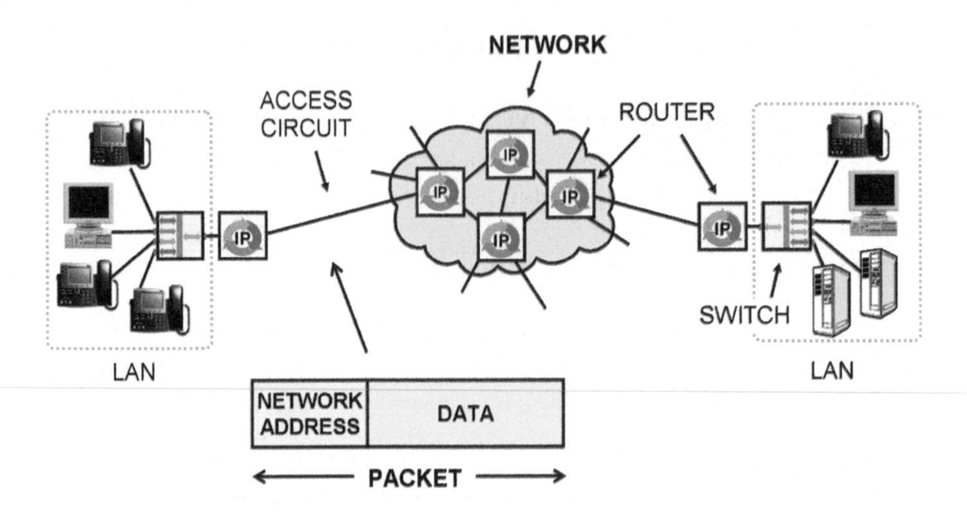

FIGURE 182 NETWORKS ARE MADE OF HIGH-CAPACITY LINKS CONNECTED BY ROUTERS OR SWITCHES. ACCESS CIRCUITS CONNECT THE USERS TO THE NETWORK.

Networks always have two points of view: from the user's point of view, how does the user indicate to the network where the data is to be delivered? This information usually takes the form of a *network address*.

From the network's point of view, if it receives data to be sent to a particular network address, how does it actually decide which route to take to reach that destination address?

14.6.1 Packet-Switched Networks

The most widely-deployed type of network used to be a *circuit-switched* network, the traditional Public Switched Telephone Network (PSTN). To place a call, the caller tells the network the address of the person to whom they wish to connect - their telephone number - then a route is chosen and then trunk circuits are switched in and reserved to form an end-to-end path for the duration of the call.

This is now replaced with a *packet-switched* network, where trunks are not reserved for the duration of a communication session, but rather voice, video, Internet traffic or anything else is segmented and placed in packets that are transmitted into the network on an as-needed basis, and relayed from one router to the next, interspersed with many other users' packets until it is delivered to the far end. This is also called *bandwidth on demand.*

Every destination on the network is assigned a network address. To transmit data to a destination, the address of the desired destination is placed in the header at the beginning of the packet, and each router uses the destination address to determine the next hop.

The router implements the routing by taking the packet from an incoming circuit (or more precisely, the incoming broadcast domain), and transmitting it out on a different circuit. Routers perform essentially a relay function. Knowing **which** circuit to move the packet to is the routing part of the story. This whole process is called *packet switching*.

The most popular protocol for assigning network addresses and formatting packets is IP, the Internet Protocol, developed by the Department of Defense and is now maintained by the Internet Engineering Task Force (IETF). It is, of course, the protocol for network addresses and packet format on the Internet, and also used for networks not directly connected to the Internet... and has become the only standard for packets worth learning about.

14.6.2 Routing Table Updates

The devices that perform the routing of packets must have a way of making route decisions. In general, they use routing tables and look up the routing for each packet based on the destination address.

The routing tables are kept updated using protocols like Open Shortest Path First (OSPF) and Border Gateway Protocol (BGP). There are other proprietary protocols also available for this purpose.

14.6.3 MPLS

It should be noted that in their core, IP networks generally have Multiprotocol Label Switching (MPLS) implemented to allow the management of traffic and transmission characteristics. MPLS is an implementation of *virtual circuits,* where a path for packets is predetermined and programmed into the routers by equipment at a Network Operations Center. In this case, MPLS replaces OSPF and IP addresses for routing.

MPLS is covered in detail in Chapter 17.

14.7 Transport Layer: TCP and UDP

14.7.1 Reliability

With IP, there is no guarantee that a packet will be transmitted, when that might happen, or how often it might happen. There is no guarantee that if a packet is transmitted, it will be received.

In fact, in IP, there is no way for the transmitter to know whether it was received or not. Nada.

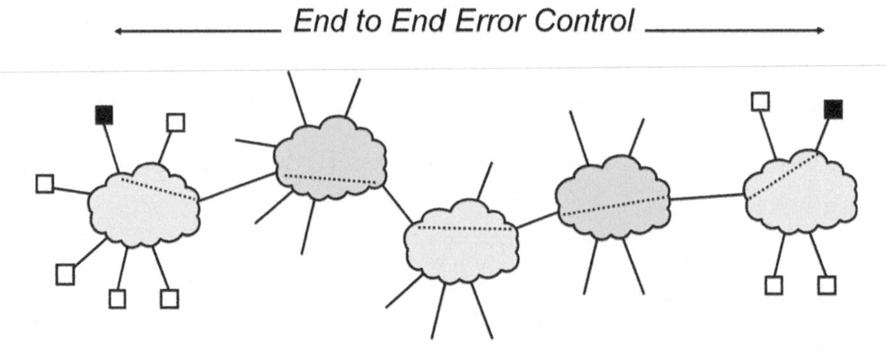

FIGURE 183 THE TRANSPORT LAYER IMPLEMENTS ERROR CHECKING END-TO-END BETWEEN THE SENDER AND RECEIVER

So, what if some packets are, indeed, not delivered? This means segments of the message are missing... so a protocol is required to deal with the problem.

Dealing with missing data at the receiver is one of the two functions performed by Layer 4, the transport layer.

Before a segment of a message is put in an IP packet by the sending device, it is passed to its Layer 4 software, which puts a sequence number and error check on the message segment, then passes it to the IP software.

At the far end, its Layer 4 software checks the sequence number and error check on the message segment. If a message segment fails the error check, or is missing, there are two basic strategies: retransmit the missing segment, or interpolate it.

The most popular transport protocol is the Transmission Control Protocol (TCP), which provides sequence numbers, error checking and retransmission of data that is received with errors or not received at all.

The TCP software at the sender puts a sequence number and error check on the segment, and the TCP software at the receiver normally returns a message to the sender acknowledging successful receipt.

If the sender's TCP software does not receive this acknowledgment, it automatically retransmits the segment. The result is 100% error-free communication.

This is for file transfers, including email messages and web pages.

For live, streaming communications, like Voice over IP and video over IP, there is no time to perform retransmission of bad data, so a different transport protocol, the User Datagram Protocol (UDP) is used instead.

UDP implements error checking, but not retransmission. Instead, the receiver might *interpolate* the missing data – fill in the gaps – using prior and subsequent data values to guess what the missing one was.

14.7.2 Port Numbers

Another important function of the transport layer is to identify the application that is sending the message and the application it is intended for on the far end.

There is usually more than one application using an Internet connection on a computer or a phone; for example, email and browser both running.

When a packet arrives at the computer, how does the computer know whether this packet is for the email application or for the browser?

Every application is given a number called a *port number*. The first two bytes of the Layer 4 header are a field where the port number of the source application is populated, and the port number of the destination application is populated in the next two bytes of the header.

FIGURE 184 THE PORT NUMBER IDENTIFIES THE SENDING AND RECEIVING APP

This information at the beginning of the layer 4 header is used by the far end computer to determine where to direct the data – which application this data is for – on the far-end computer.

The near- and far-end computers are called terminals, endpoints or *hosts*.

The IP address of the host concatenated with the port number of the application is called a *socket* in UNIX and IP. It is called the *transport service* in the OSI model.

This allows segments of messages to be moved reliably from a particular application on one host to a particular application on a different host.

14.8 Session Layer: POP, SIP, HTTP

The remaining layers are referred to as the *upper layers*.

The first stop on the upper layers is Layer 5, the session layer.

Once we can get the data to the destination, the next question would be, are we **allowed** to send data to that computer? This is the function of the session layer.

The session layer manages communications sessions between applications, including initiation, maintenance, sometimes restoration and certainly termination of information transfer sessions.

14.8.1 Password Authentication

Establishing a session is often implemented by "logging on" to a remote system with a username and password. An agreement on how the password chosen by the user is transmitted to the far-end computer during account creation.

One can imagine the complexity when it is desired that the password be encrypted before transmission, so that it cannot be intercepted and re-used... how to transmit the decryption key for the password without encrypting **it**?

14.8.2 Authentication Servers

Another area of development in session establishment is *authentication servers*. Without an authentication server, there are two basic choices for remembering user names and passwords on servers:

1) Use different usernames and passwords for every server you access. The question is, how does one remember all these user names and passwords? Perhaps recording them in a file called "user names and passwords.doc" in your My Documents folder? That does not sound very secure!

2) Use the same username and password on every server. This exposes you to a serious security risk: that your username and password will be stolen from one of those servers by an intruder or a technician, then re-used to log in to your accounts on other servers.

An authentication service, like Google Accounts or Log in with Facebook, allows you to only have one username and password, which allows you to access many services. You log in the authentication server and it provides credentials to the server you are logging in to, without revealing the username and password to the third party.

If a particular service, like the control panel for your web-based email wants extra protection, it will ask you to log in again – but the username and password you type in are not validated on the email server, they are passed to the authentication server for verification.

This way, you don't have to store a user name and password on every server. Just one. Google, Facebook and many other companies implement forms of authentication services.

14.8.3 Password Caching

Other strategies for password management involve *caching* or storing a copy of the password for each site somewhere in the cloud, and retrieving the correct password for a particular login page as needed.

This function is implemented by browsers, and by add-ons to browsers.

14.8.4 Cookies

After you log on to a server, it would be nice if it remembered what you were doing last time you logged on... restore your previous session. A method for session restoration used on the Web is *cookies*.

When your browser uses HTTP to request a file from a web server, the web server replies with the file – but first an instruction to your browser to store a cookie for the server's domain using the Set Cookie instruction.

The cookie is one or more name-value pairs and the server's domain name in plain text, saved in a small file in a folder on your hard drive.

These name-value pairs could be your username and password for the server, to be used to log you on transparently later on. In this case, the cookie might be userid=yaright; password=fuggedaboudit; domain=forgetit.com.

Every subsequent time you request a file from that domain, your browser automatically supplies the name-value pair as part of the HTTP file transfer request.

The example above has the problem of storing your password in a plain text file on your computer, and giving it out to pretty much anything that asks for it.

Additionally, since the session information is stored on the client, the designer of the system would have to account for obvious issues for users with more than one client computer.

To avoid these problems, Google sets a cookie that is one name-value pair, an encrypted code that identifies the user, and the cookie information for their applications is stored on their servers.

A problem is privacy. One well-known Internet web page banner advertising company was caught defining their cookie in such a way that every client computer returned ALL of its cookies to them.

They were accused of using this trick to – unethically – gather data for data-mining to determine where you had surfed.

They were **literally** stealing cookies from children (!).

14.8.5 Client-Server Sessions

An example of a standard session layer protocol is the Post Office Protocol (POP), an agreement on how your computer logs on to a mail server to check for new e-mail messages then downloads them.

When setting up a POP-type email account, such as in Outlook, it's necessary to start the client software, then configure your user name and password and the name of the POP server.

Then, when you click "send and receive", your POP client attempts to log on to that POP server using that user name and password.

If it is successful the server indicates how many messages there are and then it uses the file transfer protocol to download the email messages – which are data files – one at a time, from the server to the client.

If the transfer is interrupted in the middle of a message, the next time the POP client runs, the transfer resumes from the beginning of the message – so POP implements session state and session restoration as well.

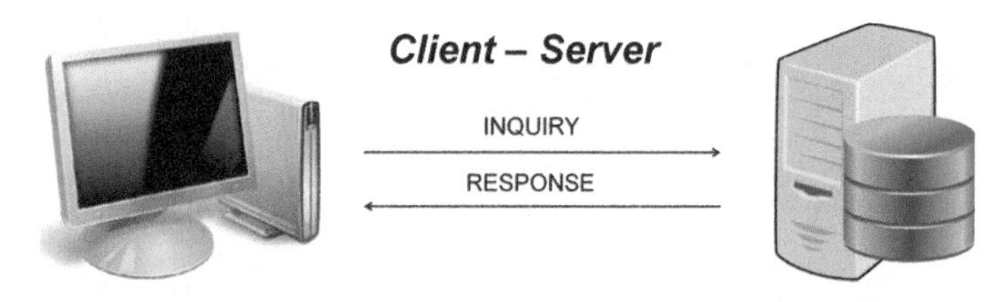

FIGURE 185 CLIENT-SERVER COMMUNICATIONS

One could argue that the Hypertext Transfer Protocol (HTTP) is a session-layer protocol; this is the protocol used to initiate a download from a web server by a browser.

The session only lasts for the transfer of all of the files referenced in one web page, and there is no authentication – but an example of a session establishment nonetheless.

14.8.6 Peer-Peer Sessions

Instead of a client-server session, a VoIP phone call is a client-client or *peer-to-peer* session.

To set up a phone call, it is necessary to communicate the IP address and communication port numbers used by a phone to the other phone. The two phones subsequently transmit IP packets from one phone to the other.

In standards-based VoIP systems, the Session Initiation Protocol (SIP) is used to establish VoIP phone calls.

SIP includes *proxy servers* that act as intermediaries between the caller and called party so that the caller only finds out the called party's IP address if the called party wants to take the call.

If the caller "picks up", the result is a session between two telephones, where the two telephones are *peers*, meaning they are equals.

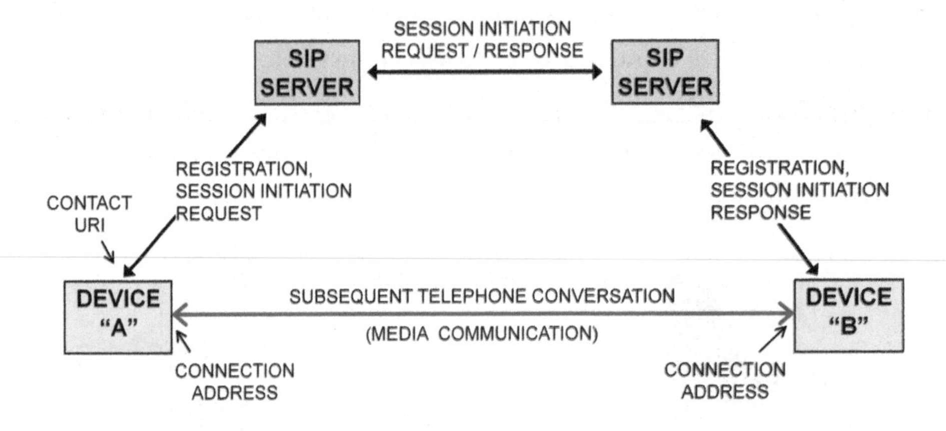

FIGURE 186 SIP USES PROXY SERVERS TO SET UP PEER-TO-PEER OR "CLIENT-CLIENT" VOIP PHONE CALLS

14.9 Presentation Layer: ASCII, Encryption, Codecs

Layer 6, the presentation layer is very important: this layer is responsible for coding application-layer messages into 1s and 0s.

Examples of presentation layer protocols include keystroke-coding protocols like ASCII and Unicode, transformation protocols like MIME, data compression, encryption and codecs.

14.9.1 Character Coding

The American Standard Code for Information Interchange (ASCII) was primarily defined for teletypes, i.e., printed English characters, which can easily be supported within a limit of 128 codes, and so specified codes that were 7 bits long.

As computers standardized on octets, more languages with different characters were required to be supported, and parity checking was abandoned, sets of 8-bit codes were required. ASCII formed the basis of two 8-bit code sets that are now in wide use: ISO-8859-1 and Windows-1252.

Unicode and its Unicode Transformation Format (UTF) may end up becoming universal standard codes for character sets. Unicode defines a huge superset of characters, including Kanji characters used in Asian languages, and methods of representing them, called transformation formats. The most popular is UTF-8, typically using two bytes per character.

14.9.2 E-Mail Coding

Another example of a presentation layer protocol is MIME: The Multipurpose Internet Mail Extensions. This is a protocol for transforming or transcoding messages consisting of 8-bit bytes (like an image, spreadsheet or computer program) into messages consisting of 6- or 7-bit bytes that look like ASCII text for backwards-compatibility with email systems based on 1970s-era UNIX computer technology that only supported 7-bit ASCII coding.

A MIME header is placed at the beginning of the transformed message so that the far-end can apply the reverse transformation to re-create the message in its original 8-bit bytes.

This is transforming images into what look like giant plain text messages to email them, and is no longer necessary, but is still almost universally implemented for backwards-compatibility with old systems.

See Chapter 7 for more detail on ASCII and MIME.

14.9.3 Codecs

Codecs for voice and video digitization are presentation layer protocols.

The G.711 standard from the ITU specifies voice coding at 64 kb/s – carried in IP packets or DS0 channels. There are many other voice coding standards, most of which use fewer bits per second.

The H.264 standard, specified in Part 10 of the MPEG-4 standard for video coding is used for HD video.

All of these are protocols for coding messages into 1s and 0s.

14.9.4 Data Compression

Another example of a presentation layer protocol is data compression like WinZip, implementing the ITU standard V.42bis.

This is again a method of representing information in 1s and 0s – just using fewer 1s and 0s to represent our information.

14.9.5 Symmetric Encryption: Private Key

Encryption is a presentation layer protocol. There are two basic methods of encryption: *symmetric key* encryption and *asymmetric key* encryption.

Symmetric key encryption is also called *private key* or secret key encryption, as there is a single key (a binary number) that both encrypts and decrypts the file, so the key is kept private.

Popular methods include the Advanced Encryption Standard (AES) using the Rijndael ("rhine-doll") algorithm.

14.9.6 Asymmetric Encryption: Public Key Encryption and Digital Signatures

The other type of encryption is called asymmetric key encryption. What this means is that there is a key pair; and what key A encrypts, key B can decrypt… and what key B encrypts, key A can decrypt.

This is used in two different ways: for secure communications, called *public key encryption*, and for authentication, called a *digital signature*.

For secure communications, a *key pair* is generated. One of the keys is made public, available on a public key server, and the other key is kept private.

To communicate securely, the sender creates a message then uses the receiver's public key to encrypt it and transmits the encrypted message. The receiver uses the private key to decrypt the message.

This avoids the problem inherent with symmetric or private key encryption for communications, which requires transmitting the key, exposing it to potential eavesdropping. With public key encryption, the decryption key is never transmitted.

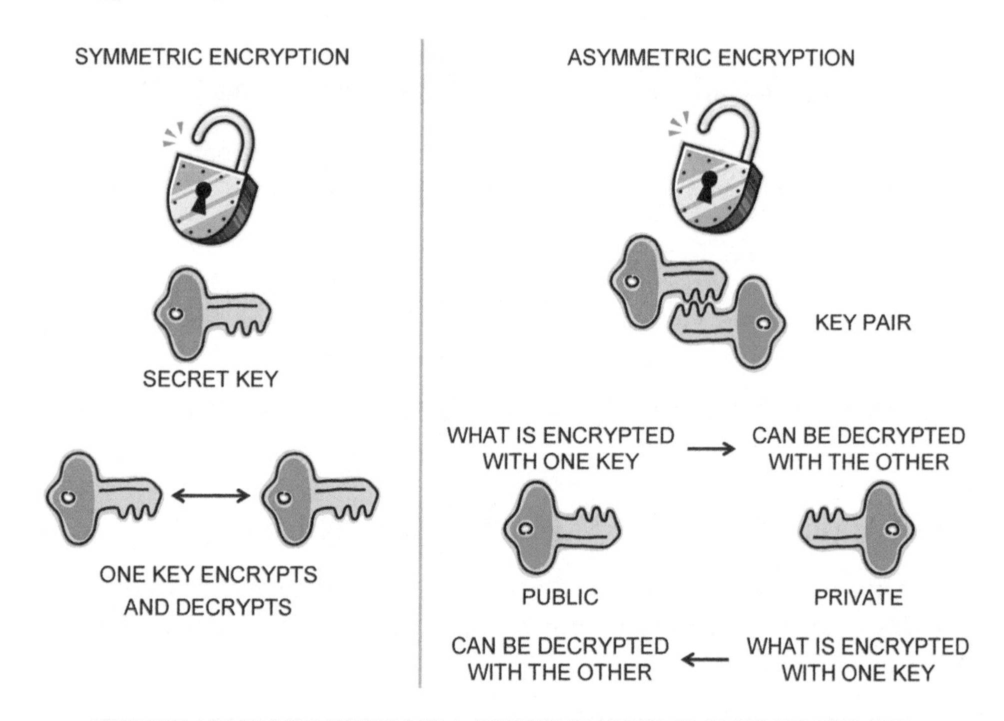

FIGURE 187 ENCRYPTION IS A PRESENTATION LAYER PROTOCOL

Algorithms developed by Rivest, Shamir, and Adelman (RSA) for public key encryption are widely used. These are so good, if you use 1024-bit long keys, it's estimated that it takes the National Security Agency more than milliseconds to decrypt your messages.

They don't like this idea, and had software that implements RSA declared a weapon so it is controlled. Spooky. Servers implementing free public RSA must be located outside the USA.

On the other hand, people who want to kill you for religious reasons use it to try to hide their communications.

14.9.7 Example of Separability of Layers

Encryption is a very good illustration of the independence and separability of layers in the OSI model.

Higher and lower layers know nothing of the encryption process. Higher layers have simply selected a secure communications option, and know nothing of the details of how this is accomplished.

They pass messages to the presentation layer, which performs the encryption on the transmitting side, then decryption on the receiving side before passing the message up to the application layer at the receiver.

Lower layers know nothing of the encryption process – they are just tasked with moving 1s and 0s just like any other data.

14.9.8 Example of Peer Protocol

Encryption is also a very good illustration of the idea of peer protocols: having the same plan on both systems at each layer.

If the sender uses one protocol for encryption and the receiver is using a different protocol, there is not going to be any communication.

14.10 Application Layer: SMTP, HTML, English ...

The application layer, Layer 7, is the highest layer in the protocol stack, and includes the specification of the messages to be exchanged and often, Human-Machine Interfaces (HMIs).

The application layer transmits and receives *messages*. The application layer protocol is an agreement on the format of the message and how it is going to be exchanged.

14.10.1 Email

An easy example of an application layer protocol is e-mail, and an easy example of that is the Simple Mail Transfer Protocol (SMTP), a standard for exchanging email between computers.

SMTP specifies the three upper layers; it describes the Layer 5 functions of how mail transfer sessions are initiated and terminated, how pieces of the file will be sent and acknowledged, and for coding Layer 6 and message format Layer 7 references RFC 5322 "Internet Message Format".

RFC 5322 specifies US-ASCII coding for Presentation Layer 6, and finally, to get to the part of interest here, the specification of the structure and format of a mail message, which is Layer 7.

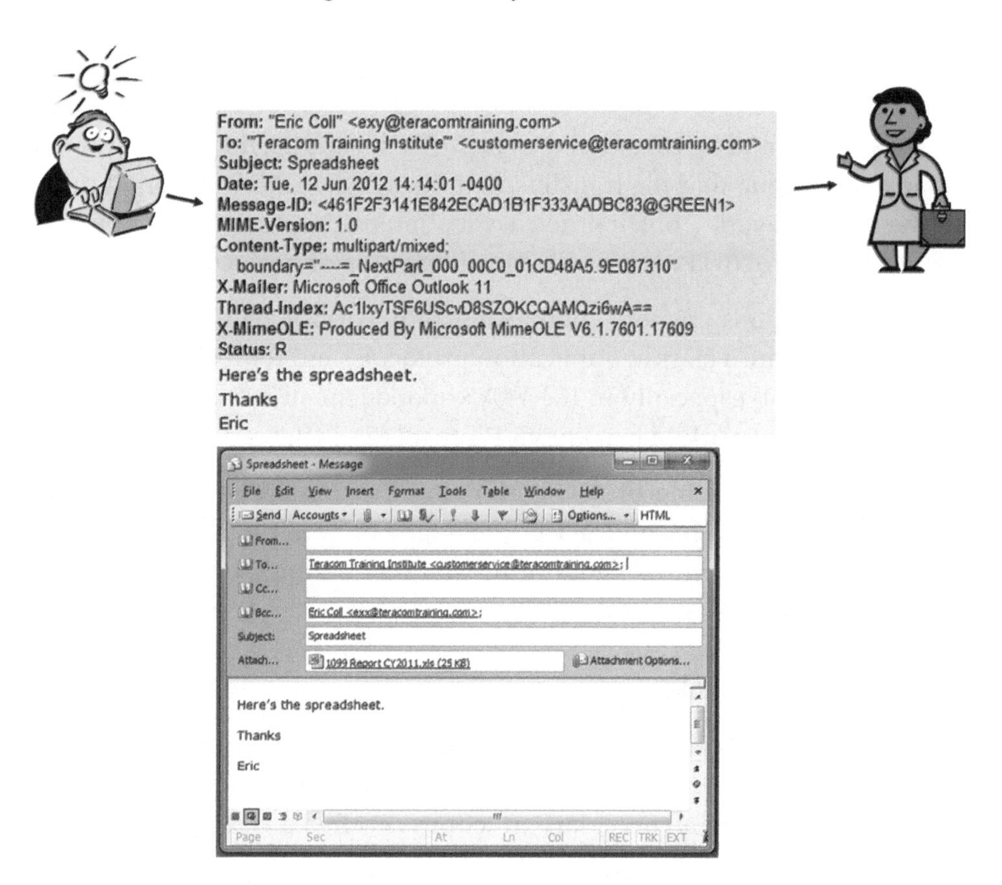

FIGURE 188 SIMPLE MAIL TRANSFER PROTOCOL

RFC 5322 Internet Message Format tells you a message is made of lines. It tells you a line is a sequence of characters coded into US-ASCII that ends with 13 (Carriage Return) 10 (Line Feed). It tells you how the lines are grouped: header and body, separated by a blank line.

RFC 5322 defines the structure and content of the header. It tells you the format of a header line is field-name:field-body. It tells you what the header field-names are. It tells you what the allowed field-body values are. It tells you how to format the time, and the allowed values. It tells you how to format an address.

It doesn't tell you what message to write in the body.

14.10.2 More Application Layer Examples

English is an application layer protocol: its syntax rules define the format of messages and its vocabulary is the allowed content of messages.

HTML is also an application layer protocol: it specifies the structure, syntax and vocabulary of messages colloquially referred to as web pages.

File transfers could be considered as being in the application layer, though some might argue that file transfers and file systems are actually all three of the upper layers... but it's not a very interesting argument. The File Transfer Protocol (FTP) is an example of this type of protocol.

Remote operations: remote monitoring and control of devices from a central station are a class of application protocols, and represent a growing market segment, especially in the WAN management arena. An example is the Simple Network Management Protocol (SNMP).

This is also a messaging protocol, allowing the transfer of status inquiry and response messages and alarm messages between network elements like routers and a central monitoring station running software like HP Openview.

14.11 Protocol Stacks

It is necessary to choose particular protocols to implement each layer. The protocols are usually drawn one on top of another, so the collection of protocols to perform all necessary functions is called a *protocol stack*.

The same protocol is required at each layer. This is called a *peer protocol.*

The peers communicate, with their communications *encapsulated* inside other protocols' data units to be carried to the other system.

Figure 189 provides a visual summary of the material discussed in the previous pages, and is used to illustrate how information travels down through the protocol stack on the left, through the network equipment in the center, and back up the protocol stack on the right.

14.11.1 Example: Web Surfing

The protocol stack when surfing the web is: application-layer messages formatted following HTML, coded into 1s and 0s using ASCII at the presentation layer, retrieved from a server using HTTP at the session layer, communicated reliably between server and browser using TCP at the transport layer, in network layer IP packets, in link layer MAC frames, on a LAN cable for the last three feet.

14.11.2 Voice over IP

The protocol stack for a VoIP telephone call is: application-layer messages formatted using English, coded by the presentation layer into 1s and 0s using the G.711 codec.

The session is set up between the telephones using SIP, communicated using best efforts using UDP at the transport layer, in network layer IP packets, in link layer MAC frames, on a LAN cable for the last feet.

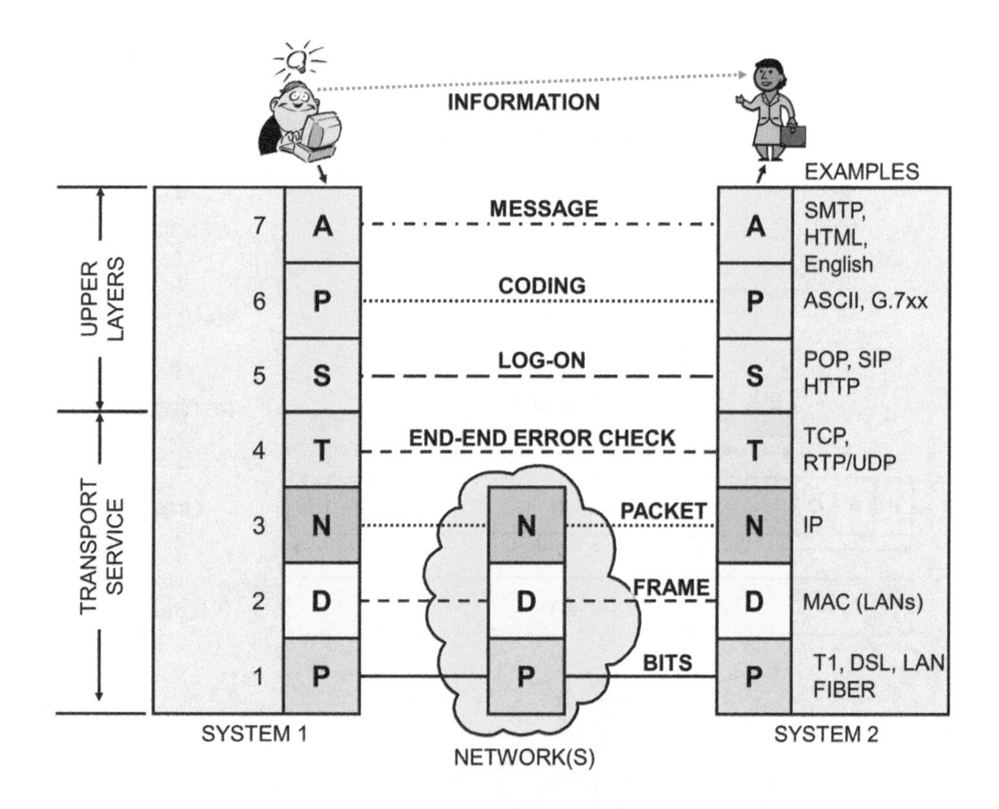

FIGURE 189 PROTOCOL STACKS AND PEER PROTOCOLS

14.12 Protocol Stack in Operation: Ukrainian Dolls

To actually communicate between systems, protocols covering all seven layers, stacked on top of each other, have to be implemented on both the sending system and the receiving system... and it must be the same protocols at each layer implemented on each system.

14.12.1 Communications Flow

Communications begins with a person having information - a thought - to communicate, starting at the top of Figure 189. To communicate this information to another person, the user creates a message and enters it into the system via software implementing the application layer protocol.

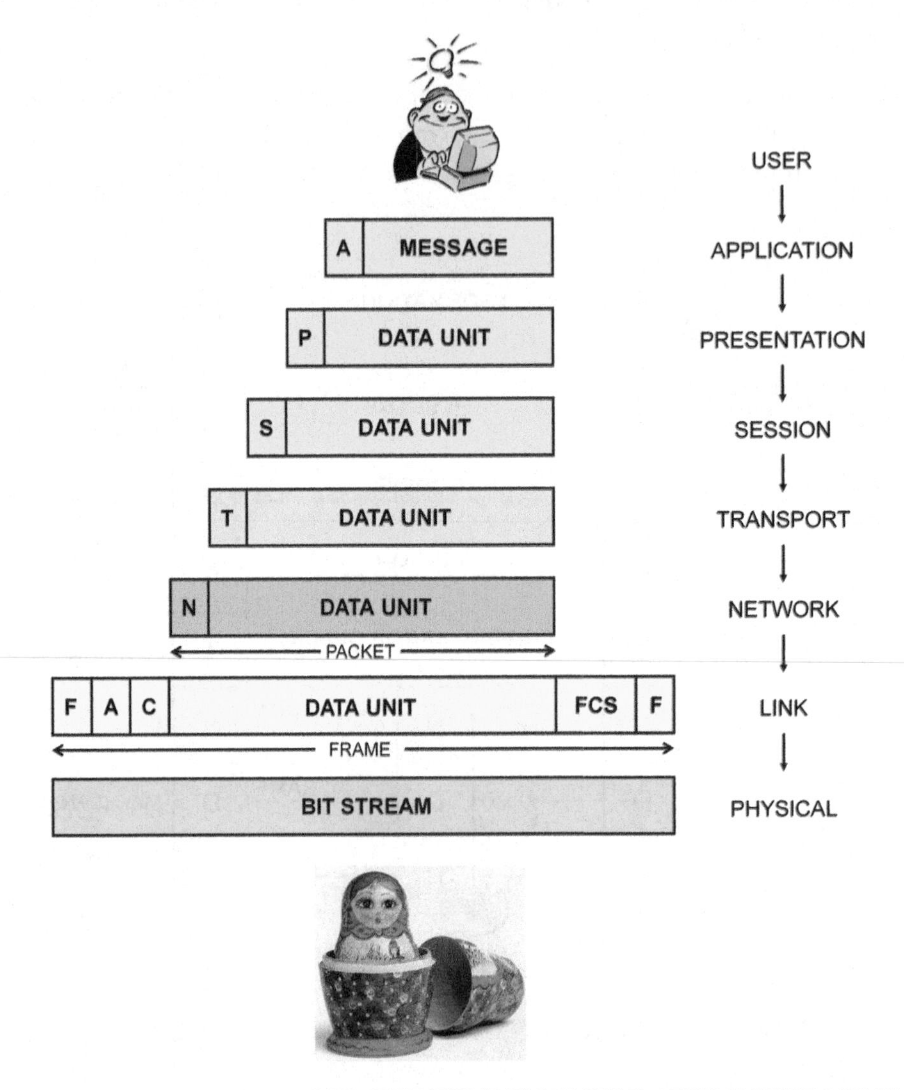

FIGURE 190 EACH LAYER'S OUTPUT IS ENCAPSULATED IN THE NEXT LAYER'S PDU

The application layer protocol encapsulates the message with application layer control information such as which person the message is to, then hands the result, called the application layer Protocol Data Unit (PDU) to the next layer down, the presentation layer.

The presentation layer will code the message into 1s and 0s, and add a header with control information such as an indication of which coding scheme has been used, and pass the result to the session layer.

The session layer might put some audit information on the front, like which client it came from and any session authentication information that's necessary, then give it to the transport layer.

The transport layer is responsible for identifying the source and destination applications, and for end-to-end error checking, so it will take what it gets from the session layer, and put the source port number and destination port number, a sequence number and error check on it, and give this transport layer protocol data unit to the network layer to transmit to the far-end host.

The network layer will take the incoming transport layer PDU and put that into a network layer PDU – called a packet – with the network address of the final destination on the front of the packet.

The packet goes into the data link layer PDU – called a frame – with the MAC address of the destination on this particular circuit on the front of the frame, for transmission via a physical port.

The frame is then transmitted one bit at a time over the physical layer: one bit at a time over the LAN cable, over an airlink, over a fiber.

The physical layer on the next system receives the bits and passes them up to the data link layer protocol software, which performs an error check on the frame, looks at the MAC address on the frame, and compares it to the MAC address hard-coded into its LAN interface, and if they are a match, indicating this is the desired receiver, it extracts the payload from the frame (which is a packet) and passes it up to the network layer.

The network layer software will look at the address on the packet and use that as the basis of making a route decision. If it is going to route the packet somewhere, the way it implements the route decision is to take the packet and put it back in a frame and change the destination MAC address, (because now it's going to a different destination on a different broadcast domain), recalculate the frame check sequence and then transmit it out on a different circuit or different broadcast domain.

Eventually the packet will arrive at the far-end network layer software, which will see that the destination IP address on the packet is the same as its own IP address, so that will extract the data from the packet and give it to the transport layer on the far-end computer.

The transport layer will check the error check that its peer (on the originating computer) put on the information, and if it fails the error check, discards the received segment.

If it passes, the transport layer extracts the payload from the transport protocol data unit, and passes it to the software application on the far-end computer indicated by the destination port number in the layer 4 header.

The received codes are passed to the presentation layer on the far end, which will decode what it receives and pass the result to the application layer, which will recreate the original message and display it to the person at the far end via a Human-Machine Interface.

14.12.2 Segmentation at Each Layer

At each stage, the protocol might segment the data unit it receives from a higher layer and transmit a number of smaller data units to its peer protocol on the opposite system, which reassembles them back into the original size to hand back up to the higher layer.

14.12.3 Nested Headers

By passing segments of data to a lower layer, which performs its function, adds a header and passes the result to a yet lower layer, the protocol data units of each layer end up being nested one inside another inside another like Ukrainian nesting dolls.

The innermost, smallest doll is a segment of the application-layer message.

At the bottom of the protocol stack, all of the headers added by the layers are in place, one after another as illustrated in Figure 190. The result is a lot of overhead – all those headers – but also the ability to make the best choice for protocols at many different levels independently.

14.13 Standards Organizations

Since standards are such a good idea, we write lots of them!

Many different organizations with different perspectives and agendas have become involved. Out of the resulting myriad choices, particular protocols become standards in the actual sense of the word through popularity - the choices most popular in the market, sometimes referred to as the "thundering herd".

14.13.1 ISO

The International Organization for Standardization (ISO) defined the OSI Reference Model that we examined in detail.

It's important to keep in mind that the OSI Reference Model does not tell us **how** to do all of these functions - it tells us **what** we have to do, and gives us a structured way of discussing what we have to do so we can discuss separate issues separately, and not get things jumbled up.

In addition to the reference model, ISO does publish particular protocols, such as the data link protocol HDLC. These OSI-published protocols enjoy varying degrees of actual industry use: slim to none. This is a side issue to the OSI Reference Model and the concept of open systems.

14.13.2 DOD and IETF

The US Department of Defense (DOD) published specifications for a suite of protocols including the Internet Protocol (IP) and Transmission Control Protocol (TCP).

These are now maintained by the Internet Society, through the Internet Advisory Board (IAB) and the clique called the Internet Engineering Task Force (IETF) that publishes Internet standards called Request for Comments (RFCs).

14.13.3 ITU and Bellcore

Lest we forget! The PSTN was the world's biggest network; it made the Internet look tiny in comparison in the previous millennium. Eventually, the Internet and the Public Switched Telephone Network will be the same thing. In the meantime, there are standards specific to the PSTN.

The Comité Consultatif International de Téléphone et de Télégraphe (CCITT), now officially called the Telecommunications Standards Sector of the International Telecommunications Union (ITU-T) is an international treaty organization, with strong European telephone company influences. This organization publishes many standards, including the V. series of modem standards, the X. series of data network access standards, and the I. series of digital telephone network standards.

The former Bellcore (Bell Communications Research), now called Telcordia and originally part of Bell Labs, publishes standards for the North American public telephone network.

14.13.4 TIA and IEEE

Industry organizations include the Telecommunications Industries Association (TIA), which is a subgroup of the Electronic Industries Association (EIA), which publishes the old RS-232 standard for modem cables connections, and the newer TIA-568 standard for building wiring.

The Institute of Electrical and Electronic Engineers (IEEE) publishes standards for how to build LANs on TIA-568 cabling, the 802 series of LAN standards. "Ethernet" is 802.3 and 802.2 together.

14.13.5 3GPP

The increasingly inaccurately-named Third-Generation Partnership Project publishes standards for mobile communications in Releases, which are ... released every couple of years.

14.13.6 ANSI

In addition to these organizations with specific areas of interest, there are national organizations such as the American National Standards Institute (ANSI) that try to coordinate standards at a national level.

Sometimes, in an attempt to coordinate two similar but not identical competing standards from different groups, and come up with a unified standard, ANSI ends up creating a third standard that then competes with the two existing "standards".

15 Ethernet, LANs and VLANs

LANs were commercialized in 1979 by DEC (now HP), Xerox and Intel with a product called *Ethernet*. LANs became popular for connecting computers, printers and file servers in-building during the 1980s.

Ethernet then escaped the building, and is now used in the form of Optical Ethernet to implement today's telecom network, in the form of point-to-point fiber links between routers in different locations.

One of the original requirements for a LAN was to connect computers in an office to a shared laser printer.

This requirement generalized to connectivity between devices for the sharing of all kinds of resources, including hardware resources: hard disks and surveillance cameras; information resources: centralized databases; software resources: network address configuration programs, communication resources: WAN circuits, and countless other examples.

Ethernet, and its many subsequent updates in the IEEE 802 standards, is now almost universally used on fiber, twisted-pair copper, and over the air to implement the Layer 2 links that make up the telecom network.

In doing so, Ethernet achieved one of the long-sought goals in the telecommunications business: the same technology used in the network core, on the network access circuit, and at the customer premise.

15.1 LAN Basics

15.1.1 Bus Topology

The original design for a LAN used a *bus topology* and coaxial cable. *Topology* is the way the system looks viewed from the top, its layout.

Bus comes from electrical power distribution systems: a power distribution bus is a thick bar conducting electricity, for example, the bar in a circuit breaker panel to which all of the circuit breakers are attached.

In LAN terminology, this term was borrowed to mean a cable running down an office building hallway connecting a floor full of PCs. As illustrated in Figure 191, all of the PCs, or workstations, terminals, devices or simply *stations* were connected to the bus.

15.1.2 Broadcast Domain

The bus implements a multi-drop circuit: anything any station transmits is received by all of the other stations.

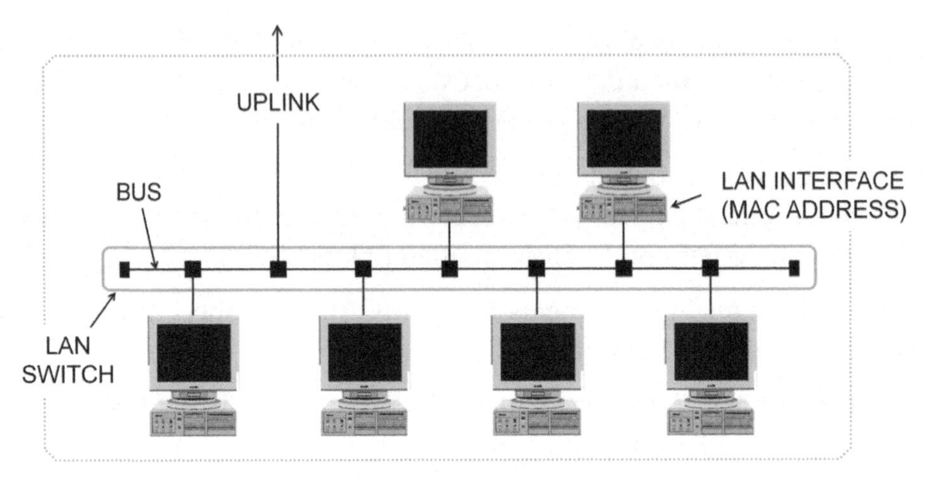

FIGURE 191 BROADCAST DOMAIN

For this reason, this group of stations is said to form a *broadcast domain*: any station has the possibility of communicating directly with any other station in the broadcast domain without the need of other equipment or protocols.

This has obvious implications for both security and performance.

The bus has been replaced with an *Ethernet switch*, also called a LAN switch and Layer 2 switch. Ethernet switches are covered in detail in Section 15.4.

15.1.3 Balanced Configuration

LANs are balanced configurations with combined stations, which means there are no controllers - all stations are equal on the LAN.

Multidrop distribution networks like cable TV are *unbalanced configurations*, with controllers and controlees.

15.1.4 Collision Domain

With the advent of LAN switches (Section 15.4) and two-way simultaneous communications, devices are no longer connected to a bus cable, but rather devices are connected to a LAN switch with point-to-point LAN cabling like the kind in Figure 195.

There are only two stations on each cable: the device and the hardware port on the switch it's plugged into, with the full-duplex connection, there can be no collisions, so discussion in this section is not relevant.

On wireless LANs, where everyone uses the same radio band, (and on bus cables in the Flintstones era), it is relevant, as *access control* is required.

In addition to a broadcast domain, Figure 191 also illustrates a *collision domain*, meaning there is the possibility that the transmissions of two stations will collide: both transmitting on the circuit at the same time.

Since anything any station transmits on a bus is received by all of the other stations, only one can transmit at a time.

If two transmitted at the same time, with the electricity that the two stations are producing added together, receiving stations would not be able to understand what they heard, and nothing would be communicated.

An *access control mechanism*, a protocol for determining if a particular station may begin transmitting is required for the bus architecture of Figure 191, so that only one station transmits at a time.

The designers of Ethernet chose Carrier-Sensing Multiple Access with Collision Detection (CSMA-CD), a *contention-based* access control protocol, where stations contend for the use of the physical connection on a first-come, first-served basis. In essence, the CSMA-CD protocol is:

1) Listen for a carrier signal, to hear if any other station is transmitting.

2) If you hear another station transmitting, wait.

3) When there is silence, transmit a frame containing the intended receiver's MAC address, your MAC address, a block of data and error checking.

4) Listen while you transmit.

5) If you hear something different happening, this means another station started transmitting, so stop transmitting and return to Step 1.

Access control for Wi-Fi is OFDMA beginning with 802.11ax Wi-Fi 6.

15.1.5 MAC Address

As illustrated in Figure 191, anything any station transmits is received by all stations: this is a multidrop circuit, a *broadcast domain*.

Since all stations receive everything, when transmitting data, a station has to indicate the address of the desired receiver on the link: an indication of **which** station on this link should react to the data; in other words, indicating for which station the data is intended.

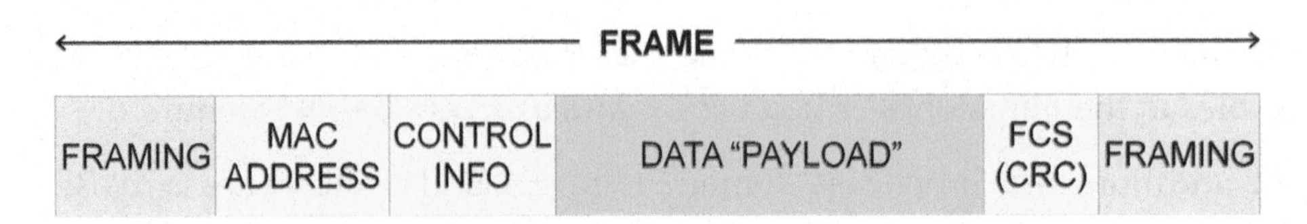

FIGURE 192 MAC FRAME

The software that implements the addressing is described in the standards documents as being part of the Media Access Control sublayer, and so the link addresses are called *MAC addresses*.

Every LAN interface is given a hard-coded 48-bit MAC address by its manufacturer. The first three bytes of the address identify the manufacturer and the last three bytes are a serial number.

15.1.6 Communication of MAC Frames

The mechanism for communicating to another station on a LAN is to transmit a frame with the MAC address of the intended receiver in this broadcast domain in the destination address field of the frame.

The frame is transmitted, all stations receive it, perform the CRC error check (which protects the address), then compare the destination address on the frame to their own MAC address.

If the MAC address on the frame is not the same as a station's MAC address, it is supposed to ignore the frame.

If they are the same, then the station knows it is the intended receiver and processes the frame, extracting the payload from the frame and passing it up to the next higher-level software.

15.2 Ethernet and 802 Standards

The LAN was invented at the Xerox Palo Alto Research Center (PARC) in Silicon Valley in California – along with the mouse and the windows graphical user interface.

And people say Xerox never does anything original!

LAN technology was commercialized in 1979 by a consortium of three companies: Digital Equipment Corporation (DEC), now part of HP, Xerox and Intel.

It was branded *Ethernet,* presumably by the marketing department, evoking the idea of communicating via *the Ether,* the fabric of space itself.

However, instead of the Ether, thick yellow coaxial cables were used.

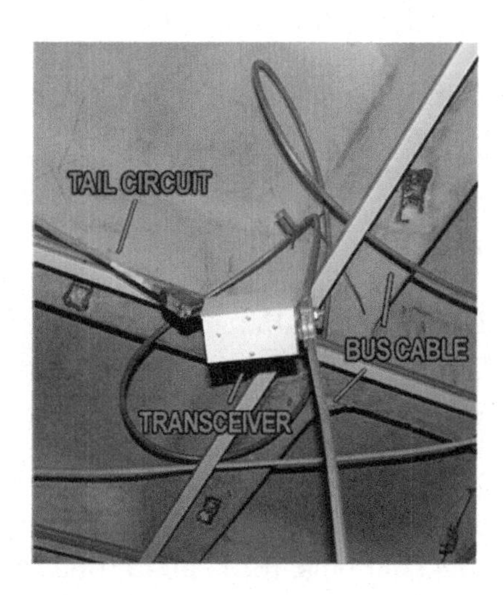

FIGURE 193 ETHERNET BUS

15.2.1 IEEE 802 Standards

The Institute of Electrical and Electronic Engineers (IEEE) subsequently formed standards group 802 in February of 1980, and developed a number of standards for LANs.

Standards 802 and 802.1 described the overall architecture and MAC addressing scheme.

802.2 described Logical Link Control, the protocol for exchanging frames between stations.

802.3 described physical coax cabling, signals and framing in a way that was almost identical to Ethernet.

15.2.2 Ethernet vs. 802.3

Other than a difference in the address format, the two were identical. Both Ethernet and 802.3 used the bus topology, provided 10 Mb/s communications and used the CSMA-CD access control protocol.

Eventually, the market adopted the "open" 802.3 standard and Ethernet failed as a commercial product.

We now use the term *Ethernet* to refer to the 802.3 standard, which copied the design of Ethernet then stole its market. The people who invented Ethernet must spin in their graves each time this happens.

15.2.3 Token Ring

The IEEE also developed standards for other LAN architectures, notably 802.5 which was IBM's Token Ring product.

This involved passing frames neighbor-to-neighbor around a ring, and passing a token or permission to originate a new frame around the ring. Token Ring is now obsolete legacy technology.

15.2.4 Baseband LAN

Ethernet is a *baseband* system: a station uses the entire capacity of the bus when transmitting. The CSMA-CD access control mechanism is used to decide if a station can transmit on the bus at any given time.

IBM attempted to commercialize an in-building communication system very much like modems on Cable TV, where there was a wide frequency bandwidth and multiple channels. IBM called this a *broadband* LAN.

IBM's product no longer exists – there is no such thing as a broadband LAN, and all LAN technologies are "baseband" LANs, hence the designation BASE in the 802.3 standard.

15.2.5 10BASE-5

In the initial design, to connect a station, a transceiver was physically attached to the coaxial cable bus and a short tail circuit run from the transceiver to the station's Ethernet card.

Stations communicated by broadcasting frames with the MAC address of the source and desired destination at the beginning of the frame. Anything a station transmits is received by all other stations.

The original design is referred to as 10BASE-5, since it provides 10 Mb/s, implements a single baseband channel on the bus, and the maximum length of a cable segment is 500 m.

15.2.6 10BASE-2

The first improvement was to reduce the cost of the bus cable and transceiver. A thinner coaxial cable was specified, and the transceiver function moved to the adapter card inside the PC instead of being a separate device.

This was referred to as a Thinwire Ethernet or 10BASE-2, as the maximum cable segment length is 185 m with the thinner cable. It is sometimes still used to run Ethernet over existing in-building coaxial cable TV wiring.

15.2.7 10BASE-T

The next improvement, 10BASE-T, implemented Ethernet using point-to-point twisted pair cables connected to a passive hub to replace the bus. The maximum length of twisted pair cable is 100 m.

15.2.8 100BASE-T

100BASE-T is 10BASE-T ten times faster, on Category 5 unshielded twisted pair, employing two of the pairs for data with a 3-volt, 3-level Manchester line code. The other two pairs are either unused, or sometimes used to deliver power to terminal devices.

Cable categories are covered in the next section.

15.2.9 1000BASE-T

1000BASE-T is Gigabit Ethernet, implementing two-way simultaneous transmission over all four pairs of a LAN cable in parallel.

The bit stream is divided into four and 250 Mb/s is transmitted over each pair, using bandwidth of approximately 100 MHz on each of the pairs.

For those who like details: the line coding is 5-level Pulse Amplitude Modulation (PAM) with 4-dimensional 8-state Trellis Forward Error Correction encoding, pulse shaping and signal equalization, Near-End Cross Talk (NEXT) cancellation and echo cancellation using digital signal processing. Available in quantity 1 for less than $10!

In theory, the next step is 10 Gb/s on Category 6 copper cables.

15.2.10 Optical Ethernet

Optical Ethernet, that is, signaling MAC frames point-to-point by flashing a light on and off on a fiber, begins with 1000BASE-SX and 1000BASE-LX Gigabit Ethernet over fiber, the SX being short wavelength (850 nm) and the LX being long wavelength (1550 nm) with a specified range of 5 km.

Optical Ethernet is covered in Section 10.5.

At time of press, the high end for Optical Ethernet is 100GBASE-ER4, 100 Gb/s Extended Range, signaling the bits on four wavelengths in parallel with a range of up to 40 km. This will increase in the future.

15.3 LAN Cables and Categories

LANs for the most part run over cables inside buildings.

The term *cable* is often used to mean "bundles of wires". Connectors or terminations may also be included as a package.

15.3.1 Unshielded Twisted Pair (UTP)

Historically, copper wires have been used for two-wire telephone access circuits, called loops. Pairs of copper wires are also used for LAN cables.

Copper is used because it is a good conductor of electricity, inexpensive, pliable, corrosion-resistant, and easy to extrude into long, thin wires.

The two wires are twisted together to reduce pickup of noise, and so are often referred to as a *twisted pair.*

The wire may be *solid* or *braided,* the latter being more expensive to manufacture but better resistant to breakage.

15.3.2 Shielding

A *shield* may be placed around individual pairs, and/or around the entire bundle of wires in a cable.

The shield is a metal foil or mesh that prevents both the ingress and egress of electro-magnetic energy, which causes interference on copper wires.

Unshielded Twisted Pair (UTP) is often used, as adding shielding to reduce noise also reduces the frequency response.

15.3.3 TIA-568 LAN Cable Categories

The most widely-followed standard for LAN cables is TIA-568, published by the Electronic Industries Association and its Telecommunications Industry Association sub-group.

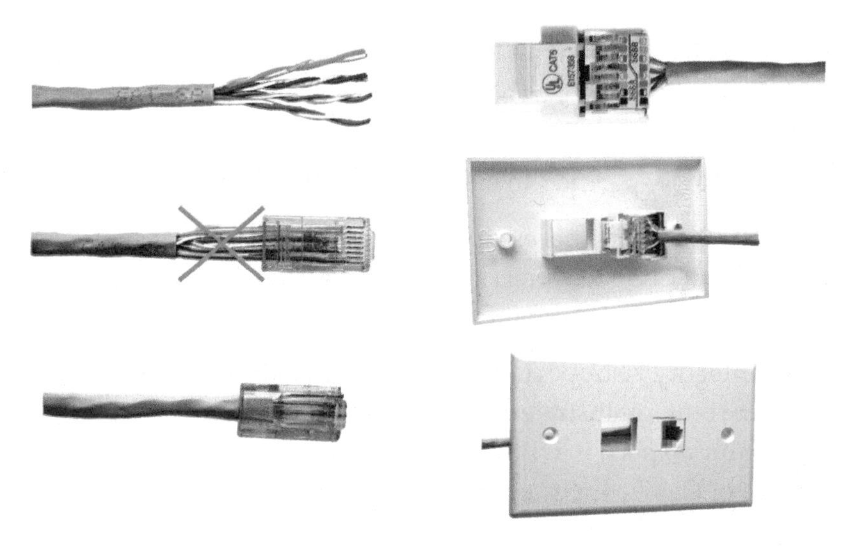

FIGURE 194 CATEGORY 5E LAN CABLE AND CONNECTORS

Bulk cable is terminated on keystone connectors that snap into the back of a cover plate. Patch cables, illustrated on the left, can be made by crimping RJ-45 connectors on bulk cable.

Telecommunications Systems Bulletin TSB-67 adds the requirements and methods for field testing installed cable systems. Taken together, these are the authority how to design and install a structured cabling system.

TIA-568 defines *categories* of twisted-pair cable for different line speeds:

- TIA-568 Category 1 cable is existing telephone cabling, also called Rusty Twisted Pair (RTP).
- Category 2 cable was 25-pair multiconductor cables for old key telephone systems that had buttons to press to access different lines.
- Category 3 cable was for 10 Mb/s Ethernet on twisted pair, 10BASE-T.
- Category 4 cable was specified for 16 Mb/s token ring.
- Category 5 cabling was for The Future at up to 1000 Mb/s.

Categories 1 through 5 are no longer installed.

Category 5 (*Cat 5*) cable was supposed to support Gigabit Ethernet, but in practice turned out to be missing the specification of some required transmission characteristics.

Enhanced Category 5 (Cat 5e) was subsequently specified to guarantee the operation of Gigabit Ethernet on twisted pair, 1000BASE-T.

Whether a cable can be certified as conforming to a standard is often dependent on the consistency and placement of twists during manufacturing.

Category 6 cable is specified to support 10 Gb/s on twisted pair.

At 1 Gb/s, it becomes necessary to specify the frequency bandwidth supported on the twisted pair, along with all of the other transmission characteristics, to enable signaling at these line speeds.

In theory, Category 7 supports 100 Gb/s on twisted pair. This is in the same league as current mainstream fiber-optic transmission systems, so one could probably expect it will be a while before there is any significant deployment of Cat 7 copper wires.

15.3.4 TIA-568A vs. TIA-568B

There are two specifications for which wires in the cable go to which pins on the connectors: TIA-568A and TIA-568B. There is no difference between the two in terms of performance – but it is necessary to pick one of the two configurations and use it consistently on every jack, every patch panel, every patch cord and every connector.

TIA-568B is the most popular choice. Holding a male Category 5e connector in front of you with the retainer clip facing you and the metal contacts on the top, pin 1 is on the left. The wires are color-coded in a standard way, using white, orange, green, blue and brown.

The TIA-568B connections are:
Pin 1 – white/orange
Pin 2 – orange
Pin 3 – white/green
Pin 4 – blue
Pin 5 – white / blue
Pin 6 – green
Pin 7 – white / brown
Pin 8 – brown

This pinout must be used consistently, as the design of both the connector and the cable and their performance measured in transmission characteristics such as crosstalk, insertion loss, echo and other metrics are based on particular signals being on particular wires.

15.3.5 Maximum Cable Length and Cabling Architecture

All categories specify cables with four pairs (eight wires) and a maximum length of 100 meters.

This means the maximum run length of the cables – including runs through risers, poles, conduits – is 100 m (330 feet).

To be conservative, devices would be connected to a switch located in a wiring closet within a radius of perhaps 200 feet.

These wiring closet switches could be connected to centralized Ethernet switches on each floor and/or connected to a router in the communications room, possibly using fiber.

15.3.6 Difference Between Categories

The difference between the categories rests in guaranteed transmission characteristics of the cable, including specifications for Near-End Crosstalk (NEXT), Attenuation to Crosstalk Ratio (ACR), supported frequency bandwidth, all of which affect the maximum possible information transfer rate, and hence what kind of devices can be successfully attached to each end of the cable.

One of the main factors in getting a cable certified to meet the TIA-568 category is quality control, particularly in the consistency of the twisting and placement of the pairs.

Two pairs will be twisted at a particular number of twists per inch, but offset by half a period to minimize crosstalk between the pairs. The other two pairs will be twisted at a different rate that is not a multiple of the other, and similarly with the twists exactly not lined up.

How well and how consistently this is accomplished during the manufacturing process determines how successful the manufacturer will be in having the cable certified as meeting the standard.

15.3.7 Which Category to Use

When determining which category of cable to use, life cycle and labor cost are determining factors.

For a patch cable connecting a DSL or Cable Modem to a device inside a residence, where we have an expectation that the line speed will not exceed 100 Mb/s in the foreseeable future, then Cat 5 patch cables may be used.

For an extra ten cents, a Cat 5e patch cable would allow the continued use of the cable were the line speed to increase above 100 Mb/s, as it inevitably will at some time in the future.

Since the labor cost is usually far greater than the cable, it is strongly recommended to install cabling inside walls with capacity greater than immediate needs, and twice as many cables as what the conventional wisdom dictates.

Two Category 6 cables to each work area would be the Cadillac solution.

Two Category 5e cables to each work area would be well positioned for the future. One Category 5e cable to each work area is the minimum.

One Category 5 cable to each work area would probably be viewed as a mistake ten years down the road.

15.4 LAN Switches: Layer 2 Switches

Since the invention of Ethernet and its standardization, the technology has evolved and improved in several different ways. One main improvement is the Ethernet switch, also called a LAN switch or *Layer 2 (L2) switch.*

This device replaces bus cables and hubs, providing dramatic improvements in performance plus the possibility of implementing improved traffic management and security through Virtual LAN (VLAN) technology.

15.4.1 Hardware

In concrete terms, a Layer 2 switch is a small dedicated-purpose computer with anywhere from two to hundreds of LAN hardware ports, an internal bus, memory, and software performing the switching function and possibly additional VLAN-related functions.

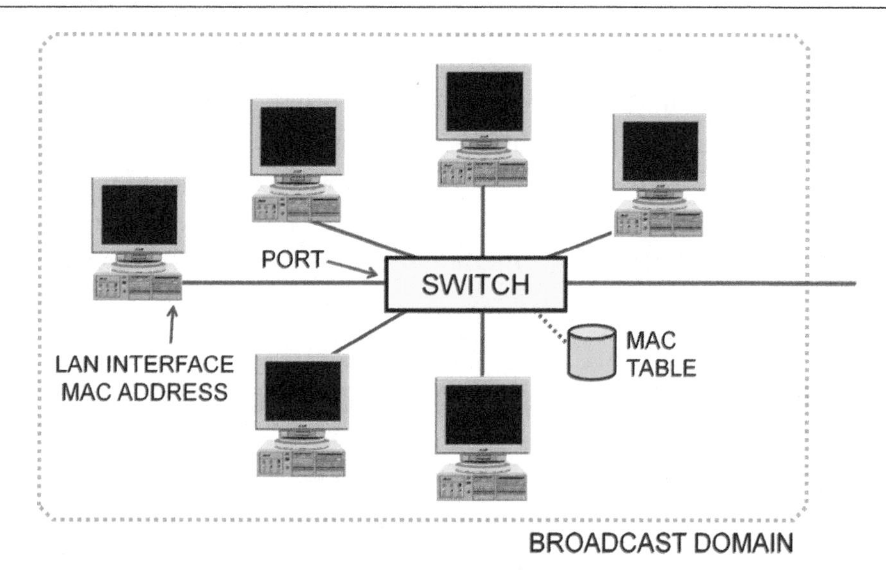

FIGURE 195 LAN SWITCH

Each hardware port is an Ethernet jack, and should support 1000 Mb/s full-duplex. Expensive switches have Optical Ethernet ports.

15.4.2 Purpose and Operation

The essential function of a Layer 2 switch is to examine the destination MAC address on each incoming frame, determine which hardware port a device with this MAC address is connected to, then relay the frame to that hardware port, to the device with the destination MAC address.

To determine the MAC address of the computer connected to a particular hardware port, the processor reads the sender MAC address on frames and stores this information in what might be called a *MAC table.*

15.4.3 Buffers

Since there is typically only one processor and one internal bus connecting these hardware ports inside the switch, small amounts of memory called buffers are provided for each port to allow stations to send frames simultaneously.

The frames are stored in the buffer then relayed to the appropriate port(s) by the switch's processor, normally on a first-come, first-served basis, or in a prioritized order in the case of an expensive switch implementing prioritization protocols.

15.4.4 Frame Forwarding

In normal operation, the processor relays a frame from one port to one other, and it does this in a lightning-fast manner, since it only reads the destination MAC address, does a lookup in the MAC table then forwards the frame to the indicated port. It does not receive the whole frame and perform an error check; the destination computer performs error recovery.

15.4.5 Broadcast Domain Defined by Switch

In exceptional circumstances, the Layer 2 switch **broadcasts** the frame to all the hardware ports.

This happens when there is no entry in the MAC table for the destination MAC address. It also happens when the content of the destination MAC address field explicitly instructs the switch to broadcast the frame by populating the destination address all 1s. This occurs when the sending computer is running the Address Resolution Protocol, for example, attempting to discover the MAC address of the computer that owns a particular IP address.

Because there is the possibility the switch will send a copy of the frame to all of the hardware ports, all of the computers connected to a Layer 2 switch are in a ***broadcast domain:*** any station has the possibility of communicating directly with any other in the broadcast domain without the need of other equipment or protocols.

15.5 VLANs

15.5.1 Broadcast Domains Defined in Software

VLANs are essentially a software trick, implemented by the switch, to define broadcast domains in software, for the purpose of traffic management.

A basic LAN switch does not implement VLANs. All of the devices physically connected to the basic LAN switch form a broadcast domain; there is no way of preventing one device from communicating with another.

A more sophisticated switch supporting VLANs allows an administrator to identify specific hardware ports as belonging to a particular VLAN group, identified by a 12-bit number.

In the simple example illustrated in Figure 196, the ports for the IP phones producing voice packets are defined to be in VLAN 1 and the ports for the desktop computers are defined to be in VLAN 2.

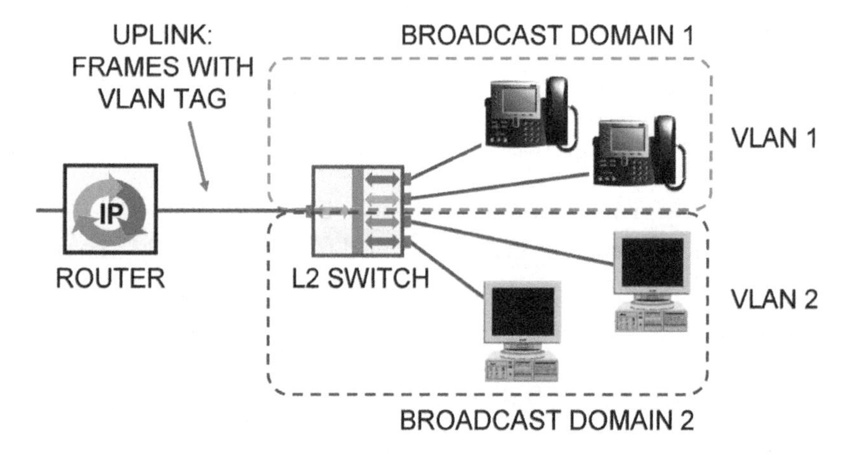

FIGURE 196 VLANS IMPLEMENTED BY A LAYER 2 SWITCH

Once this is set up, the processor will only forward frames between hardware ports that are in the same VLAN, and if it is necessary to make a copy of a frame to send to "all" ports, a copy is **only** sent to the ports in the same VLAN and not to any others.

15.5.2 Routing Between VLANs

In the example illustrated in Figure 196, the port labeled "uplink" is the connection leading to the rest of the network, i.e., to a router. This port is defined by the administrator to belong to both VLAN 1 and VLAN 2.

Communication from a device in VLAN 1 to a device in VLAN 2 can implemented by transmitting a packet in a frame from the device in VLAN 1 to a router via the uplink port, whereupon the router could transmit the packet in a frame back to VLAN 2.

The purpose of this architecture is to prevent direct communications between devices in different VLANs, and allow communications only through an external router... where rules can be entered by an administrator specifying if the communication between the VLANs is allowed or denied.

15.5.3 Header Tag

To indicate to the device on the uplink port which VLAN a frame is originating from, an extra Tag Header conforming to the 802.1Q standard is added to the frame immediately following the address fields, and the VLAN ID is populated in the Tag Header.

Tagging the frame with the VLAN ID allows the definition of VLAN groups that span multiple physical switches.

Additionally, the Tag Header includes the Tag Protocol Identifier identifying the frame as a tagged frame, and following the 802.1p protocol, can optionally carry a three-bit number indicating the priority of the frame for Quality of Service mechanisms.

15.5.4 Traffic Management and Network Security

VLANs are a powerful low-level tool for traffic management and network security. It allows the grouping of devices into separate broadcast domains so that devices in one VLAN cannot communicate to devices in a different VLAN, a measure against attacks launched from infected Windows computers against a VoIP system, for example.

It is also an essential tool used to separate customers of a carrier who are using a shared facility.

By putting each customer's hardware ports in a unique VLAN, traffic from different customers will be interspersed on a shared circuit, but the customers cannot communicate to each other nor receive copies of other customers' traffic.

16 IP Networks, Routers and Addresses

This chapter could equally be called "Layer 3".

In this chapter, we cover networking, which is Layer 3 of the OSI model, including routers, packets and network addresses. We'll understand how the network is built by connecting circuits with routers, and trace the flow of a packet from end to end.

The standard method of formatting packets and assigning network addresses is of course IP, formerly known as the Internet Protocol, so this chapter could also be called "All about IP".

16.1 Definition of Network

The definition of *a network* includes the requirement to make route decisions: not broadcasting the data to every station, but instead forwarding, switching or *routing* the data from one circuit to the next to the next to eventually deliver it to a particular destination.

A packet network is constructed of point-to-point circuits connecting routers in different locations. The routers physically move packets from one circuit to a different circuit, a forwarding or relay function.

Knowing **which** circuit to relay a packet to is the routing part of the story.

Packet networks incorporate two main ideas: *packet switching* and *bandwidth on demand*.

Packet switching means forwarding IP packets from one router to another, and using IP addresses, routing tables and algorithms to determine which circuit to forward them to.

Bandwidth on demand means giving many devices access to a circuit, and giving each the **possibility** of transmitting a packet, but not reserving capacity for any particular device. If a device does not have anything to transmit, another can use the available capacity.

Since the term *bandwidth* is used to mean transmission capacity, this is called a capacity on demand or bandwidth on demand strategy.

16.2 IPv4 Address Classes

To send information from a machine in one broadcast domain (Section 15.1) to a machine in a different broadcast domain, it is necessary to have a router relay the information from one broadcast domain to another.

16.2.1 Packets and Network Addresses

The mechanism for this is to give each machine a unique **network** address. The information to be communicated is placed in a packet and the network address of the desired destination is populated in a field in the packet header.

The packet is then encapsulated in a MAC frame with the router's MAC address, and broadcast by the sending machine on its broadcast domain, which includes the router.

The router receives the frame, extracts the packet and uses the contents of the destination address field in the packet header as an input to making a route decision.

The decision is implemented by the router transmitting the packet in a frame on a different outgoing broadcast domain, usually one that contains the next-hop router.

IP version 4 is a standard method of formatting packets and network addresses, specifying 32-bit-long network addresses called IPv4 addresses or simply IP addresses.

IPv4 is the current network address standard for the Internet. IPv6 is coming.

16.2.2 Historical Network Classes

In the beginning, the Internet was the Inter-net, a protocol for addressing machines that were on different pre-existing networks that used other packet and network addressing schemes.

A method of assigning IP addresses to machines on these pre-existing networks was necessary.

To make routing tables efficient, it was desirable to associate a contiguous range or block of addresses with a pre-existing network. The block would ideally have as many addresses as there were machines in that network.

The developers decided to standardize on three typical sizes of networks, which they called *classes* of networks: big, medium and small, or Class A size networks, Class B size networks and Class C size networks respectively, and so three standard sizes of blocks of addresses that would be assigned: Class A blocks, Class B blocks and Class C blocks.

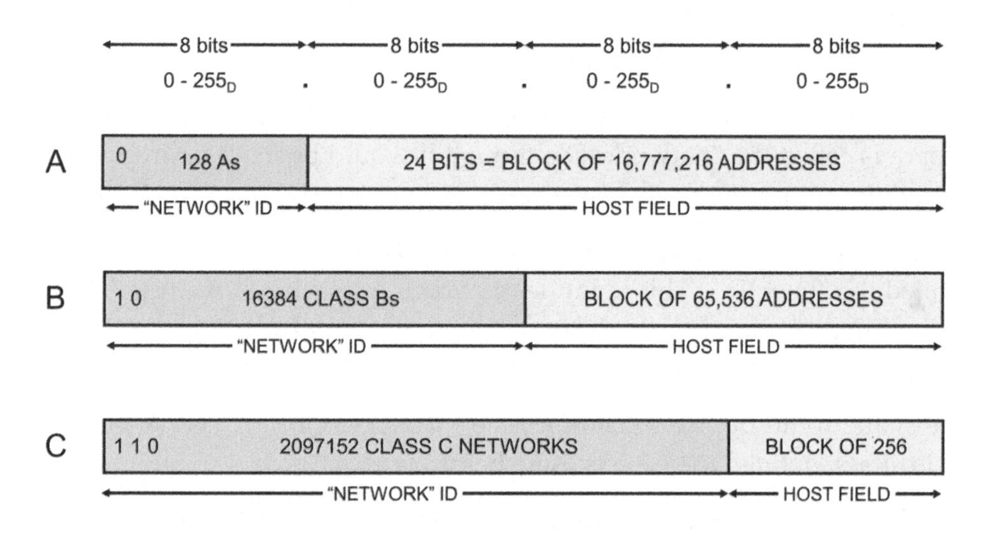

FIGURE 197 IPV4 ADDRESS CLASSES

Today, the inter-net addressing protocol has taken over, and the "other" packet and network addressing schemes have disappeared, so the terminology of A-sized networks, B-size networks and C-size networks being connected by the inter-net is dated.

Class C blocks are still often used on private networks, as they are the simplest small address block size in dotted-decimal notation.

The term "address class" is still used; today, it might best be interpreted to mean "block of IP addresses".

16.2.3 Class A, B and C

A Class A address space is a contiguous block of 16,777,216 IP addresses. Since this was such a large block size, there weren't very many of them available: 128 Class A blocks.

A Class B address space is a contiguous block of 65,536 addresses. Since it is a smaller block size, more were available: 16,384.

A Class C address space is a contiguous block of 256 IP addresses. 2,097,152 Class Cs were available.

IPv4 addresses are 32-bit binary numbers. Knowing that $2^8 = 256$, $2^{16} = 65,536$ and $2^{24} = 16,777,216$, one can see that the classes or blocks were defined so that they lined up with the byte boundaries in the address space.

16.2.4 Network ID and Host ID

The first byte of a Class A address identifies a block of 16,777,216 addresses.

In keeping with the original idea of the inter-net, this part of the address space is called the "network ID", though it would be more appropriate to call it the "block ID" or "address range ID" or "prefix" today.

The remainder of the Class A address, the last three bytes or 24 bits, was called the "host ID". This identifies a particular machine. A Class A block of addresses can be used to sequentially number 2^{24} or 16,777,216 machines.

To make route decision-making easier, the prefix of a Class A block of addresses, defined to be 8 bits long, begins with a 0.

A Class B block begins with 10. The first two bytes are the "network ID" or prefix and so would be the same for all machines in a Class B block. The last two bytes or 16 bits can be used to sequentially number $2^{16} = 65,536$ machines.

A Class C block begins with 110. The first three bytes of the address are the prefix and so the same for all machines. The last byte or 8 bits are used to number 2^8 or 256 machines.

Addresses beginning with 111 were originally reserved for "escape to extended addressing mode", then divided into two parts:

16.2.5 Class D and E

Addresses beginning with 1110 are multicast addresses, sometimes referred to as Class D. The division of address space between "network" ID and "host" ID is not defined.

Addresses beginning with 1111 remain reserved for some unknown use, and are sometimes called Class E.

16.2.6 Dotted-Decimal Notation

IPv4 addresses are 32-bit binary numbers – but writing 32-bit binary numbers on pieces of paper or computer screens, or speaking them between people is unwieldy if not impossible.

Hexadecimal, a numbering system based on 16s, is a good short form for binary numbers, as it is simple to convert to binary and segments the address into groups of four bits.

Unfortunately, those that came up with IPv4 decided to use decimal as a short form for binary numbers, and came up with an awkward notation called *dotted-decimal,* where the 32 bits are divided into four groups of 8, then the groups of 8 are converted independently to decimal, yielding addresses written like 232.155.166.1.

Using decimal as a short form for binary numbers is awkward, since it is difficult to convert between decimal and binary.

16.3 Subnets and Classless Inter-Domain Routing

For more flexibility, blocks of IP addresses are no longer restricted to one of the three sizes. As illustrated in Figure 198, IP addresses are now organized into random-sized groups of IP addresses called *subnets*.

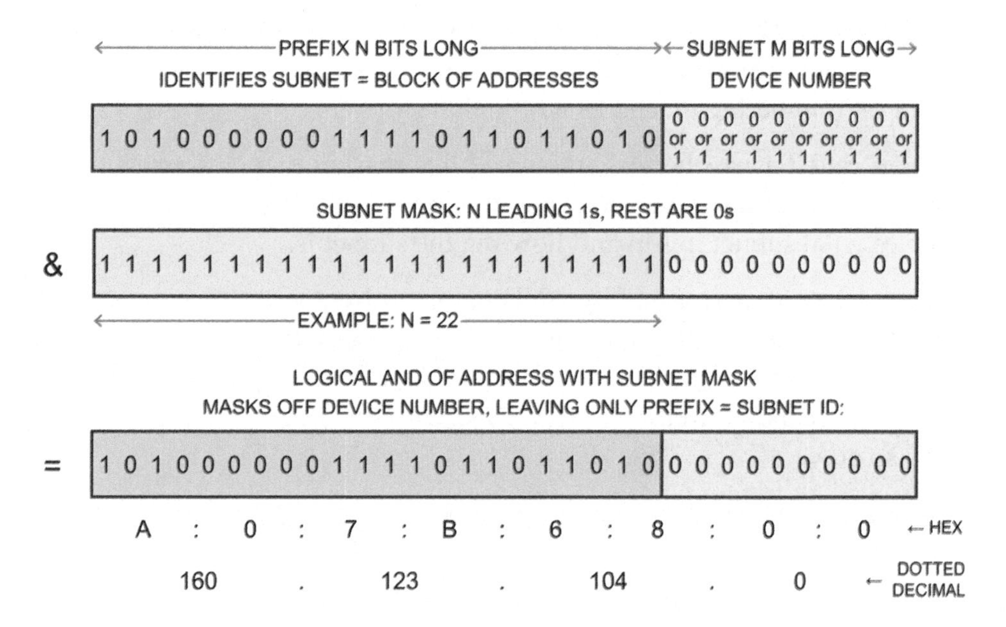

FIGURE 198 SUBNET AND PREFIX

This was originally called Classless Inter-Domain Routing (CIDR).

This allows the assignment of blocks of addresses more appropriately-sized than "Class B" or "Class C" to an organization, and allows an organization to manage their subnet, i.e., the network addresses assigned to that organization, as a group of smaller subnets.

As illustrated in Figure 198, a subnet is identified by the value of its *prefix*, which are the most significant bits of the address. All devices that are given addresses within that subnet will have addresses that start with the prefix.

The remaining part of the address, after the prefix, is the subnet, and the addresses in the subnet can be used for device IP addresses.

In the example shown, the prefix is 22 bits long. Since IPv4 addresses are 32 bits long, that leaves ten bits for numbering machines, meaning that this subnet has $2^{10} = 1024$ IP addresses available to assign to devices.

When a device is first powered on, it does not know which of the 32 bits in its address are the prefix identifying the subnet, and which of the 32 bits are used to number machines within the subnet.

As we will see in a subsequent lesson, at the lowest level of routing, it is essential for a device to know where this dividing line is, so it can identify which subnet it is in, and which subnet the destination is in.

The dividing line is indicated with a *subnet mask,* a number with 1s identifying the bits that are the prefix, and zeros indicating the bits that are used for numbering machines within that subnet.

This is called a *mask* because it is used like masking tape, to mask off the bits used to number the machine and leave only the prefix, telling the device what subnet it is in and how big the subnet is.

The masking is performed by ANDing an address with the subnet mask. Anything AND 0 = 0; anything AND 1 = whatever it was. It is applied bitwise, i.e., the first bit in the mask is ANDed with the first bit in the address, second bit in the mask ANDed with the second bit in the address and so on.

Since the mask has 0s in the positions that are used to number machines within the subnet, ANDing an address with the mask turns all of the bits that are used to number machines into 0s and leaves the bits that are the prefix unchanged.

In the example shown, the subnet mask is 11111111111111111111110000000000, which can be written as 255.255.252.0 in dotted-decimal, or FF:FF:FC:00 in hex.

ANDing any address in the subnet with the subnet mask yields the prefix 1010000001111011011010 followed by ten 0s: 10100000011110110110100000000000, which can be written as 160.123.104.0 in dotted-decimal or A0:7B:68:00 in hex.

To convey both the size of the subnet and its prefix, it is usually written as the prefix padded out with trailing zeros to make a 32-bit number, followed by /n, where n is the number of 1s in the subnet mask.

In the example of Figure 198, one would write 160.123.104.0 /22 in dotted-decimal or A0:7B:68:00 /22 in hex to identify the subnet.

16.4 DHCP

IP addresses may be static or dynamic. *Static* means that the address assigned to a machine generally does not change.

Dynamic means that an IP address is assigned to a computer on demand, for a fixed lease period. The computer may be assigned a different address each time it demands one.

Addresses are assigned to a computer using the Dynamic Host Configuration Protocol (DHCP).

16.4.1 Dynamic Addresses for Clients

Dynamic addresses are acceptable for a machine running client software, since the way things are organized is that the client initiates communications with a server, and includes its return address (the source IP address) in every packet sent to the server.

16.4.2 Static Addresses and DNS for Servers

To communicate to a server, it is necessary to find out the numeric IP address of the server before the client can communicate to it. That is often accomplished through the Domain Name System (DNS), essentially tables where the IP address of a server can be looked up.

To avoid having to frequently update those tables, servers are generally assigned static addresses.

16.4.3 DHCP Client – Server Communications

The system administrator provisions a DHCP server, configured to assign IP addresses within a specific block (within a subnet) to clients. Computers are loaded with DHCP client software.

Communications between the DHCP client and server are effectively application-layer messages, coded into ASCII and carried in UDP protocol data units, which are carried in IP packets, which are carried in MAC frames.

The desired recipient of the messages is indicated as being the DHCP on a machine by populating in the UDP header destination port = 67 for messages to the server and destination port = 68 for messages to the client.

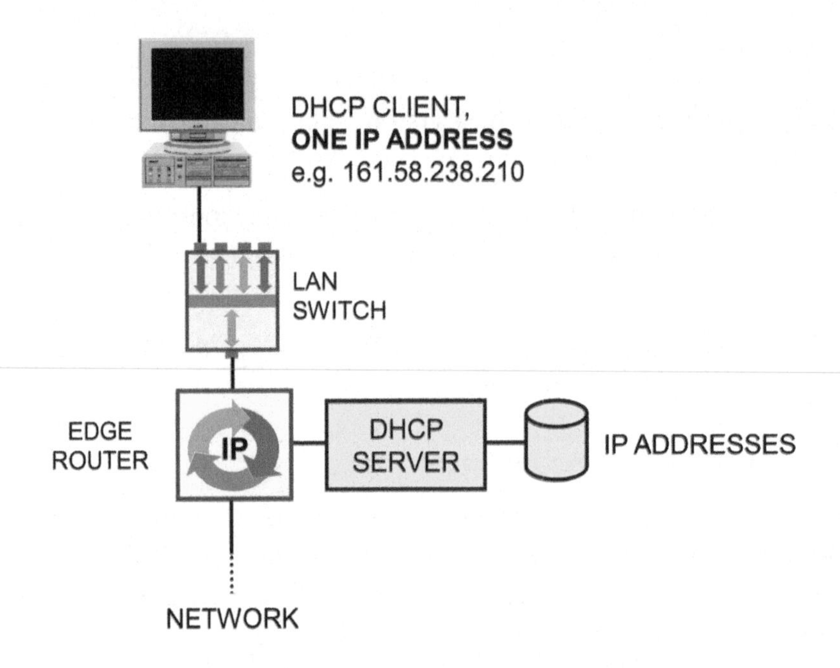

FIGURE 199 DHCP CLIENT AND SERVER

The messages are "broadcast", which means that the destination IP address is all 1s and destination MAC address is all 1s. The actual addresses are used for source MAC and IP addresses, except that the client uses "0" as its IP address, since of course the whole point of the exercise is to get an IP address.

16.4.4 DHCP Message Exchange

Each computer will run a DHCP client when it starts, generating a DHCP Discover message.

Any DHCP server that receives it, and there may be more than one, will respond with a DHCP Offer message, with an offered IP address and a lease time.

The client will answer with a DHCP Request message to confirm its selection of an offered address, then the server will complete the cycle with a DHCP ACK that usually includes other configuration information such as the IP address of the default gateway (CE router), the IP address of one or more DNS servers, the lease time, and the subnet mask of section 16.3.

There are several variations on the basic process, all of which are enumerated in the relevant standards document, RFC 2131.

16.4.5 Lease Expiry

On expiry of the lease time, the DHCP client must begin the discover process anew.

If a DHCP client runs while still holding a valid lease, it will request to be assigned the same IP address.

If there are many clients constantly running DHCP (an ISP's customers, for example), then it is likely that a different IP address will be offered by the server each time a computer runs its DHCP client.

The lease time may be configured by the system administrator to any value. This function might be useful to help manage situations where there are more clients than addresses.

If the DHCP server reaches the end of its configured range of addresses, it attempts to re-assign previously-assigned addresses to new requestors, beginning with those previously-assigned addresses for which the lease has expired. Before re-assigning the address to a different machine, the server might optionally *ping* the address to determine if it is still in use.

16.4.6 DHCP to Assign Static Addresses

Even though it is the "dynamic" host configuration protocol, DHCP is also used to assign static addresses to machines.

This is accomplished with a table in the server, configured by the system administrator, which relates MAC addresses to IP addresses.

Whenever a computer with a MAC address contained in the table asks for an IP address, it will always be assigned the IP address specified in the table.

This allows the assignment of static addresses to computers from a centralized management system (the DHCP server), conveyance of other information like default gateway and netmask, and eliminates the need for any human involvement (and its associated errors) in configuring computers.

In Windows, you can see the IP address currently assigned to a computer, as well as its LAN card MAC address by opening the Network Connections folder and viewing the detailed "status" of the LAN card.

If under "properties" of the TCP/IP protocol the choice "obtain a network address automatically" is selected, the DHCP client is run at startup and when "repair this connection" is clicked.

16.5 Assigning Subnets to Broadcast Domains

While this is a relatively advanced topic, and not necessary for a basic understanding of IP networks, this concept is at the heart of how route decisions are made. It is an essential strategy to enable routing at the lowest level (Section 16.6.7), without having to have one entry for every machine in the routing tables.

Grouping machines in broadcast domains, defined in software by VLANs or defined in hardware by simple Layer 2 switches, is one of the basic building blocks of network security.

Machines in different broadcast domains means that they cannot communicate packets directly; they must ask a router to relay the packet from one broadcast domain to the other.

The router then becomes a point of control where rules determining what is allowed to be relayed where can be implemented to compartmentalize the network and control the flow of traffic.

To facilitate routing, standard practice is to assign all the machines in the same broadcast domain to have IP addresses within the same subnet.

Typically, an organization will use private addresses (Section 16.8.3) inside the building. Since the private address space is large, and it does not cost anything to use them, Class C sized subnets, i.e., groups of up to 256 machines, are often defined. For example, one subnet could be 192.168.1.0 /24, a second could be 192.168.2.0 /24 and a third 192.168.3.0 /24. For situations where there are groups of more than 256 machines, like Wi-Fi at an airport, larger subnets would be used.

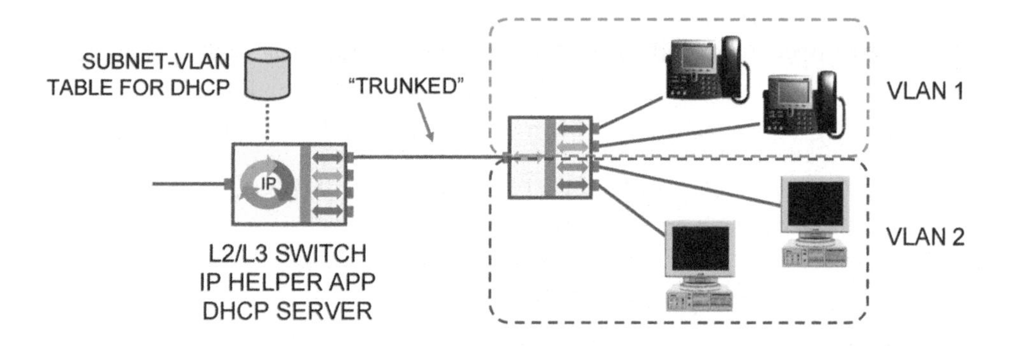

FIGURE 200 ASSIGNING SUBNETS TO BROADCAST DOMAINS

Devices within the same broadcast domain are assigned IP addresses in the same subnet by a DHCP server. This is implemented on a device that could be called an L2/L3 switch, or more simply a router – but a router with special features.

This device has both Layer 2 LAN/VLAN switching plus Layer 3 routing capability, plus a DHCP server and some helpers. Since the functions are in the same device, the inter-function communication happens at computer bus speeds. These are typically expensive, high-end devices such as Cisco Catalyst switches.

Assigning machines to VLANs requires configuring the Layer 2 switch to associate a VLAN ID with each hardware port.

Assigning an IP address to a machine, based on which VLAN the machine is in, can be done automatically using Dynamic Host Configuration Protocol (DHCP) running on the router plus an "IP helper application", and a table which we might call the "subnet-VLAN table" where someone has defined and typed in the prefix of the subnet associated with each VLAN.

When a machine runs its DHCP client, it will send a DHCP request in the form of a DHCP discovery packet in a frame. When passing through the Layer 2 switch on the right, the frame will be tagged with the VLAN ID.

In Figure 200, the link between the router and the final Layer 2 switch on the right is labelled as "trunked", meaning it carries frames from more than one VLAN, and meaning that an extra field with the VLAN ID is populated in the frame header by the Layer 2 switch on the right.

When the frame is received by the DHCP server running on the router, the IP helper application will query the subnet-VLAN table using the VLAN ID. This will return the subnet ID and subnet mask. The DHCP server will in turn assign an IP address within that subnet to the machine.

This automates the process, so that machines in a particular VLAN are all assigned IP addresses in the same subnet.

16.6 IP Network: Routers Connected with Point-to-Point Circuits

We begin with the simplest example of a network illustrated in Figure 201: three locations connected in a ring with point-to-point circuits.

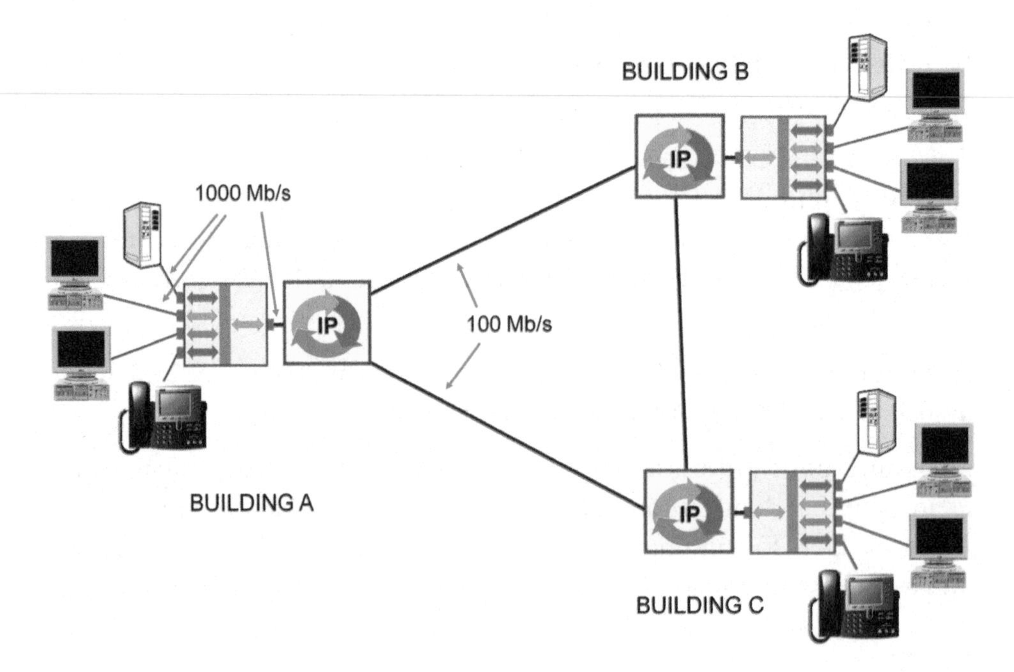

FIGURE 201 SIMPLEST EXAMPLE OF A NETWORK

This is called a *private* network, since there are connections only between those three locations, and the circuits that implement the connections are not shared with others. It is also in fact how modern telecom networks are built: carriers' point-to-point fibers connecting routers in different cities.

The inter-building or Wide Area Network (WAN) circuits could be implemented with point-to-point fibers, point-to-point radio links or full-period "dedicated line" services like T1 from a carrier.

This is useful as the simplest framework for understanding how circuits, routers, routing, IP packets and IP addresses, MAC addresses and MAC frames, copper and fiber work together to implement a packet network.

In the next chapter, the story is made more realistic – and more complex – replacing the dedicated lines with packet-switched bandwidth-on-demand services from a carrier.

16.6.1 Broadcast Domain at Each Location

In the example of Figure 201, at each location there are a number of terminals or devices – VoIP telephones, desktops, servers, plus a router, all connected to a LAN switch.

All of the devices connected to the LAN switch in building A are in the same broadcast domain. This allows the communication of MAC frames between the devices in building A and the router in building A.

16.6.2 Edge Router at Each Location

The router in building A has **three** Ethernet jacks for terminating circuits. One of them is connected to the LAN switch in building A. The other two are connected to circuits leading to the routers in other buildings.

By internally moving packets from one jack to another, the router moves packets between buildings.

In Windows, the building A router is called the *default gateway* by building A devices. In any device's routing table, the IP address of Router A is listed as the *default route.*

If a device wants to send a packet outside its broadcast domain, its only hope is to send the packet on its default route – to that gateway – for onward forwarding.

That gateway, more generally called the *edge router* for building A, is the only device in building A that connects to other broadcast domains.

Each broadcast domain, i.e., each building, is assigned a unique range or block of IP addresses, called a subnet.

The terminals and the router in each broadcast domain are assigned an IP address within the subnet for that broadcast domain.

16.6.3 Default Gateway

The terminals are informed of the IP address of the router in their subnet, which is their default gateway to other broadcast domains.

Informing the devices of the IP address of their edge router allows the devices to communicate packets to it for onward forwarding.

16.6.4 Packet Creation

To communicate VoIP from the telephone in building A to the telephone in building C, telephone A first has to find out the IP address of telephone C, usually using the SIP protocol as described in Section 8.4.

Once the conversation starts, telephone A creates IP packets addressed to telephone C containing snippets of digitized speech.

16.6.5 Packet Transmission from the Source

To send a packet from telephone A, there are only two choices: send the packet directly to telephone C, or if that is not possible, send the packet to its default gateway, router A, for onward forwarding.

To determine which of these two possibilities to use, telephone A first determines if it can send the packet directly to telephone C.

By definition, that would require telephone A and C to be in the same broadcast domain. Since each broadcast domain has been assigned a subnet, by definition, that requires telephone A and C in the same subnet.

Telephone A can determine the answer by applying the subnet mask using the logical AND operation to its own address, and to the address of telephone C, then comparing the result.

If they are the same, the two telephones are in the same subnet, and thus in the same broadcast domain, and so telephone A can transmit the packet directly to telephone C.

In this example, the result will not be equal, allowing telephone A to determine that telephone C is in a different subnet, which means in a different broadcast domain, and so by definition, telephone A knows it cannot communicate the packet directly to telephone C.

It must instead send the packet to the router in building A (which **is** in the same broadcast domain) for onward forwarding.

Once telephone A has decided the destination is router A, it transmits the packet to router A by putting the packet in a MAC frame with destination MAC address that of router A, then representing the bits that make up the frame one at a time on the copper wire LAN cable plugged in to the phone by putting electrical voltage pulses on the wires.

16.6.6 IP to MAC Address Resolution Protocol (ARP)

Since telephone A has been informed of router A's IP address, it can determine router A's MAC address by asking the router what it is, using the Address Resolution Protocol (ARP).

Using ARP, telephone A transmits a packet addressed to router A in a frame with all 1s as the MAC address, an instruction that a copy of the frame should be sent to all devices in the broadcast domain.

After router A replies to telephone A with its MAC address as the source address in the frame header, telephone A can address frames to router A.

16.6.7 Packet Routing

Upon receiving the MAC frame and extracting the packet from it, router A will physically forward or relay the packet from the LAN in building A to a circuit that can get to building C.

Determining **where** the packet should be relayed is the *routing*.

Networks are built with redundant connectivity for service availability reasons: more than one way to get from A to C. The router in building A is connected to two circuits that lead to building C.

Router A must decide which circuit to forward the packet on.

Router A has a routing table, which has entries relating ranges of IP addresses (subnets) to the IP address of a device that can forward a packet there, and the cost.

Cost is usually measured by number of hops, i.e., circuits to traverse.

The routing table is populated by entries manually typed in by a technician, by the routers communicating with each other in the background, or by a central control system in a Network Operations Center.

In this case, the routing table will end up being populated with two entries for subnet C, which contains telephone C:

- All of the devices in building C are reachable by going to router C, and the cost is one hop, and
- All of the devices in building C are reachable by going to router B, and the cost is two hops.

The router picks the least-cost route, and forwards the packet to router C.

16.6.8 Overbooking & Bandwidth on Demand

Beside the question of routing is a different discussion: performance.

In the example of Figure 202, each device has the possibility of transmitting packets to the LAN switch then to the router at 1000 Mb/s, and onward to other buildings at 10 Mb/s... but none of those bits per second are reserved for any particular device, either on the LAN or the WAN.

Statistically speaking, most of the time, telephone A does not transmit anything. Occasionally, it will transmit a packet in a frame to the LAN switch then router A over the LAN at 1000 Mb/s.

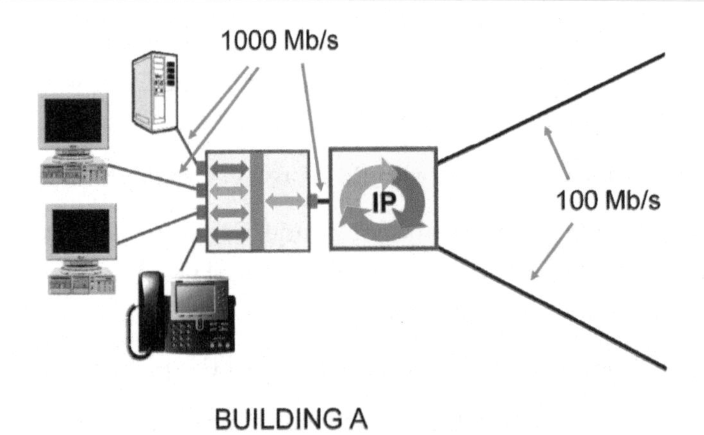

FIGURE 202 OVERBOOKING LAN AND WAN CIRCUITS

The router will relay the packet to a jack that has a 10 Mb/s dedicated line to another router, and transmit it at 10 Mb/s to the other router. Occasionally, a different device will transmit a packet to router A.

If many packets arrive at router A on the 1000 Mb/s LAN cable from the LAN switch, the router may have to temporarily store or *buffer* the packets before being able to forward them on a WAN circuit at 10 Mb/s.

The same problem exists (to a much lower degree) at the LAN switch, where there are four devices with 1000 Mb/s connections sharing a single 1000 Mb/s connection to the router.

If the overload persists, the oldest packets in the buffer, still waiting to be transmitted, get **overwritten** by the newest incoming packets. When a packet is overwritten in a buffer, it disappears. These would show up in any measurement of packet delivery rate or *dropped packets.*

16.7 Routers and Customer Edge

In this section, we take a closer look at the router in Building A of Figure 203, and how it is configured to both implement the network and control network traffic.

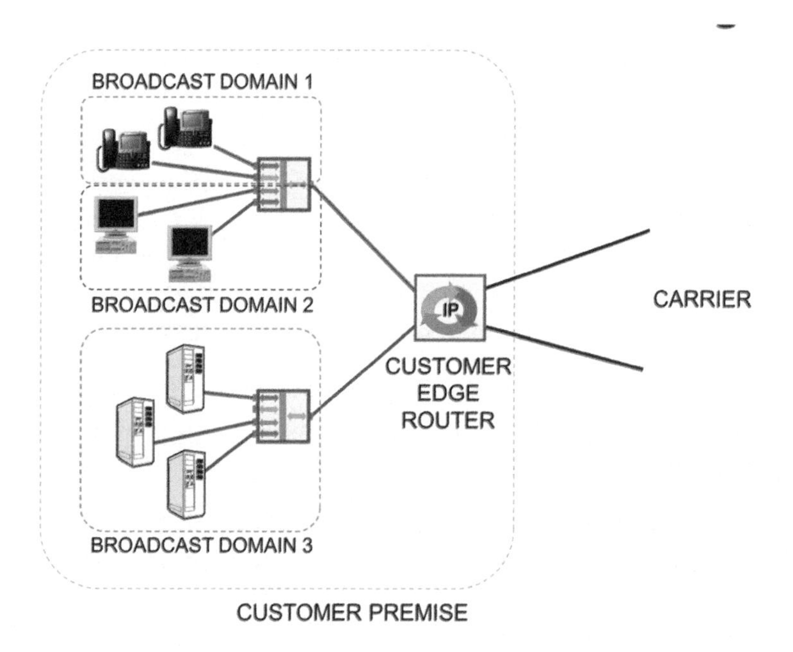

FIGURE 203 CUSTOMER EDGE ROUTER

16.7.1 Customer Edge Device

The router in Figure 203 is the connection between the LAN and the WAN. It is connected to in-building Layer 2 switches on one side, and to carrier circuits on the other side.

From the point of view of a carrier that might be providing the point-to-point links, it defines the edge of the customer's in-building network, and so is called the *Customer Edge (CE)* device by carriers.

This device has also in the past been called the *premise router*, the customer premise router, and is called the "default gateway" by Windows computers.

As illustrated in Chapter 12, the Customer Edge can be implemented as a $20 stand-alone device, or included in the same device that houses a DSL or Cable modem or fiber termination for home or small office use, or with industrial-strength versions from companies like Cisco for larger offices.

16.7.2 Router Connects Broadcast Domains

In the configuration illustrated in Figure 203, the CE router belongs to four broadcast domains: the two VLAN-defined broadcast domains on the upper LAN switch, the hardware-defined broadcast domain on the lower LAN switch, and the WAN circuit is a fourth broadcast domain.

Without a router, these four broadcast domains are like individual standalone circuits. The router implements the network by implementing the possibility of communications between the broadcast domains.

16.7.3 Routing

The router examines the destination address field in the Layer 3 header (Network Layer header) on a packet, and uses this value along with information in its routing table to determine where to forward the packet.

The routing table essentially lists ranges of addresses (subnets) against the address of a device that can relay a packet to any address in that subnet, and at what cost.

The "answer", result of the route calculation is the address of the *next hop*, in other words, the device to which the packet should be forwarded to get to the destination address.

The next hop address is resolved to a broadcast domain, then to a hardware interface, then the packet is physically transmitted in a MAC frame via pulses of voltage or light to represent the value of the bits.

16.7.4 Denying Communications

In addition to implementing the network by implementing the possibility of forwarding packets between broadcast domains, a router also acts as a point of control, *denying* communications.

This is part of basic network security. The objective is to compartmentalize the network, allowing communications only between machines and/or applications when there is a legitimate and desirable reason to do so.

In practice, this is implemented by denying all communications by default, then permitting communications between specified machines and/or between specific applications.

16.7.5 Packet Filtering

Permitting communications to specific machines is implemented with rules based on source and destination network addresses in the Layer 3 header, and is called packet forwarding. When denied, it is called packet filtering.

16.7.6 Port Filtering

Permitting communications to specific applications is implemented with rules based on the source and destination port number in the Layer 4 header.

The port number is essentially an identification of the computer program running on a machine. This is referred to as port forwarding and filtering.

16.7.7 Firewall

Packet or port filtering alone is **not** a firewall.

Packet or port filtering is a low-level traffic management tool that is the first stage in a firewall.

Stateful Packet Inspection (SPI) firewalls bring in the beginning of a message - contained in a number of packets - and examine the **content** of the packets to determine the application being carried in the packet, and apply permit / deny rules based on that.

A packet or port filter bases its permit/deny decision only on the address or port number on the packet. It does not look inside the packet to see what the content is.

Hence, a properly-configured packet or port filter restricts communication of packets to destinations that have a legitimate and desired use – but allows all communications, including attacks, to reach those destinations.

If the traffic does not come from a trusted source, it is necessary to examine the content of packets permitted through a packet filter to make a final permit or deny decision.

All firewalls built into Customer Edge devices have both the packet filter and SPI functions.

16.8 Public and Private IPv4 Addresses

16.8.1 Public Addresses

Generally speaking, to obtain an IP address that is valid on the public IP network (the Internet), it is necessary to rent it from an Internet Service Provider (ISP).

16.8.2 Regional Internet Registries

The ISP is either in turn renting addresses from an upstream ISP, or renting addresses from its Regional Internet Registry (RIR), which, in turn is allocated addresses by the top-level Internet Assigned Numbers Authority (IANA). The RIR that rents blocks of addresses to North Americans is the American Registry for Internet Numbers (ARIN).

Others include AFRIC (Africa), APNIC (Asia/Pacific), LACNIC (Latin and Central America), and RIPE (Europe).

FIGURE 204 INTERNET ADDRESS AUTHORITIES

There are no more public IPv4 addresses left to be allocated by IANA to the five Regional Internet Registries.

The Regional Internet Registries have some blocks of IPv4 addresses still available, but the supply is dwindling, and so the policies for being able to rent blocks of addresses are stringent.

An ISP has to already be efficiently using $2^{12} = 4096$ addresses rented from an upstream ISP before ARIN will consider allocating them their own block.

An end-user has to demonstrate a need for a block of addresses, such as multi-homing where they have more than one ISP for availability reasons, and must prove they will use a minimum block of 4096 addresses efficiently.

The cost for a block of 4096 addresses from ARIN is $2,250 per year, or about 50 cents per address per year.

ISPs resell these addresses as addresses bundled with end-user Internet access service. Web hosting providers resell these addresses as addresses bundled with a hosting plan.

Providers also rent additional static addresses at costs like $2 per month per address… a markup of 2400%, and a very lucrative business.

16.8.3 Unassigned or Private Addresses

However, the Internet Society didn't give all of the IP addresses away. RFC 1918, "Address Allocation for Private Internets" defines three contiguous blocks of IPv4 address space that are not used, and not valid, on the public IP network (the Internet).

10.0.0.0 - 10.255.255.255 (1 CLASS A)
172.16.0.0 - 172.31.255.255 (16 CLASS B)
192.168.0.0 - 192.168.255.255 (256 CLASS C)

FIGURE 205 IPV4 "PRIVATE" ADDRESS RANGES

These addresses are officially called *unassigned addresses* and usually referred to as *private IP addresses*. Sometimes they are called non-routable addresses, though this is not very accurate; routers can route them, just not on the Internet.

Using private addresses in-building allows the use of IP and all of its associated protocols and services for in-building communications without having to pay anyone for a block of rented addresses.

While it would be theoretically possible to use any IP addresses on a private network not connected to the Internet, it is recommended to use addresses in the ranges defined in RFC 1918.

Presumably, network equipment would be configured by default to know these addresses are not valid on the Internet, and so would be better suited to handle them if and when the private network is connected to the Internet.

But – it is necessary to have a legitimate public IP address to be able to receive anything from the Internet.

A popular solution is to use private addresses in-building, pay for one public IP address for external communications, and connect the two worlds with a Network Address Translator (NAT).

16.9 Network Address Translation

In the previous section, we covered private IP addresses, and why these were preferable to use on an in-building network. We also noted that if any of the users on the private network want to receive packets from the Internet, a public IP address is required.

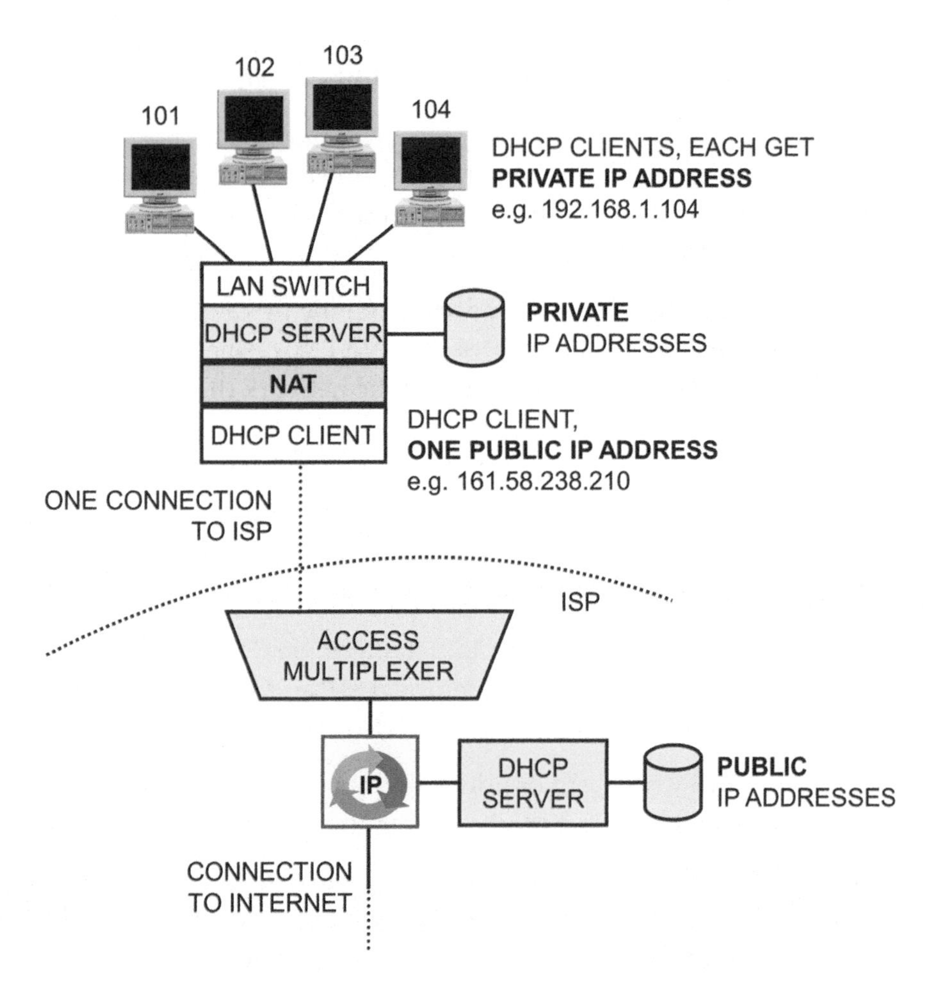

FIGURE 206 NETWORK ADDRESS TRANSLATION

To enable Internet communications for all users in-building without having to rent a public IP address for every user, a *Network Address Translator* (NAT) may be used.

16.9.1 Network Address Translator

A Network Address Translator is a software program running on the Customer Edge device.

It has a DHCP server, connected to the in-building network and configured to assign private addresses to the machines in-building, and a DHCP client, connected to the ISP, which obtains a public IP address from the ISP.

16.9.2 Outbound

When a computer on the private side initiates communications with a server, it populates the source IP address field in the packet header with its private address and the destination IP address field with the public IP address of the server.

The packet is then transmitted in a MAC frame to the computer's "default gateway", which is the Customer Edge router, where the NAT function is performed.

The NAT changes the source IP address from the private IP address of the sender to the public IP address of the NAT, i.e., the CE router, then transmits the packet in a frame on the public network (the Internet).

16.9.3 Inbound

The Internet server uses the source address in the packet it receives as the destination address to answer back to the client. Therefore, it will send the response back addressed to the NAT.

When the NAT receives the packet, it changes the destination IP address on the packet from the Internet to the private IP address of the appropriate computer, then transmits the packet in a MAC frame to the computer.

One question that arises is: how does the NAT know what computer on the private network a packet received from the Internet is intended for?

It turns out that the NAT uses the Layer 4 header to keep track of things. The Layer 4 header (TCP or UDP) begins with two octets that are called the "source port" then two octets for the "destination port".

These fields are used to indicate which application on a computer the message is being sent from and to.

The NAT selects an arbitrary "fake" port number to identify a computer on the private network, and records this port number against the private address in a table.

When a packet is transmitted to the Internet, the NAT records the actual source port number then changes the source port value to the "fake" port number.

When the reply from the server is received from the Internet, it has the "fake" port number in the destination port field of the Layer 4 header. The NAT uses this to look up the correct private IP address and correct port number and enter those values in the destination address and destination port number fields, thus relaying the incoming packet to the correct computer on the private network.

16.9.4 Advantages of NAT

NAT provides a number of advantages:

1. A NAT allows multiple computers in-building to share a single Internet address and Internet connection.

2. A NAT provide a truly "always-on" connection to the Internet. Services like DSL and Cable modem described as "always on" are always connected at the Physical Layer. They do not provide "always on" at the Network Layer, since DHCP must be run every time the attached device restarts to get a public IP address.

The NAT runs DHCP to get the public IP address; so, if the NAT is not powered off, the site will always have a public IP address assigned, and thus a connection to the Internet always ready for immediate use.

3. A NAT shields machines from attacks from the Internet. Since a private IP address is not reachable from the Internet, there is no way for a machine on the Internet to initiate communications to a machine on the private network. The only device exposed to the Internet is the NAT.

Normally, the NAT is not running on a computer running Windows, so attackers have a greatly diminished chance of finding a vulnerability to exploit compared to a computer running Windows.

16.9.5 Implementation

Most ISPs now provide the CE router with NAT function integrated in a device that includes the DSL, Cable, fiber or wireless modem.

16.10 TCP and UDP

Communicating over the Internet means your ISP giving your packet to another ISP, and **hoping** that ISP actually transmits it onward to another ISP, and so on, to eventually get your packet to the receiver's ISP.

There are no guarantees that a packet will be transmitted, when that might happen, nor how often that might happen. Packets may be corrupted, overwritten or discarded at intermediate nodes, and never delivered.

This is called an *unreliable network service.*

To make communication over an IP network reliable, users must run transport-layer protocols that communicate end to end.

The most popular is the Transmission Control Protocol (TCP), which performs retransmissions of missing and errored data during file transfers to assure integrity.

Another choice is the User Datagram Protocol (UDP), which is used for "best efforts" transmission of individual packets, and does not do re-transmits.

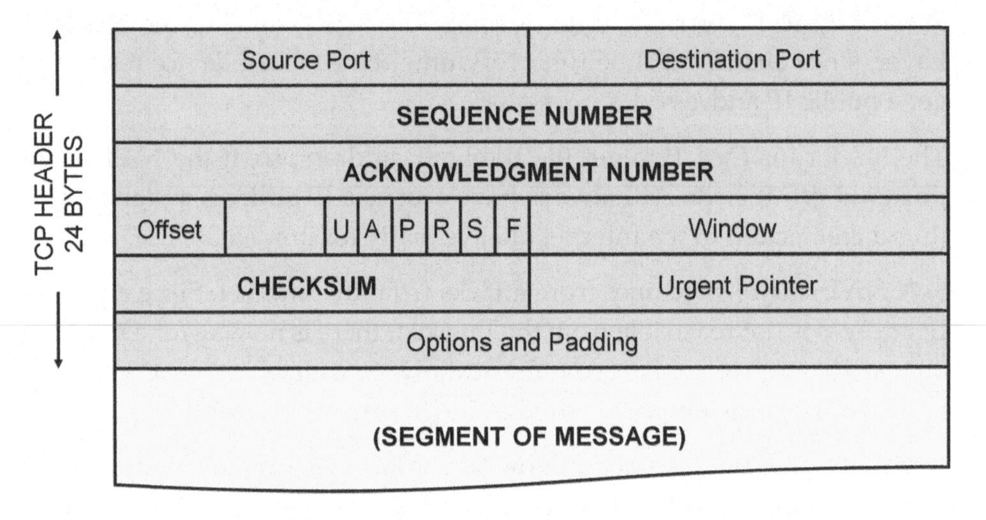

FIGURE 207 TCP PROTOCOL DATA UNIT

Before a segment of data is passed to IP, TCP adds a header to the segment with an error check and sequence number, and starts a timer at the sender.

The receiver's TCP checks the error check and sequence number. If the data is corrupted, the receiver discards the data, and after a period of time, the timer started by the sender's TCP will expire, and the sender will automatically retransmit the segment.

If the data is received without error, the receiver's TCP sends an acknowledgment back to the sender's TCP and the sender stops retransmitting.

TCP turns the underlying unreliable connectionless IP network into a reliable transport service for use by upper layers. TCP is for file transfers – when it absolutely, definitely has to get there, and if it doesn't, the missing piece(s) will be retransmitted.

This is not very useful for voice and video. During a live, streaming communication session, there is no time to retransmit missing pieces. For these applications, a different transport-layer protocol, the User Datagram Protocol (UDP), is employed.

UDP is similar to TCP, but only implements port numbers and error checking. UDP does not implement sequence numbers or retransmission of missing or errored data. It provides *best-efforts* transport.

16.11 IPv6

The main limitation of IPv4 was a shortage of network addresses. Though having a 32-bit address space, yielding 2^{32} or 4.3 billion addresses, the assignment of large blocks of addresses, particularly Class A, caused a rapid exhaustion of available addresses.

Additionally, mechanisms for security and traffic management were not provided or not well supported by IPv4, requiring the development of additional protocols, headers and overhead to perform these functions.

RFC 2460 Internet Protocol, Version 6 (IPv6) emerged from a pack of contenders to be the eventual replacement for IPv4. The improvements that IPv6 offers over IPv4 are expanded addressing capabilities, header simplification, improved support for extensions and options, support for traffic management and support for data integrity and data security.

16.11.1 Expanded Addressing Capabilities

The main improvement is expansion of the address field from 32 bits to 128 bits, expanding the available address space to $2^{128} = 3 \times 10^{38}$ addresses (340,282,366,920,938,463,463,374,607,431,768,211,456 to be exact) ... enough to allocate a block of 560 trillion trillion addresses to every person on earth.

16.11.2 Header Simplification

Some IPv4 header fields have been dropped or made optional, to reduce the bandwidth cost of the IP header, and to reduce the number of operations – and thus time – required to forward a packet.

16.11.3 Improved Support for Extensions and Options

A flexible mechanism for adding to the IP header with variable-length *extension headers* has been implemented. This allows optional implementation of error detection, source authentication and encryption as standardized services at the network layer that would be available to all applications.

16.11.4 Support for Traffic Management

Fields in the header allow identification of priority, and identification of a packet as belonging to a flow of packets, that is, a sequence of packets originating from the same source and going to the same destination and intended to receive the same forwarding treatment. Currently, this would be an MPLS label number. These capabilities can be used to implement traffic management and prioritization as Quality of Service mechanisms.

16.11.5 IPv6 Packet Format

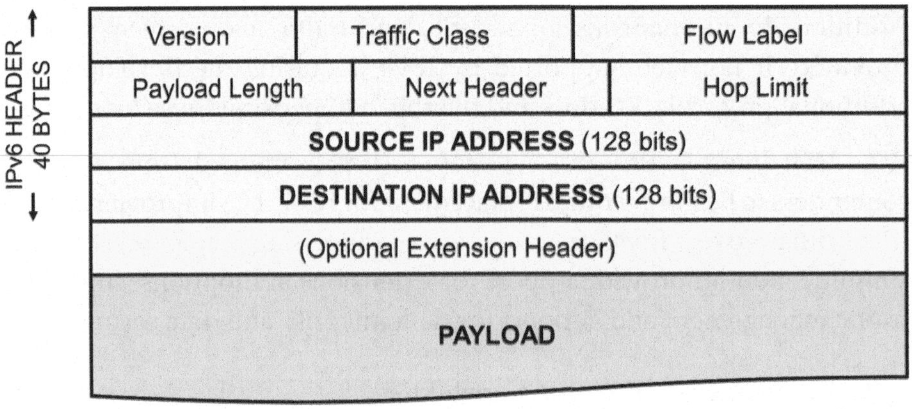

IPv6 HEADER 40 BYTES

Version | Traffic Class | Flow Label
Payload Length | Next Header | Hop Limit
SOURCE IP ADDRESS (128 bits)
DESTINATION IP ADDRESS (128 bits)
(Optional Extension Header)
PAYLOAD

FIGURE 208 IPV6 PACKET

The IPv6 header is 40 octets long, and includes the following fields:

Version field: 4 bits, indicating the version of IP. This would be "6".

Traffic Class: 8 bits, indicating a "priority" or precedence for this packet. This field can be populated by the originator of the packet, or by subsequent network equipment. This could be used to support differentiated Classes of Service for different applications.

Flow Label: 20 bits that can be used to identify the packet as belonging to a group or class of packets which should receive the same forwarding treatment on the network. This would typically be an MPLS label.

Payload Length: 16 bits, containing length of the payload immediately following this header, which includes any optional extension headers.

Next Header: 8 bits identifying the type of header following. In the simplest implementation, the IP packet will be encapsulating a transport layer protocol data unit, such as that output by TCP. In that case, the header immediately following the IP header would be the TCP header.

Hop Limit: 8 bits, populated by the source with a number between 1 and 255. This number is decremented by each device that forwards the packet. When it reaches zero, the packet is discarded. This prevents endless forwarding of packets in loops.

Source Address: 128 bits, identifying the originator of the packet.

Destination Address: 128 bits, usually identifying the final destination on the network. If a Routing extension header is present, the destination address field will contain the address of the next router through which the packet must travel.

The IPv6 header could also be followed by one or more IPv6 extension headers, which can include a Hop-by-Hop Options header, Routing header, Fragment(ation) header, Destination Options header, Authentication header and/or an Encapsulating Security Payload header.

16.12 IPv6 Address Allocation and Address Types

IPv6 addresses identify *interfaces*. An interface is typically an integrated circuit driving a wired or wireless LAN connection on a device.

The notation /n is used to mean the first n bits in the address.

16.12.1 Internet Registry Identification

The first 12 bits of the address identifies the Regional Internet Registry.

In North America, ARIN's policy is that the first 32 bits of the address identifies a block allocated to a Local Internet Registry, most of the time a big ISP.

16.12.2 Sites and Global Routing Prefix

The first 48 bits of the address is called the *Global Routing Prefix* and identifies a *site*. Most of the time, this will be an ISP's data center, though it might be a university campus or large organization's building.

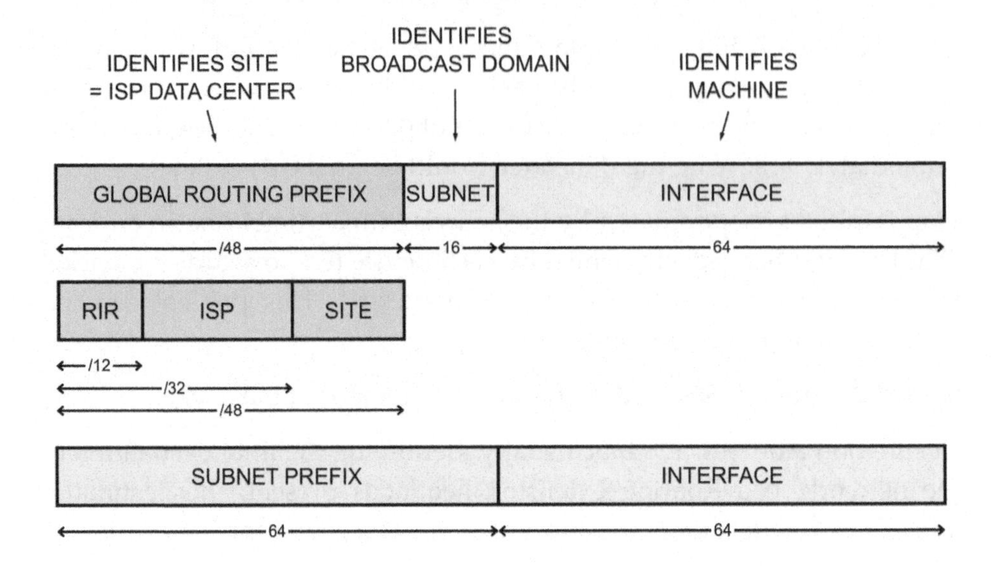

FIGURE 209 IP VERSION 6 ADDRESS STRUCTURE

16.12.3 Interface ID

The last 64 bits of the address is called the *Interface ID*, and could be the updated version of a MAC address called EUI-64, or a random number for privacy reasons.

The Interface ID identifies the integrated circuit running a LAN connection – wired or wireless – on a device. For consumer equipment with one LAN connection like a PC or smartphone, it effectively identifies the device.

16.12.4 Subnet ID

Between the 48-bit Global Routing Prefix, which essentially identifies buildings or campuses, and the 64-bit Interface ID, which identifies LAN connections is 16 bits called the *Subnet ID*.

The Subnet ID can be used to implement a hierarchy of addresses assigned to end-users and/or subnets at a particular end-user.

16.12.5 Allocation

Residential users generally do not have multiple subnets, so in the case of an ISP's site, this 16-bit field can be used to assign one subnet, that is, one /64 block to 65,536 customers per site.

In this case, all IP addresses at the residence (the end-site) would have the same first 64 bits, and the last 64 bits would be IDs of interfaces at the residence. Every light switch, light bulb, every electrical socket, both slots in your toaster ... everything will have an IP address in the future.

Large government and corporate end-sites would normally have more devices and multiple subnets (broadcast domains) to be compartmentalized for network security reasons, so they might be assigned multiple subnets, for example a /56 block from the ISP's site.

In this case, the end-user in the office building would employ the lower 8 bits of the subnet field to identify up to 256 subnets (broadcast domains) at their end-site.

On each of these, the first 64 bits of the IP address would be the same for all devices on the subnet (in the broadcast domain), and the last 64 bits are the Interface ID. The ISP could service up to 256 of this kind of customer from one /48 site block at the ISP's data center.

Customers of an ISP that have sites bigger than a large corporation or government building include... smaller ISPs. For this type of customer, a /48 block would allow the downstream ISP to resell the /64 block residential and /56 block corporate / government scenarios just described.

16.12.6 Subnet Prefix

The first 64 bits of the address, containing the Global Routing Prefix and the subnet ID are called the *subnet prefix,* which identifies a broadcast domain.

The second 64 bits of the address identify a device in the broadcast domain.

16.12.7 IPv6 Address Types

RFC4291 "IPv6 Addressing Architecture" is the authoritative reference for the discussion in this section.

Three main IPv6 address types are defined: *unicast, anycast and multicast.*

A unicast address identifies a single interface. A global unicast address is basically a valid Internet address – that may or may not be directly reachable from the Internet, for security reasons.

Both anycast and multicast addresses identify a set of interfaces. A packet addressed to an anycast address is delivered to the nearest interface in the set, while a packet addressed to a multicast address is delivered to all of the interfaces in the set.

The address 0 is called the *unspecified address,* used as the source address of an interface in the process of acquiring an address using DHCP, for example. The address 1 is called the loopback address, and is used by an interface to reference itself.

Addresses beginning with 1111:1101 (FD_H) are called *unique local addresses,* used in the same way as IPv4 private addresses. These addresses can be routed on a private network, but are not valid on the public Internet.

Addresses beginning with 1111:1110:10 ($FE8_H$ - FEF_H) are *link-local unicast addresses*. They end with a 64-bit interface ID, and are valid only on a single broadcast domain, for functions like neighbor discovery. Routers are not allowed to forward packets addressed to these addresses to a different broadcast domain.

Addresses beginning with 1111:1111 (FF_H) are multicast addresses. All other addresses are global unicast addresses, i.e., addresses for the public Internet. Anycast addresses are taken from unicast address space.

One other type of address worth noting is the IPv4-mapped IPv6 address. This begins with 80 zeros, then 16 ones ($FFFF_H$), followed by a 32-bit IPv4 address. This is a method for transition from IPv4 to IPv6 and may end up being the way that a "legacy" IPv4 addressing scheme is accommodated on an IPv6 network.

17 MPLS and Carrier Networks

17.1 Introduction

Packet networks incorporate two ideas: packet switching and bandwidth on demand. In this chapter, we examine how these principles are implemented by common carriers, i.e., organizations that build networks and carry many users' packets over common facilities.

17.1.1 Overbooking

To recap: *packet switching,* also called *packet forwarding* and *routing,* means relaying user data in packets from one circuit to a different circuit, or to be exactly precise, from one broadcast domain to a different broadcast domain. The network address is used to determine where to send the packet next at each intermediate router.

Bandwidth on demand means giving many devices access to a circuit and giving each the **possibility** of transmitting. If a device does not have anything to transmit, another device can use the available capacity.

This allows the implementation of *overbooking* or *oversubscription,* where the total of the incoming line speeds is greater than the outgoing line speed.

The appropriate level of overbooking can be calculated based on the historical demand statistics, how often the devices actually transmit data – regardless of what access line speed they have – and so overbooking is also called *statistical time-division multiplexing.*

17.1.2 Congestion, Contention and Packet Loss

When the demand exceeds the available capacity – more packets being sent in to a router than can be sent out – the network is said to experience congestion. At a router, packets are stored in temporary memory called buffers while waiting to be transmitted.

Under heavy load, the buffers can fill up. Then, in the case where a new packet arrives before the oldest one in the buffer can be transmitted, the new packet over-writes the oldest one in the router's buffer memory.

When a packet is over-written, it disappears. This is also called a dropped packet, non-delivered packet and *packet loss*.

For applications like email and web pages, packet loss is typically not a problem; the TCP software re-transmits the missing data in a new packet from the source. The user might only notice the page taking longer to load.

For live telephone calls and television programs carried in IP packets, there is no time to retransmit missing data, so packet non-delivery and excessive delay can result in poor voice quality on phone calls, and serious pixelation or block-averaging distortions on video.

17.1.3 Class of Service (CoS)

Performance is usually defined as packet delivery percentage, maximum delay and maximum variability in delay.

Specific performance thresholds are called *Classes of Service (CoS)*, and are part of the service contract between the carrier and their customer called a *Service Level Agreement (SLA)*.

One Class of Service could be defined for delay- and packet-loss-sensitive applications like voice and video, with guaranteed high packet delivery rate and low delay.

Another CoS could be defined for delay- and packet-loss-tolerant applications like web pages and email, with lower guaranteed packet delivery rate and longer delays.

If there is congestion at a router, packets with the higher CoS are transmitted at the expense of packets with a lower CoS that are delayed or dropped.

In addition to managing performance guarantees, traffic management is also required for network load balancing and recovery from equipment failure and cut lines.

17.2 Carrier Packet Network Basics

A carrier builds a packet network by obtaining expensive, high-capacity routers and placing them in buildings in different cities.

These buildings might be called switching centers, toll centers or Central Offices when the carrier is also the local telephone company; other times the buildings might be called POPs or data centers.

The carrier then connects these routers with point-to-point circuits at their own expense. This forms the carrier's network *core* or backbone.

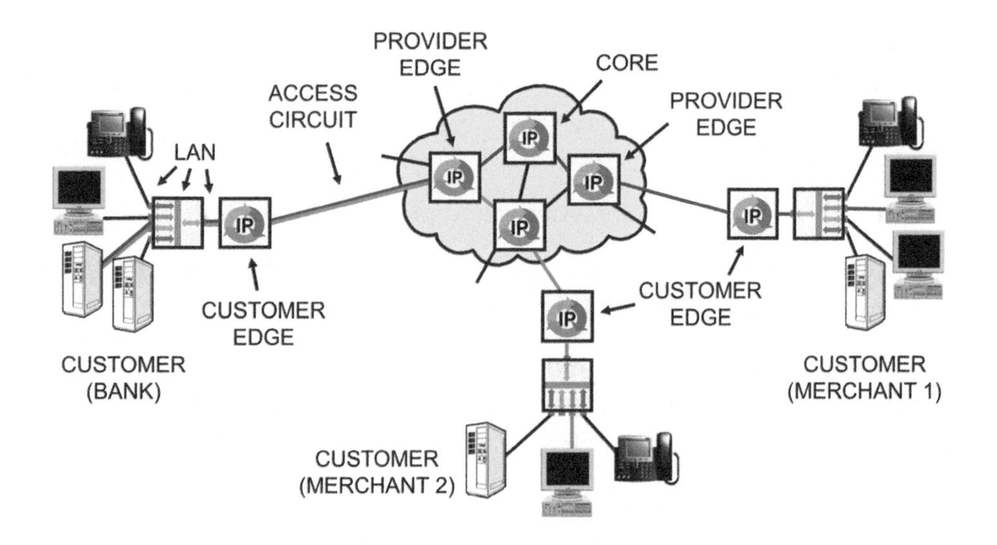

FIGURE 210 CARRIER PACKET NETWORK

The connections are implemented with capacity on a fiber, either owned by the carrier or leased from a third party.

Redundant connections are made to ensure high availability: a minimum of two connections are required at each location to protect against cut lines.

The cheapest way to implement two connections at each location is to connect neighbor-to-neighbor to form a ring. In practice, additional "shortcuts" will be implemented where traffic warrants.

17.2.1 Provider Edge (PE) and Customer Edge (CE)

This network core is front-ended with carrier edge equipment, often called the *Provider Edge (PE).*

The Provider Edge router is connected to the *Customer Edge (CE)* router with a physical dedicated access circuit.

The Customer Edge router is customer premise equipment situated between the customer circuits and the access circuit. It performs functions that can include acting as a point of control for traffic as part of network security, a data concentration function and conversions between frame formats, physical mediums and line speeds.

The Provider Edge is equipment owned by the carrier, performing similar functions: a point of control for network security, data concentration, media and format conversions. Some of the PE functions may be implemented in provider equipment deployed at the carrier premise. This is discussed in an upcoming section.

17.2.2 Access

The access circuit is generally a dedicated physical connection.

Current choices for access circuits include copper-wire DSL and cable modem technologies, Optical Ethernet and wireless.

In days past, copper-wire technology including 56 kb/s digital service and 1.5 Mb/s T1 were popular choices. Higher-end services might have used SONET on fiber at rates like 150 Mb/s (OC3) or 500 Mb/s (OC12).

17.2.3 Advantages of Packet Networks

There are a number of significant advantages to the use of packet network services from carriers instead of dedicated lines or circuit-switching.

First, there is no circuit set-up delay as with a dial-up modem; with a packet-switched service, the possibility of communicating to the network is maintained constantly over the access circuit, so communications can begin anytime without delay simply by transmitting a packet from the Customer Edge to the Provider Edge router. The packet is delivered to the far end CE sometime later.

Second, users can send packets addressed to many different destinations interspersed on a single access circuit, thus communicating 'simultaneously' to many destinations with only one access circuit at each location.

Since there is a monthly charge per access circuit or *port* in this business, this is a large cost advantage compared to dedicated or circuit-switched services, which require a separate access circuit for each simultaneous connection.

The third advantage is cost.

For all types of services, there is a flat rate per month for the access. For dedicated lines, there is an additional mileage charge per month. Circuit-switched phone calls are billed per minute.

Packet services are (in theory) billed per packet. Since the users normally will be doing nothing, it will be cheaper to pay per packet than pay all the time – which is a "dedicated line" actually is.

Whether or not a packet service is billed per packet, or flat rate, or flat rate up to a bandwidth cap then per packet for overage is a business decision on the part of the carrier.

In any case, the cost for packet network service is less than the cost of dedicated line or circuit-switched services, because the network circuits are overbooked and so the cost to the carrier is lower.

17.3 Service Level Agreements

A Service Level Agreement (SLA) is a contract between the customer and the service provider.

It is the core technical specification of the relationship between the customer and the service provider... specifying what the customer is going to get when transmitting packets over the carrier's network.

The service provider agrees to provide specified transmission characteristics, on condition that the customer stays within a specified traffic profile.

A particular set of guaranteed transmission characteristics is often referred to as a Class of Service (CoS). It would typically include specification of the minimum packet delivery rate, the maximum end-to-end delay and the maximum variability in delay.

17.3.1 Traffic Profile

A *traffic profile* is the specification of the bandwidth – the number of bits per second that the user will transmit over time. It is the "statistics" in statistical multiplexing.

The traffic profile would specify the maximum average bit rate measured over a specified period of time, the maximum peak or *burst* rate, how much data the user is allowed to transmit at the burst rate, and how often bursts are allowed to occur.

17.3.2 Contract

The contract is: as long as the customer traffic is less than the restrictions of the traffic profile, the service provider is obliged to provide the specified Class of Service, i.e., meet the specified transmission characteristics.

If the provider does not meet the specification, then in many cases, the customer can get a partial rebate of service cost for the month.

17.3.3 Business Decisions

Whether the service provider requires the customer to specify a traffic profile, and the details of traffic profiles are a business decision on the part of the carrier.

Some carriers, perhaps to distinguish themselves from the competition, do not enforce traffic profiles. Or perhaps more accurately, they only offer one traffic profile, which is "transmission at full access line speed 24/7".

If the customer has a physical access to the provider's network at 10 Mb/s, then they are allowed to transmit 10 Mb/s all the time – and still require the carrier to meet the Class of Service guarantees.

Other carriers only sell SLAs with traffic profiles; in other words, they will only guarantee a Class of Service if the customer guarantees they will restrict their traffic below set limits.

Historically, this has been the case because of limited resources – because the carrier network is heavily overbooked, and the carrier must restrict the incoming traffic to be able to meet the CoS guarantees.

Their network is in no way capable of supporting all customers transmitting at full line speed 24/7 at the same time. The carrier's network was purposely designed to support lower traffic profiles, to reduce the cost, giving the customers high apparent network speeds at a lower cost.

An easy example of this today is residential Internet access. To reduce the cost to customers, the access network in a neighborhood is heavily overbooked.

When a customer signs up for Internet access at 25 Mb/s for $60/month, they are not paying for a traffic profile that allows them to transmit and receive 25 Mb/s 24 hours per day, 7 days a week, 365 days per year.

They are paying for a residential user traffic profile, which is "receive at 25 Mb/s in short bursts once in a while".

Two-way 25 Mb/s Internet access guaranteed to work at full speed 24 hours per day, 7 days a week, 365 days per year is available – but that costs $500 per month not $60 per month.

For business customers, traffic profiles were required in the past for the same reason: the network was overbooked.

Today, if there is practically infinite bandwidth underlying all parts of the networks between business customer locations, then there is no technical need for traffic profiles.

17.3.4 Enforcement: Out of Profile Traffic

Traffic is metered, and traffic exceeding the agreed traffic profile is said to be *out of profile*.

The question is how to treat out of profile traffic. If the network is heavily overbooked, customers might need to be forced to conform to their agreed traffic profile if the network is going to meet the Class of Service standards for all customers... and not have to give refunds to everyone every month.

In this case, out-of-profile traffic might be:

- Assigned to a lower Class of Service, and so becoming more likely to be dropped by some equipment downstream,
- Temporarily delayed to bring short-term overages into compliance with the traffic profile (called *traffic shaping*), or
- Discarded at the input to the network (*traffic policing*).

There is a business reason why a carrier would continue requiring traffic profiles after the technical need for them has diminished: revenue.

Out-of-profile traffic of course could be, and is, billed as "overage". From a service provider business point of view, the service provider is keeping the cost of service offerings low for everyone, and charging extra to those that transmit more than everyone else (and making extra revenue).

From a customer point of view, this might be seen as a hidden cost that is only discovered after a contract is signed.

17.3.5 Abusive Applications

Residential customers that run the BitTorrent file "sharing" application transmit and receive more traffic than regular users, and can face traffic policing and/or charges for overage.

Some go as far as to claim that *net neutrality* is required, and in their case, they believe net neutrality means they should not be subject to traffic caps, traffic policing (throttling), or extra charges, instead pay the same as their neighbors – even though they are using far more bandwidth.

It must be noted that the activity these customers are usually undertaking is reception and distribution of intellectual property for which they have not paid the copyright holder.

From a technical point of view, these users are essentially asking their neighbors to subsidize what the copyright holder would likely consider to be criminal activity.

With the rise of streaming video: Netflix, YouTube and others, the impact of traffic generated by BitTorrent on network performance is decreasing, as streaming video requires as much or more bandwidth.

17.4 Provider Equipment at the Customer Premise

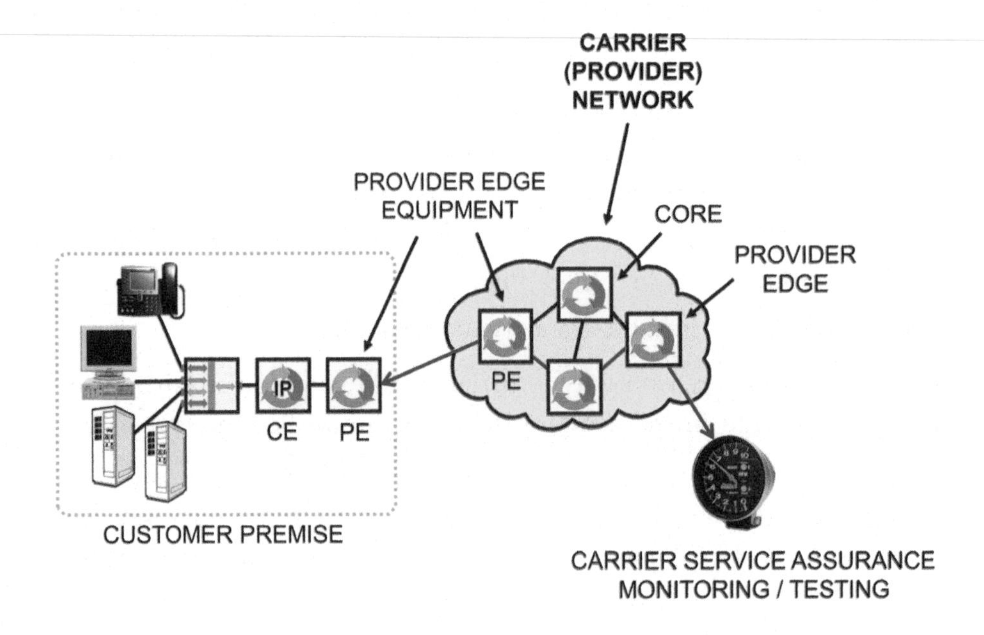

FIGURE 211 PROVIDER EQUIPMENT AT THE CUSTOMER PREMISE

The Provider Edge (PE) is equipment owned by the carrier, performing a number of functions: a point of control for network security, data concentration, media and format conversions.

As we will see in this chapter, one of the "format conversions" the PE may implement is affixing and removing virtual circuit IDs – called *labels* in MPLS. It turns out that the best place to perform this function is at the customer premise – so it is not unusual to see at least part of the Provider Edge function in equipment deployed at the customer premise.

The PE equipment at the customer premise can also be the traffic meter and traffic policing device described in the previous section.

This PE equipment at the customer premise is the *ingress device*, called a Label Edge Router in MPLS and the upcoming sections.

Having equipment at the customer premise also allows the service provider to use it as a remote test head, so the provider can perform service level assurance: monitoring, tests and troubleshooting between a centralized test system and the equipment at the customer premise.

17.5 Virtual Circuit Technologies

Traffic management on IP networks is implemented with *virtual circuits* internal to the network, invisible to users, following a protocol called Multi-Protocol Label Switching. MPLS is a tool enabling the centralized control of routing and prioritization of different kinds of network traffic such as telephone calls vs. file transfers.

The remainder of this chapter explains the basic principles of virtual circuits, then MPLS in particular and how MPLS is used for traffic management, VPNs, service integration and traffic aggregation.

17.5.1 IP Routing vs. Centralized Control

In an IP network, the route decision for a packet arriving at an IP router is calculated using a relatively complicated algorithm that takes into account the destination IP address, the cost of different routes to the destination and other factors.

This algorithm is run for each packet, on every router in the chain. This takes a relatively long time, increasing network delay.

More importantly, it makes practically impossible the *control* of routes, prioritization and resulting traffic characteristics since each router operates independently.

The idea behind virtual circuit technology is to not run the IP routing algorithm on each packet at every router, but instead defining *classes* of traffic, predetermining the end-to-end route for all traffic in a class, and programming the route for the class into the routers from a software application in a Network Operations Center.

17.5.2 Traffic Classes

A class of traffic has the same source, same destination and should experience the same transmission characteristics, such as maximum delay and loss. They are internal to a large network. An example would be "VoIP traffic NY - DC".

To establish communications, many traffic classes are defined, then a route is determined for each class by control software in the Network Operations Center (NOC).

17.5.3 Virtual Circuits

The route is generically called a *virtual circuit:* it is the path that all of the traffic belonging to the class will follow – if any such traffic is ever transmitted in the future.

A number is used to refer to both the traffic class and the route. This number is generically called a *traffic class number* and a *virtual circuit ID*. There is specific jargon for MPLS and predecessors ATM and Frame Relay.

To implement the virtual circuit in the network, the routing table in each router in the chain that makes up the virtual circuit is populated with an entry that specifies the next hop for the class. This completes the setup.

17.5.4 SVCs and PVCs

In some technologies, there are additional buzzwords describing two flavors of virtual circuits: Permanent Virtual Circuits (PVCs) and Switched Virtual Circuits (SVCs).

Switched Virtual Circuits are set up in a manner similar to making a phone call: your network equipment asks the network to establish a connection to some destination, the network sets it up without further human intervention, you communicate, then the connection is released when you have finished communicating.

The difference between an SVC and a phone call is that full-time capacity is not reserved in the network for an SVC... it is just a path, a route, a possibility. With a phone call, 64 kb/s are reserved in the network during your communication session, whether you're using them or not.

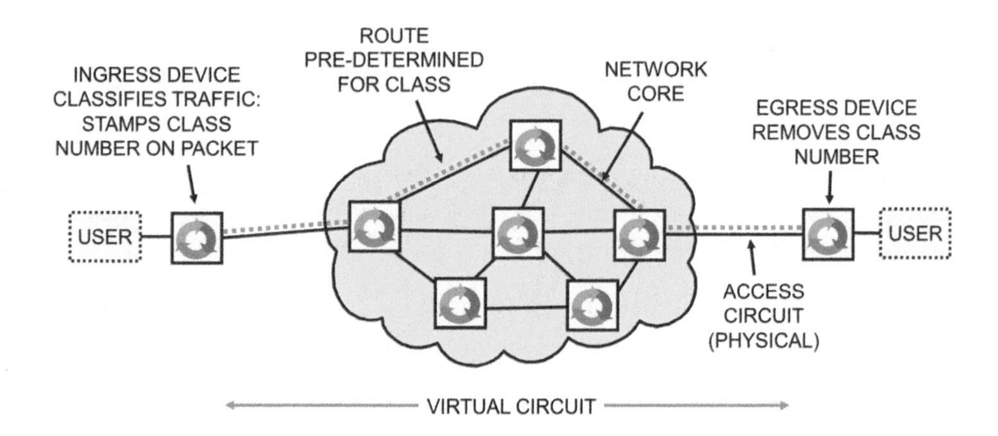

FIGURE 212 VIRTUAL CIRCUITS

Permanent Virtual Circuits are exactly the same as Switched Virtual Circuits, except that they are set up and never released.

The set-up process for virtual circuits is in many cases a manual operation, performed by a technician sitting at a control console that commands the network routing equipment.

For this reason, in practice all virtual circuits tend to be Permanent Virtual Circuits... set up and left set up. As long as the customer pays their bill.

17.5.5 Ingress Device: Packet Classification

Later, when an IP packet is actually presented to the network for transmission, a piece of equipment called the *ingress device* analyzes the traffic to determine what class it belongs to: it *classifies* the incoming packet.

Once it is decided, the ingress device stamps the class number in a field before the IP address in the packet header.

Subsequent routers in the network core use the class number to determine the next hop treatment, not the destination IP address.

In many cases, the ingress device is part of the Provider Edge (PE), meaning that the entire story of virtual circuits and traffic classes is internal to a carrier's network and invisible to the customer.

The best place to at least initially classify traffic is at the customer premise, as this allows use of local information not available in the carrier network, and allows aggregation of multiple traffic classes on a single access.

For this reason, the traffic classification function is often performed by provider (carrier) equipment located at the customer premise.

17.5.6 Forwarding Based on Class Number

Once the traffic is classified, the carrier network equipment does not use the IP packet address, but instead uses the class number to look up the next hop and possibly relative priority in its routing table.

This reduces the routing decision at each router from a complex algorithm to a simple table lookup, reducing the delay through each router and load on the router's processor.

More importantly, it provides a carrier with a mechanism for managing flows of packets end-to-end, implementing load balancing and swift service restoration after a fault, by managing the next hop entry for each class in each routing table from a centralized Network Operations Center (NOC).

17.5.7 Differentiated Services

Implementing multiple traffic classes that all go from the same place to the same place, but each associated with a different **priority** allows the implementation of *Differentiated Services*, i.e., multiple Classes of Service on the same IP network.

By classifying packets appropriately, packets carrying different content will experience different transmission characteristics; prioritizing loss- and delay-sensitive traffic like VoIP over loss- and delay-tolerant web pages.

An alternate method of implementing Differentiated Services is to have one field in the packet header for the virtual circuit ID (routing) and a second field in the header for the traffic class (priority). The IPv6 header has these two fields.

Differentiated services are covered in more detail in section 17.8.

17.6 MPLS

IP is firmly established as the standard protocol for networking… but in itself, does not have any way of implementing performance guarantees measured by characteristics like packet delivery rate and delay.

17.6.1 MPLS vs. TCP

TCP can deal with non-delivered packets, implementing communication between the source and destination for delivery confirmations and retransmission of non-delivered data.

But TCP only retransmits lost data; it does not influence the packet delivery rate or end-to-end delay, both of which are critical for telephone calls and live video over IP as well as business data services.

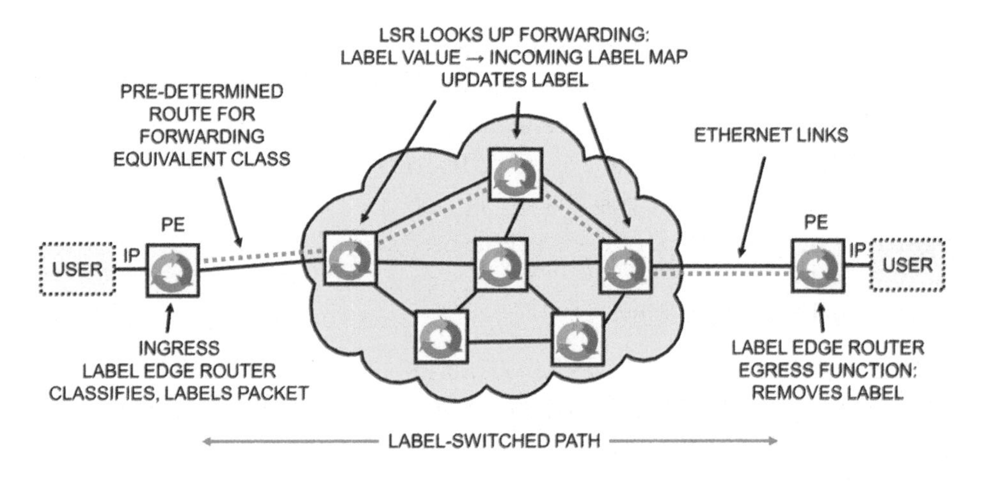

FIGURE 213 MPLS

To control packet delivery rate and delay, a traffic management system is required to manage and prioritize flows of IP packets.

Multi-Protocol Label Switching (MPLS) is used for this purpose, providing network operators with IP packet traffic management using virtual circuits.

MPLS concepts are the same as the general concepts of Section 17.5, and legacy technologies X.25, Frame Relay and ATM covered in Appendix D, but has its own particular jargon.

17.6.2 Forwarding Equivalence Class

For "traffic class", MPLS uses the term *Forwarding Equivalence Class* (FEC) to mean a group of packets that are forwarded over the same path with the same forwarding treatment.

17.6.3 Labels and Label Stacking

Instead of "virtual circuit ID", *labels* are used to identify a FEC.

In IPv4, the label is typically contained in an MPLS Shim Header, which is four bytes of extra overhead prepended to the IPv4 packet. Twenty bits are used for the label, three bits for experimental functions, one bit to indicate "last label" in a stack, and eight bits for time to live.

In IPv6, the label can reside in the "Flow Label" field in the packet header defined for this purpose.

A packet can have multiple labels, organized on a last-in, first-out basis, called the *label stack*. This allows a hierarchy of FECs, and aggregation of traffic by type (e.g., telephone calls, television, web pages, BitTorrent) so all of the instances of a single type of traffic can be managed as a single entity in the core. The processing is always based on the top label, regardless of whether any others might be "below" it.

17.6.4 Label-Switched Path

Instead of "virtual circuit", *Label Switched Path* is the term used to describe a sequence of routers that all work on a particular packet's label at the same depth in the label stack.

This is similar to the notion of the route associated with a virtual circuit. LSPs can be internal to a network. Many LSPs could end at the same egress router, to be forwarded on a common outgoing LSP… or vice-versa.

LSP route selection is the definition of the actual path through equipment and over circuits that an LSP will follow. In theory, this could be done on a hop-by-hop basis, with each LSR choosing the route to the next hop, much the same as IP routing.

In practice, particularly on carrier networks, the route for an LSP is defined by a system at a Network Operations Center (NOC), to facilitate network management functions like load balancing, route optimization and service restoration after a network fault.

17.6.5 IP User-Network Interface

MPLS is a traffic management system for IP packets implemented internally to the carrier network core, invisible to customers.

The interface between the user and the network is IP packets. MPLS labels are added to the packet by carrier equipment at the entry to the carrier network, and removed by carrier equipment at the exit from the carrier network, before the IP packet is delivered to the user. The user never sees MPLS labels.

17.6.6 Label Edge Routers

The ingress router, part of the Provider Edge function, as illustrated on the left of Figure 213 is called a *Label Edge Router (LER).* The LER analyzes an incoming IP packet, determines what Forwarding Equivalence Class it belongs to and labels the IP packet accordingly.

Since this analysis is only performed once at the ingress, the classification decision can take into account factors not available to IP routing, such as the source port or VLAN ID, and can be as complex as desired.

Subsequent devices – the LSRs – use only the label affixed by the LER to the IP packet for forwarding decisions, not the IP address.

At the destination side of the MPLS network, a Label Edge Router performs an egress router function, mainly removing the last label and its header from the packet before forwarding the packet to the user destination.

17.6.7 Label-Switching Router Operation

The routing devices internal to an MPLS network are *Label Switching Routers* (LSRs). These devices use the value of the topmost label on a packet to look up the forwarding and possibly prioritization instructions for the packet, then forward the packet.

Making the routing decision a table lookup rather than a complicated algorithm, minimizes delay through the LSR and facilitates control of routing via an external system populating the contents of the table.

In the LSR, the *Incoming Label Map* is the "lookup table", indexed by label number. The *Next Hop Label Forwarding Entry* is an entry in the Incoming Label Map that contains information on forwarding a labeled packet: the next hop, what operation to perform on the label stack, and can contain other information needed to properly forward the packet. There can be more than one entry for a given label value.

The essential function of an LSR is *label swapping*. The LSR examines the label at the top of the stack, and does a table lookup in the Incoming Label Map to get the Next Hop Label Forwarding Entry, then uses that information to encode a new label on the packet and forward it on the appropriate outgoing link with the appropriate relative priority.

The labeled packet can be forwarded to the next LSR or LER over a data link running any kind of layer 2 protocol, typically Ethernet.

17.7 MPLS VPN Service for Business Customers

The remainder of this chapter examines the practical uses to which MPLS is put by carriers.

One large part of carrier revenue is providing high-quality point-to-point communications between specific locations of a business, government or other organization. Large banks, to give one example, have budgets measured in the tens of millions of dollars per year for this type of service.

17.7.1 Private Network Service

In the 1970s, this type of service would be a *private network service,* consisting of multiple dedicated point-to-point lines between different locations of the bank. Later, the "dedicated lines" would be implemented as dedicated channels on a channelized TDM system.

In both cases, the bank's communications are private: other customers of the carrier could neither see the bank's traffic nor the bank's sites, nor can the bank communicate to other customers of the carrier over these services.

And in both cases, the carrier can sell a Service Level Agreement to the bank, guaranteeing transmission quality and service availability for a price.

17.7.2 Virtual Private Network (VPN)

A *Virtual Private Network (VPN)* means that the private network service is not, in fact, implemented with dedicated point-to-point connections – it just appears that way to the customer.

In reality, there are many users' IP packets interspersed on the circuits, hidden from each other using encryption and/or MPLS LSPs.

The term VPN is used to describe at least three different things in telecom. Two in current use are Internet VPNs and MPLS VPNs.

17.7.3 Internet VPNs

Internet VPNs (Section 5.9) are secure point-to-point communications across the Internet. What appears to the user to be a point-to-point dedicated line is implemented by encrypting packets on the sending computer, transmitting them over the Internet to a particular pre-selected and authenticated destination, then decrypting them at the far-end receiving computer.

The encryption and exchange of keys is specified in a set of protocols referred to as *IPsec.*

IPsec implements a secure VPN over the Internet – but there are guarantees of transmission quality since no single carrier controls all of the Internet circuits over which the packets travel end-to-end.

A popular application for Internet VPNs is working from home, accessing servers at work over the Internet.

17.7.4 MPLS VPN

MPLS VPNs are a different story. In the case of an MPLS VPN service provided by a carrier to a bank, the bank's traffic is not sent over the Internet between branches, it is sent over the circuits of the carrier that is selling the service to the bank.

That carrier **does** control all the circuits over which the bank's traffic travels end-to-end.

The carrier uses its MPLS traffic management system to define label-switched paths between the bank's buildings, and associate a Class of Service with each LSP. The LSP acts like a *tunnel,* carrying the customer traffic end-to-end.

This allows the marketing and sales departments to sell banks and government reliable IP packet communication services, backed up with the Service Level Agreement they require.

Multiple point-to-point IP packet communication paths connected with routers at each bank building effectively implements a private network: the bank can only communicate between the locations where the LSPs are set up, and the traffic moving over these LSPs is not visible to any other customers of the carrier. Plus, the carrier can guarantee transmission quality by prioritizing traffic on the LSP.

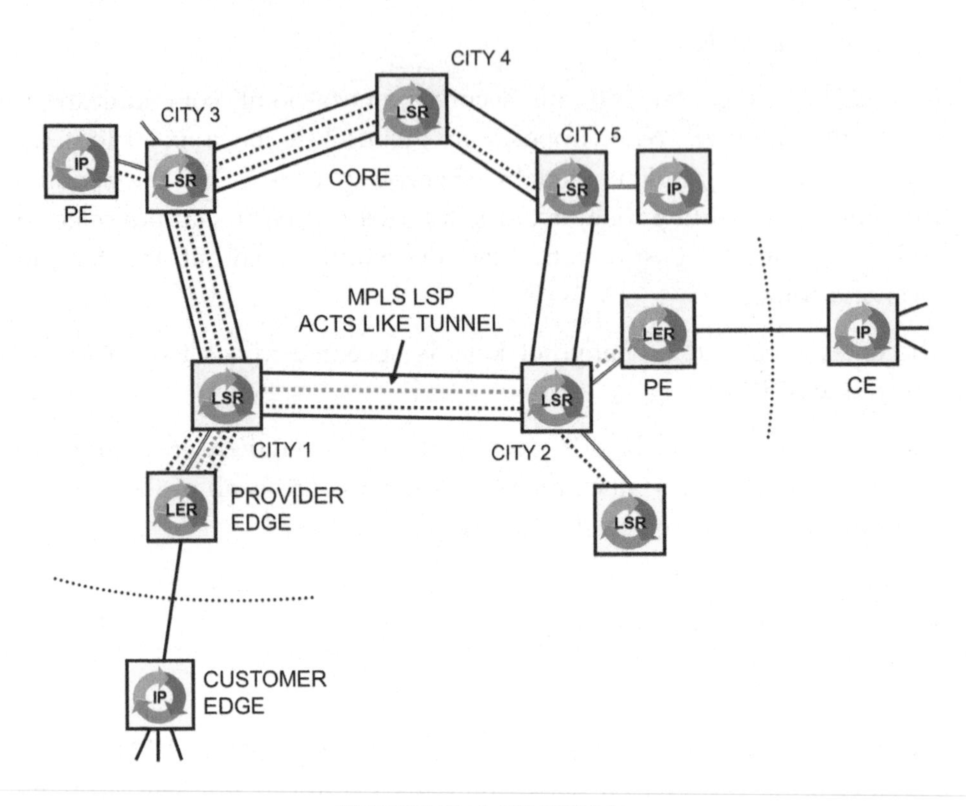

FIGURE 214 MPLS VPNS

Even though there are virtual point-to-point connections, the bank would encrypt their traffic before giving it to the carrier. The general rule in the security business is "if it is not encrypted, it has been released to the public".

To implement an MPLS VPN, a carrier defines MPLS Label-Switched Paths between customer locations in pairs. A Class of Service is associated with each LSP to implement performance guaranteed to the customer.

MPLS VPN service replaces legacy "business customer" data services like "dedicated T1s" and Frame Relay.

17.8 MPLS and Diff-Serv to Support Class of Service

Differentiated Services (DS) or *Diff-Serv* is an IP-based solution for prioritization, providing different transmission characteristics or Class of Service (CoS) for different types of traffic.

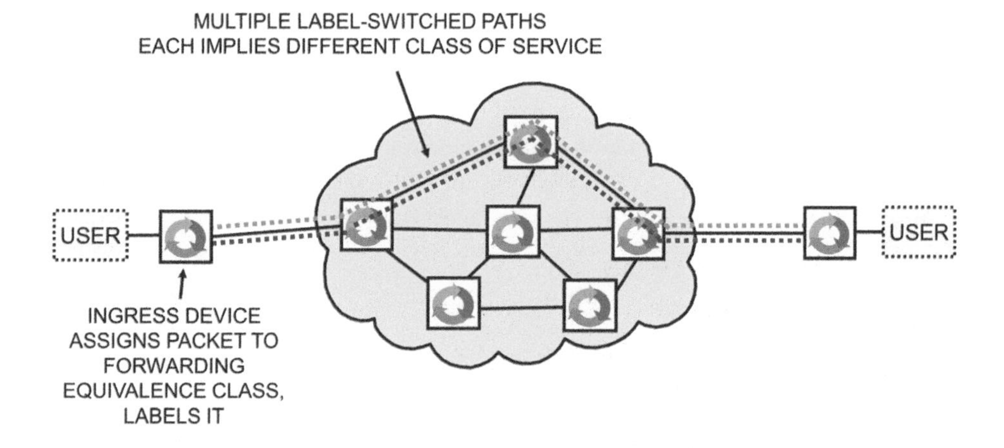

FIGURE 215 DIFFERENTIATED SERVICES

Diff-Serv provides a mechanism to classify packets as to the Class of Service (CoS) they should experience, specification of transmission characteristics like delay and packet loss, then give classified packets appropriate forwarding treatment in terms of prioritization at each hop in a DS-compliant network.

17.8.1 DS Codepoints

Packets are classified at the ingress or boundary of a network supporting DS, associating the packet with a *DS codepoint*, which is jargon to mean "Class of Service", and at the packet level, "relative priority".

In a DS router, each DS class has a Per-Hop Behavior (PHB), defining the forwarding behavior, i.e., transmission characteristics for that class.

> This only becomes meaningful when there is congestion: contention for available processing and transmission resources.

Applying PHB criteria to DS classes assigns relative priorities to packets passing through a DS router when contention occurs. The result is the ability to implement externally-observable CoS in terms of bandwidth, delay, jitter and dropped packets.

17.8.2 Assured Forwarding and Expedited Forwarding

RFC2597 Assured Forwarding and RFC2598 Expedited Forwarding contain suggestions for actual values for the DS codepoints, but appear to be largely academic exercises, defining dozens of Classes of Service.

In practice, a carrier might implement three priority levels:

1. Telephone calls
2. Television programs and
3. Internet traffic.

A more sophisticated implementation might have eight priority levels, and so eight Classes of Service, in order from highest to lowest:

1. Network control messages
2. Live telephone calls
3. Live streaming television programs
4. Live Internet web surfing
5. Video and music download
6. Email and other Internet traffic
7. Filler material like news headlines
8. Traffic from abusive applications like BitTorrent.

Since the classification is performed only at the input to a DS domain, the complex decision-making process – deciding what QoS a packet should receive – is performed once. Subsequently, each DS router has a simpler decision-making process, based on actual traffic and pre-assigned PHBs to determine relative priorities.

> This is very similar to MPLS labeling at the ingress to the network.

Considered separately, the MPLS label identifies the routing, and the DS codepoint identifies the priority for a packet. In this case, the 6-bit DS codepoint is populated in the Type of Service field in the IPv4 packet header, or in the Traffic Class field in the IPv6 packet header, and the network routers would process the label and codepoint separately.

These two ideas can be combined by implementing multiple LSPs, all going from the same place and to the same place, but each associated with a different DS priority level as illustrated in Figure 215. Then, the label on the packet is used by the router internal to the network to determine both the routing and prioritization of the packet.

17.9 MPLS for Integrated Access

MPLS labels and traffic classification can be used to combine all of the types of communications of a business or organization onto a single access circuit. This idea is sometimes called *convergence,* though *service integration* is a more accurate term. It results in a large cost savings compared to one access circuit for each type of communications.

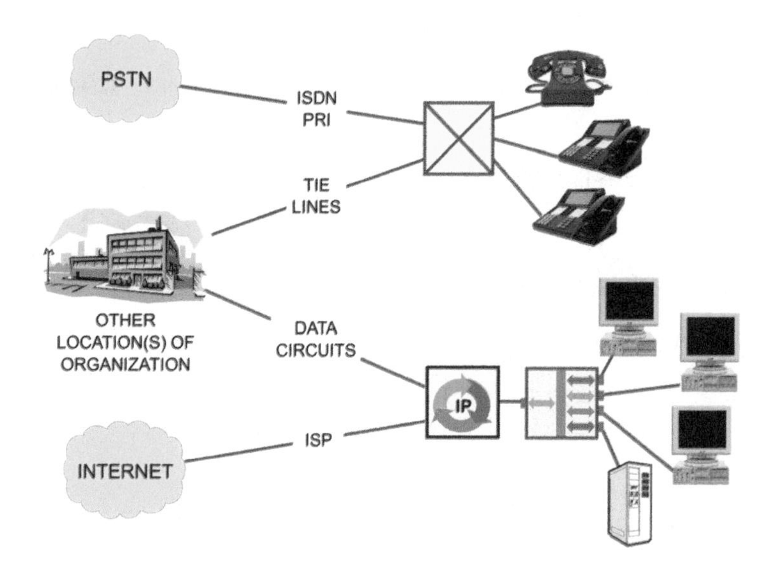

FIGURE 216 BEFORE: SEPARATE ACCESS CIRCUITS

At each location, a typical organization would communicate:

- Telephone calls to/from the PSTN,
- Telephone calls to/from other locations of the organization,
- Data to/from other locations of the organization, and
- Data, video and possibly voice to/from the Internet.

As illustrated in Figure 216, in days past, the organization would have had four physical access circuits and services – along with four bills:

- ISDN PRI over T1 to a LEC for telephone calls to/from the PSTN,
- Tie lines or a voice VPN with a custom dialing plan from an IXC for telephone calls to/from other locations of the organization,
- Dedicated T1s from an IXC for data to/from other locations of the organization, and
- DSL, Cable or T1 access from an ISP for data, video and possibly voice to/from the Internet.

17.9.1 SIP Trunking, VPN and Internet on One Access

As illustrated in Figure 217, moving to an all-IP environment, these four circuits can be replaced with one bill for one Optical Ethernet access circuit with three traffic classes, each identified with their own label number.

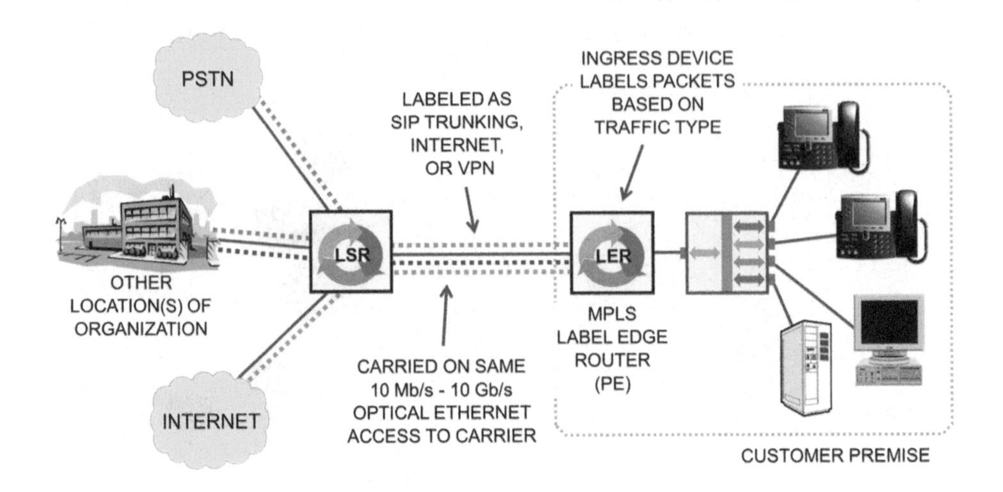

FIGURE 217 AFTER: INTEGRATED ACCESS - ONE ACCESS CIRCUIT, SEPARATE LABELS

The three traffic classes / labels would be:

- A traffic class for telephone calls. This might be called a "SIP trunking service" by the marketing department. This virtual circuit will carry VoIP phone calls to/from the carrier for communication either in native IP format to/from other locations, or conversion to/from traditional telephony for phone calls to/from the PSTN.
- A traffic class for data. This might be called a "VPN service" by the marketing department. This virtual circuit carries file transfers, client-server database communications and the like securely to/from other locations of the organization.
- A traffic class for Internet traffic. This virtual circuit carries anything in IP packets to/from the Internet.

All of this traffic is IP packets interspersed over the single access circuit. The way the traffic is distinguished is by classifying it on a piece of carrier equipment at the customer premise, traditionally called an Integrated Access Device (IAD), which in this case classifies the packet then stamps the appropriate label on each packet.

At the other end of the access circuit, the carrier uses the label to route the traffic onward and to prioritize it to assure the appropriate service level.

The result is all of the organization's traffic carried over a single access circuit, using a single technology. This is the Holy Grail of the telecommunications business, called *convergence* or *service integration*, having significant advantages in cost and flexibility.

17.10 MPLS for Traffic Aggregation

MPLS labels can be stacked. In other words, virtual circuits can be carried over other virtual circuits... or in MPLS lingo, LSPs can be carried over other LSPs.

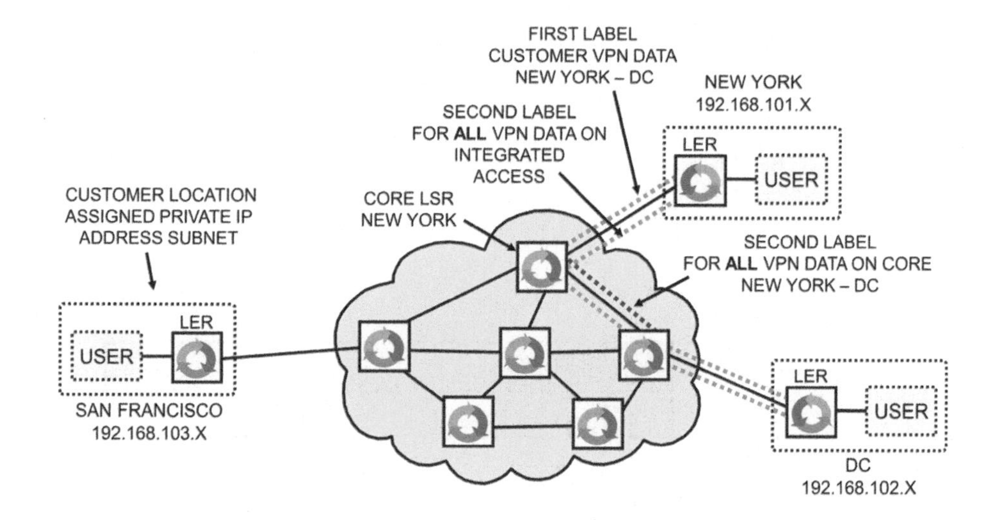

FIGURE 218 AGGREGATING VPNS FOR MANAGEMENT: LABEL STACKING

This is implemented to *aggregate* traffic so that the same kind of traffic can be managed as a single entity. This happens both on integrated access circuits and in the network core.

17.10.1 Label Stacking

Consider an example of a bank with an MPLS VPN for data between major offices in New York, DC and San Francisco. Each of these bank locations has a specific IP subnet, a unique block of IP addresses. To emphasize the fact that these communications do not go over the Internet, IP addresses in the private address space are used in the example.

For communication of data between the bank building in New York and the bank building in DC, a LSP will be established by the carrier between the two bank locations, and associated with a label number.

When the LER in New York, the ingress device, determines that an outbound IP packet contains data for the bank location in DC by looking at the destination IP address and finding it to be in the DC subnet, it will label the packet with the appropriate MPLS label for "Customer VPN Data New York-DC", the first label on the IP packet of Figure 219.

To transmit this over the integrated access, the LER will add a second label identifying it as belonging to traffic class "all VPN data on integrated access".

Other packets, like customer VPN data New York – San Francisco would also have the label "all VPN data on integrated access" added.

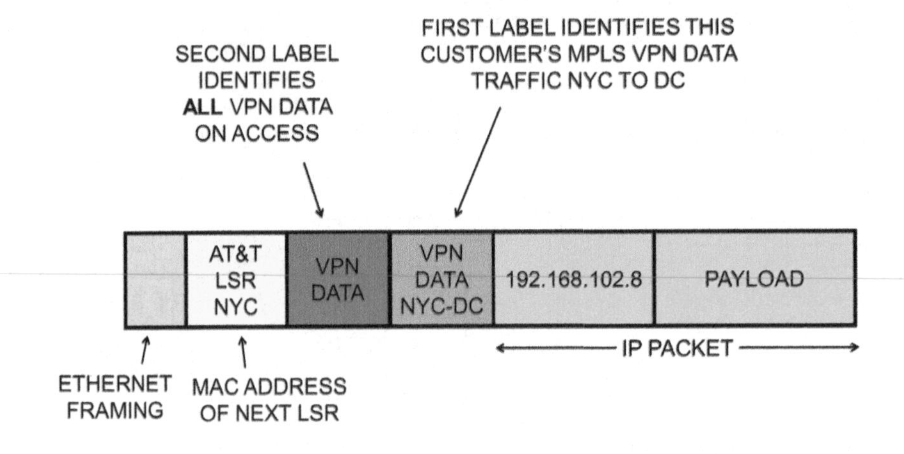

FIGURE 219 LABEL-STACKING PROTOCOL HEADERS

Then, by referring to the topmost (leftmost in Figure 219), second label, all VPN data packets on the integrated access can be managed by a control system as a single entity.

The packet at this point, with its destination IP address, the two labels, carried in a MAC frame is illustrated in Figure 219.

The carrier does the same thing on the core.

When the packet arrives at the network end of the customer's access circuit in New York, the second label saying "all VPN data on integrated access" is removed as it is no longer meaningful, and a new second label saying "all VPN data on core New York – DC" is added. Other packets, labeled as VPN data from other customers, would also have this second label added.

The result of all having the same second label is all VPN data traffic New York - DC on the core can be managed as a single entity: a single icon on a monitoring console at the NOC, configured as a single Class of Service and single route in MPLS LSRs by the carrier.

At the other end, the core LSR in DC would remove the second label and replace it with a new second label indicating data on the integrated customer access in DC.

The LER in DC, the egress device, would remove both labels and pass the packet to the customer edge router at the bank building in DC for forwarding to the machine with private IP address 192.168.102.8 in this example.

17.11 M is for Multiprotocol: Virtual Private LAN Service (VPLS)

The "M" in MPLS stands for "Multiprotocol".

In this chapter, we have referred to packets and the forwarding of labeled packets. While this is the most common use of MPLS, forwarding of labeled frames is also possible. This can be called a "layer 2 service", as opposed to MPLS VPNs, which are layer 3 services.

Carrier Virtual Private LAN Service (VPLS) moves MAC frames point-to-point using MPLS labels. From the customer point of view, the carrier appears to be a giant, nationwide LAN switch, moving MAC frames between customer locations.

Very large network operators order this service as part of a carrier redundancy strategy. When the customer pays different carriers to supply redundant connections between critical locations, the customer does not want any of the carriers involved in assigning blocks of IP addresses and routing IP packets, since the carriers support different blocks of addresses. The customer manages the IP addresses, and has the carriers provide Layer 2 and Layer 1 services.

This can be implemented with Ethernet over MPLS (EoMPLS), where the customer's MAC frame has MPLS labels pasted on the front of the frame.

LSRs do not look at anything except the label. Therefore, once a MAC frame has a label pasted on the front, it is a block of data treated the same way as a labeled packet by LSRs in the MPLS network... encapsulated in a MAC frame for forwarding over a physical circuit.

A more efficient implementation would insert the label value in the frame address field to avoid duplicate framing with a labeled MAC frame carried in a MAC frame.

That was the idea behind *Frame Relay*, an obsolete predecessor of MPLS, once a proud Section in this chapter, now relegated to Appendix D in case anyone needs to maintain a legacy system.

18 Wrapping Up

The final chapter brings together many of the concepts and technologies covered in this book with a top-down review and summary.

We begin with a discussion of technology project management and the very top of a "top-down review": requirements.

18.1 Technology Deployment Steps

When implementing new technology – and we always are – it is important to start the process at the beginning.

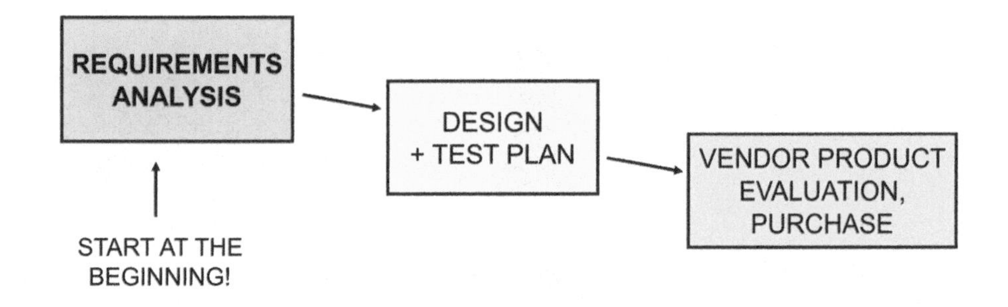

FIGURE 220 DEVELOPMENT PROCESS

First, specify the requirements, second come up with a design to meet those requirements – and a way of testing it, then third, choose vendors and their equipment to implement the design.

Unfortunately, human nature and budgeting processes being what they are, many organizations take the "backwards" path. Someone goes to a trade show in Las Vegas and two weeks later, the "solution" arrives on the loading dock, and the pointy-haired boss tells Dilbert to "plug it in".

The pointy-haired boss asked a salesperson to tell him what to do... and of course, the salesperson proposed that the pointy-haired boss buy his products.

Starting off first by buying into a product line – a particular "solution" or nifty technology – then second figuring out the design, then third realizing what your requirements really are is backwards – and guaranteed to waste a lot of time and money.

Hiring someone to enforce the requirements-before-design-before-buying-product direction will probably save far more money than their salary costs. These people are called *systems engineers* in very large technology projects, and *project managers* on smaller projects.

18.2 Requirements Specification

In a perfect world, before anything else can be done, the *requirements* for the eventual system must be explicitly identified, written down in a Requirements Specification document and signed off by the budget controllers and upper management.

FIGURE 221 START BY SPECIFYING WHAT NEEDS TO BE DONE

This is **what** the system has to do, not how it will be done. This is also called the *functional requirements.*

We must know how much money and time is available, as this places limits on what the requirements might include.

The first requirement is always connectivity: what needs to be connected and where is it located.

We must state **why** these things are going to be connected – and in a perfect world, prove with cost calculations why that is a good idea.

Identifying **who** should be connected to what and why is a necessary part of the security policy specification... as is specification of what information must be protected and to what degree.

Performance is an essential part of the requirements specification.

It is the *throughput* – the sustained end-to-end transfer rate – which is of most interest, as this encompasses the access line speed at each end plus the network connection.

It must be determined if Class of Service levels must be specified, different transmission characteristics for different types of traffic, whether a single CoS for all traffic is required, or if no CoS is required, saving money.

Service Level including *availability* requirements must also be specified. How often is it acceptable to have no connectivity, and for how long? Who is responsible for repair?

This requirement directly influences whether there is a requirement for redundant network connections and automated recovery from link congestion or failure. One might imagine the requirements for Goldman Sachs and for a gas station are different.

Security is a must. Firewalls to protect the computers and internal networks, and encryption to protect information must be implemented.

Network management is desirable for the organization implementing the network, being able to meter traffic and have alarms.

Specifying the support of open standards, for example SIP for VoIP call establishment, and SNMP for network management, instead of allowing manufacturers to use proprietary protocols should be investigated.

Allowing proprietary communications between devices locks the buyer into only using devices from that manufacturer going forward.

Flexibility is a requirement.

For the network, it must be possible to connect more locations without having loading failures.

For access circuits and services, it must be possible to increase to higher line speeds without making complete equipment changes.

Software upgradability of equipment to fix bugs and support new standards like IPv6 is an essential element of flexibility.

18.3 High-Level Design

Once the requirements document is sent out for review, modifications and finally signed by all of the people whose approval is necessary to proceed, the next step is to produce a high-level design.

The first step in designing a communication system is to identify the geographical locations to be connected.

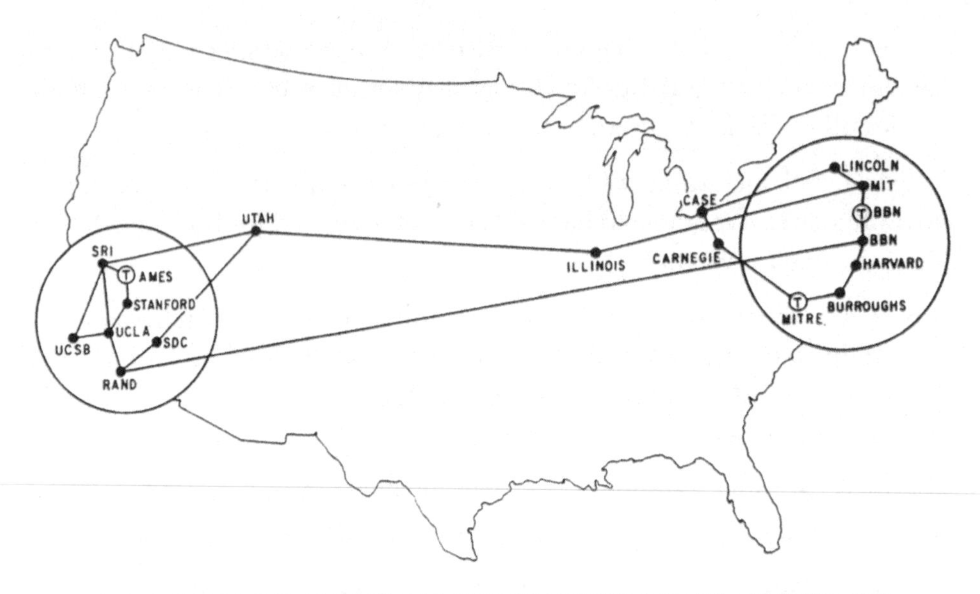

FIGURE 222 HIGH-LEVEL NETWORK DESIGN (EARLY INTERNET)

The next step is to identify the amount and type of traffic between the sites.

Then, implementing requirements for performance, availability and all the rest, a plan for connecting locations can be made.

Next, it is necessary to get costing information from service providers for the various types of services available between those locations.

This might include dark fiber between locations, Optical Ethernet, cable modem or even DSL access to MPLS VPN services, and cable modem, DSL or PON access to Internet service, along with choices for CoS if any.

With costing information for communications between those locations, and a characterization of the traffic between those locations, choices for WAN data circuits start to take shape.

At this point a design can be prepared, then vendors can be invited to submit proposals. Not the other way around.

18.4 Technology Roundup

Generic Name	Service Level	Line Speed	Physical Medium	Primary Use	Heyday
POTS	Voice grade	53 kb/s with modems	Copper	Voice	1874 →
Channel	DS0	64 kb/s	-	Anything	1958 →
56 kb/s digital	-	56 kb/s	Copper	Legacy data	1970s
ISDN BRI	2B+D	128 kb/s +	Copper	None	1980s
Fractional T1	n x DS0	64 kb/s to 1.5 Mb/s	Copper	Voice and data	1990
T1	DS1	1.5 Mb/s	Copper	Voice and data	1990
DSL	-	1 - 8 Mb/s	Copper	Internet	2000s
VDSL2	-	10 - 200 Mb/s	Copper	Internet and television	2010s
Cable Modem	-	10-1000 Mb/s	Copper	Internet, VoIP	2000s
Ethernet	-	10-1000 Mb/s	Copper	Local cabling	1980 →
Wi-Fi	-	1 - 500 Mb/s	Space	Local access	1990 →
Digital Cellular	-	9.6 kb/s to "330" Mb/s	Space	Voice and Internet	2000s
Fixed Wireless		25 Mb/s	Space	Internet	2019 →
T3	DS3	45 Mb/s	Copper	Old tx technology	1970s
Digital Radio	DS3	45 Mb/s	Space	Transmission	1980s
SONET	n x DS3	0.5 – 10 Gb/s	Fiber	Transmission	1985 →
Optical Ethernet	-	10 – 100 Gb/s	Fiber	Access and Transmission	2004 →

FIGURE 223 ACCESS AND TRANSMISSION TECHNOLOGY ROUNDUP

Figure 223 summarizes the principal technologies for physical circuit. The first column is the generic name for the technology in the telecom business.

The second column is whether there are special codes or abbreviations used to refer to the service level. This is associated with the legacy channelized digital hierarchy.

The third column lists how many bits per second the service provides. The fourth is what the technology is actually used for, and the last column is when people thought it was better than sliced bread and ice cubes.

Not included in Figure 223 is network technologies like IP and MPLS, and applications like web browsing. The content of this chart is mostly layer 1 and layer 2 technologies.

18.5 Review: Circuits and Services

It turns out that a carrier service has four components: a short access circuit between the customer and the network, a method of connection through the carrier's network, a billing plan and a service level agreement.

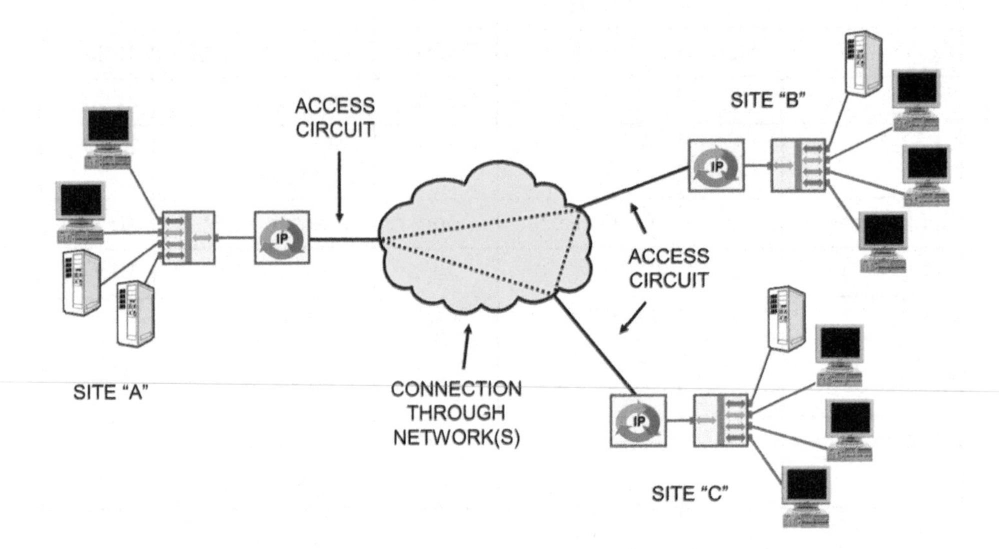

FIGURE 224 COMPONENTS OF A CARRIER SERVICE

To recap, there are many different technologies for access circuits, including:

- Plain Ordinary Telephone Service (POTS) lines
- Old data circuits at up to 56 kb/s,
- ISDN BRI at 128 kb/s,
- xDSL technology at 1 – 200 Mb/s,
- Cable modem technology at 1 – 500 Mb/s or more,
- Passive Optical Networks at 1 Gb/s or more,
- T1 digital access circuits at 1.5 Mb/s,

- Cellular and point-to-point radio,
- SONET fiber-based circuits at multiples of 45 Mb/s,
- Legacy proprietary fiber access technologies, and
- Optical Ethernet from 1 to 100 Gb/s.

Each type of access circuit must have a specific type of circuit-terminating equipment terminating the line at the customer premise to be able to transmit and receive on that circuit.

Some examples covered in this book were:

- Small Formfactor Pluggable (SFP) optical transceivers for fiber,
- LAN Network Interfaces: copper, fiber and wireless implementations,
- Modems for wireless, DSL, cable modem and POTS,
- Data Service Units (DSUs), for old 56 kb/s non-switched digital circuits,
- Channel Service Units (CSUs), used on T1 circuits,
- CSU/DSUs, used on switched-56 kb/s circuits,
- Optical Network Units (ONUs), Optical Network Terminals (ONTs), and Optical Line Terminals (OLTs) used on fiber circuits.

Network connections are made by equipment and over high-capacity circuits owned and managed by the network service provider.

Many options for connection through their networks are offered. In this book we summarized them into three fundamental choices:

- Full period: connected all the time, billed as a monthly fixed charge.
- Circuit-switched: connected on demand, billed as a monthly fixed charge for the access circuit plus a per minute usage charge.
- Bandwidth on Demand or "packet-switched": available all the time, billed as a monthly fixed charge for the access circuit plus possibly a charge based on the amount of data transmitted.

There is invariably a monthly charge for the access at each end.

The Service Level Agreement specifies the performance, cost, availability and other legal and technical terms.

A combination of access circuits with their circuit-terminating equipment, method of connection, billing plan and service level agreement make up a service.

18.6 Private Network

We finish off with a 50,000-foot view of network designs.

A *private network* solution, which consists of dedicated lines connecting locations together, could be investigated. Dedicated lines could be point-to-point fibers, or a full-period service provided by a carrier.

Simply connecting all locations together with dedicated lines in a *fully-meshed network* will usually be too expensive, so a hierarchical design is usually investigated.

Figure 225 shows a *hubbed private network* where four main locations are designated as regional centers. This architecture is also called a hub-and-spoke design.

Remote sites are connected to the nearest regional center. These connections are the spokes. The regional centers, the hubs, are connected with higher-capacity dedicated lines, perhaps using leased dark fiber.

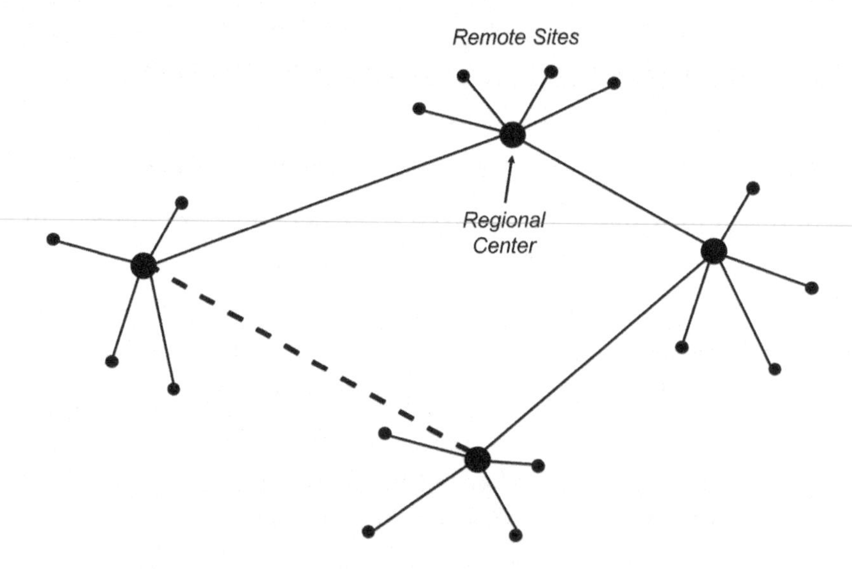

FIGURE 225 HUB-AND-SPOKE PRIVATE NETWORK

The dashed line indicates an optional connection between regional centers. This connection would provide redundancy in case of failure of one of the circuits between regional centers. With the centers connected in a *ring* as shown, traffic could be re-routed in different directions around the ring to continue communicating even if one connection is broken.

The principal advantages of a private network are very high quality of network service (full-time capacity), guaranteed performance, high security, and the fact that full-period services are available anywhere there is a telephone or cable company.

Whether a private network solution is more or less expensive than other choices depends on pricing and availability of other services such as Internet service or MPLS VPN service.

Aside from cost issues, the main disadvantage of a private network is lack of flexibility. Dedicated lines only go from point to point, so the user is responsible for ensuring that as new locations are added, connections to all points on the network can be achieved. It might take six weeks to successfully complete the addition of a new location.

A private network solution is indicated for high-security, high-capacity and high-availability requirements like those of a network connecting a bank's stock traders, their data centers, operations center and hot site backup.

Military organizations build private networks for control and availability reasons.

Google builds private networks to distribute YouTube videos between their data centers and ISPs for cost reasons.

Smaller organizations usually find that a design using carrier IP network services at each location is less expensive and more flexible.

18.7 Carrier IP Services

18.7.1 Six Main Flavors

Carriers provide six main flavors of IP packet communication services:

- Retail Internet service
- SD-WAN Internet Service
- Wholesale Internet service
- Wireless service
- MPLS VPN service, and
- SIP trunking service.

All of these services end up moving IP packets from source to destination, each with different characteristics.

18.7.2 Retail Internet Service

Internet service is the new "dial tone". In fact, the Internet and telephone network will be the same thing in the future.

An Internet connection allows the user to transmit packets to any other device on the Internet... just as POTS allows a call to any other telephone on the PSTN.

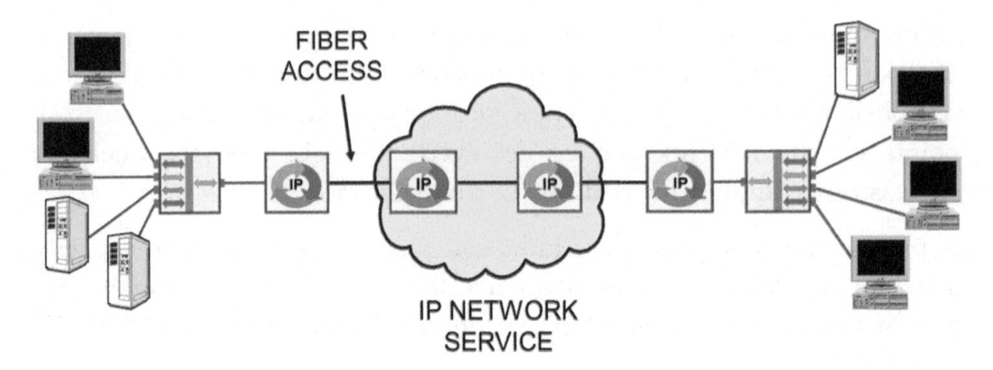

FIGURE 226 CARRIER IP SERVICES

However, unlike POTS, Internet service does not come with any guarantees. It will probably work, but anyone who has ever had a ridiculously slow Wi-Fi Internet connection in a crowded area could attest "sometimes not".

Internet service also does not come with security. One method of implementing secure connections over the Internet is IP-VPN technology, which uses encryption to provide point-to-point information security.

18.7.3 SD-WAN Internet Service

IP VPN point-to-point connections over the Internet implemented using encryption are called *tunnels*. Multiple tunnels, implemented with customer-premise equipment that can be remotely configured is called a Software-Defined Wide Area Network (SD-WAN).

Tunnels would be established pairwise between all of an organization's locations, each location having its own subnet, and routing tables matching subnets to tunnels.

There is usually no performance guarantee, as communication is over the Internet, where it is unpredictable what circuits the data will traverse. It will probably work fine… but there is no guarantee.

A guarantee requires the telephone company do it on their non-Internet circuits, called an MPLS VPN (Section 18.7.6).

18.7.4 Wholesale Internet Service: Transit and Peering

All carriers connect with other carriers to exchange Internet traffic. These services are measured and billed in the Gb/s range and are typically month-to-month or yearly contracts for a better price.

Peering is exchanging Internet IP packets with another carrier typically at similar rates in both directions, and not exchanging money. This is also known as *settlement-free.*

Selling transit services means carrying Internet packets for the purchaser of the service, relaying them between the purchaser's premise and other carriers, for eventual relay to the destination IP address. Unlike a regular Internet connection, the purchaser gets volume discounts and an end-to-end packet delivery rate guarantee.

Buying transit services means buying a very high-capacity, expensive and guaranteed Internet connection. Smaller ISPs and content providers are two examples of purchasers of transit services.

18.7.5 Wireless IP Services

Wireless IP services include both Internet service and telephone service. Starting with LTE, cellular is Voice over IP over the airlink and network. When making a connection to a PSTN address, this is connected and billed as "voice minutes" on a cellphone plan.

Cellular service also includes Internet IP packet service, billed as a "data plan". Carriers may also provide quality Wi-Fi Internet access in cities.

18.7.6 MPLS VPN Service

MPLS Virtual Private Network services are point-to-point IP packet communication services for business and government.

The IP packets can only be communicated between pre-defined locations, but the service quality, connectivity and availability are guaranteed – unlike Internet service.

The guarantee is not free. MPLS VPN Service is more expensive than Internet service.

18.7.7 SIP Trunking Service

SIP trunking replaces PBX trunks. SIP trunking services provide organizations with an in-building VoIP system connection to other organization locations and to the PSTN.

This can be configured very much like an MPLS VPN for data – with the SIP trunking service having Class of Service suitable for phone calls.

The carrier providing the SIP trunking service would carry IP packets containing voice between the organization's locations.

The carrier would also provide a gateway service to connect phone calls to POTS lines, to people with Voice over IP over Cable, and to cell phones, performing the function of an Inter-Exchange Carrier.

18.7.8 Physically Connecting

To connect to any of these carrier IP services the customer does not require special equipment, personnel or circuit types. They only need a Customer Edge device, which costs as low as $20. Most carriers give them away for residential and small business services.

For large business services, the carrier network appears as an Ethernet jack on the carrier's circuit-terminating equipment at the customer premise.

Paying for network service, proper configuration of the CE router, IP addresses and "default gateway" on the users' PCs and phones implements the "dial tone": the IP network service.

The user prepares IP packets and transmits them to the carrier, for onward forwarding to the far-end customer location.

18.7.9 Advantages

The main advantage of using carrier IP network services over private lines is the ability to communicate between all of your organization's locations with only **one access circuit** at each location.

Plus, all services – Internet, VPN, SIP trunking – provided by the same carrier can be integrated on the single access.

18.7.10 Fiber Access

Going forward, the access is fiber.

For business and government, 10 Gb/s Optical Ethernet (10,000,000,000 bits per second) will become normal "high capacity" access, just as T1 (1,500,000 bits per second) was called "hi-cap" in the 1980s and 90s.

Fiber access circuits are routinely implemented with Optical Ethernet for large customers and Passive Optical Networks in greenfields. Existing neighborhoods are being retrofitted with PON fiber at a steady rate.

Wireless and cable modem service also provide broadband last-mile access to a fiber terminal.

In the future, high-capacity or *broadband* will mean Gigabits per second then Terabits per second. The need for speed is never-ending.

18.8 The Future

We conclude with a peek at the Future.

Over the past fifty years, several attempts have been made to develop *converged networks*: networks with "dial tone" that supports all communications: speech, music, text, graphics, images and video.

For a number of reasons, convergence strategies employing ISDN and ATM were unsuccessful, did not gain critical mass and died. The third time was the charm, and packet-switched IP network service succeeded.

Beginning with grunts and gestures, progressing to verbal messages, signal fires, written messages, postal services, telegraphy, radio, telephone, television and computer networks, each advance in technology has meant communication of more information, faster and to more people – but often with the need to adapt the information to suit the communication technology being used.

Realization of a converged network is another step in the evolution of the way humans communicate, in that communication mechanisms best suited to the information will be used, rather than the other way around.

18.8.1 The IP-PSTN

The abbreviation "PSTN" has been a familiar part of the telecom lexicon for 100 years. Now its meaning is going to change.

For the past hundred years, PSTN meant *Public Switched Telephone Network:*

- Public: Accessible to anyone who pays.
- Switched: Circuit-switched service, a trunk reserved and switched onto the access lines at each end, for the duration of the call, then released.
- Telephone: Speaking at a distance.
- Network: Many interconnected nodes.

Now, "PSTN" will mean *Packet-Switched Telecommunications Network*:

- Packet-Switched: Communications segmented and encapsulated in packets that are routed using IP and/or switched using MPLS.
- Telecommunications: Transferring **any** kind of information across distance.
- Network: Many interconnected nodes.

It may be necessary to use *IP-PSTN* for a transitional period, until circuit-switching falls into the same dustbin of history as step-by-step switches.

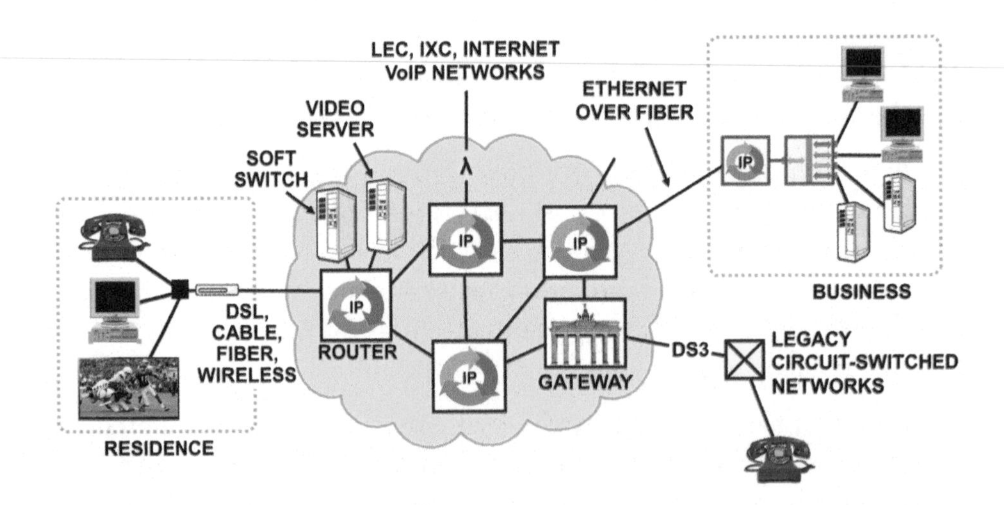

FIGURE 227 THE FUTURE

18.8.2 IP Dial Tone

This new PSTN will provide broadband *IP dial tone*: the possibility of communicating IP packets to any point on the network, at a bit rate high enough to watch guilty dog videos in HD.

"Businesses" (including government, educational institutions and the like) will use Ethernet over fiber at Gigabit speeds for access. New residences will have Optical Ethernet PON service.

Brownfields will use modems over copper wires – either DSL or cable until the outside plant is upgraded to fiber.

Wireless carriers also provide IP dial tone, in fact the majority of service in the developing world.

18.8.3 Services

Many different kinds of value-added services will be available in addition to basic IP packet communication service, some of them provided by the same company that provides the "dial tone", others provided by third parties.

The lowest level of value-added services includes service level agreements and network and information security services.

An unending list of higher-level services will be available for use on the network; some free, many not. Telephone, messaging, television, web surfing, social, radio, skyping, desktop sharing, and integrated email/voicemail are easy examples.

Internet service will become simply a few ancillary services like DNS; the capability to communicate with web servers is part of the dial tone. Telephone service will become access to a SIP server. The capability to communicate with other VoIP phones is part of the dial tone.

Many consumers will continue to pay subscription fees or pay per view to watch movies and shows. Many others will watch the shows with a slight delay while pirated copies become available, and not pay.

18.8.4 Sea Change

The change to an all-IP network supporting converged communications and broadband access is the biggest change in telecommunications since the invention of the telephone switch, and it will have a profound impact on the way people communicate.

Appendix A Modulation Techniques

A.1 Modulation of Carrier Frequencies

Pulses do not work on circuits with pass-band channels that do not support zero Hertz, like radio, CATV and POTS. A design that will work on a pass-band channel is one that employs tones or *carrier frequencies* within the pass-band.

Signals that make it through the pass-band are created by varying the *amplitude* (level or volume) of a carrier within the supported frequency pass-band, the *frequency* of such a carrier, its *phase*, or combinations thereof.

The technique of representing binary digits by varying one or more characteristics of one or more tones is called *modulation*, and the circuit-terminating equipment that performs this function is called a *modem*, a contraction of modulator and de-modulator.

The fundamental concepts are identical for DSL, cable modems, and all kinds of wireless modems including cellular and Wi-Fi... only the frequency pass-band changes.

A.2 Amplitude Shift Keying (ASK)

Amplitude Shift Keying (ASK) is the simplest technique for representing binary digits using a carrier frequency in a pass band. A single carrier frequency is defined, and one volume or amplitude is used to represent a "1" and another amplitude is used to represent a "0".

These two states of the carrier are the *signals* that will be transmitted. Signals are also referred to as *symbols*.

Note that this is not continuous *amplitude modulation* (AM) as is used on radio stations. At the radio station, a carrier frequency of something like 800 kHz is selected, and the amplitude at that frequency is varied continuously to represent the sound pressure waves coming out of the disc jockey's throat. ASK uses only two amplitudes.

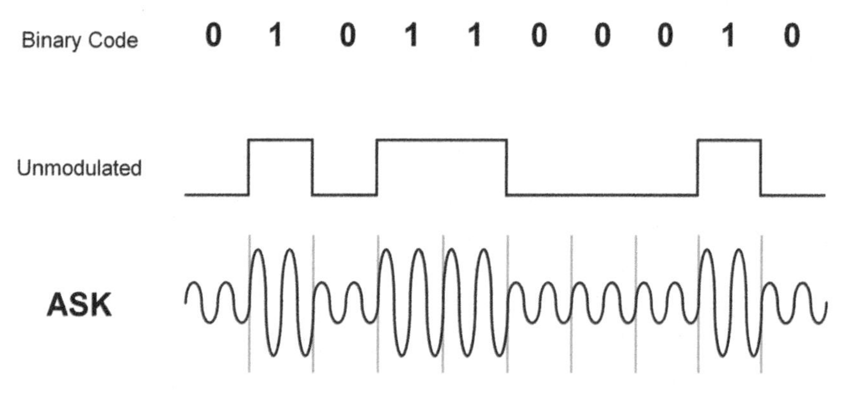

FIGURE 228 AMPLITUDE SHIFT KEYING

This technique is susceptible to noise.

Most noise, like that from microwave ovens and fluorescent lights is *additive*; noise adds to the signal. In the example of Figure 228, when transmitting a low amplitude to represent a 0, sometimes the noise adding to the signal will increase the amplitude to the point where the receiver detects a high amplitude and so in error outputs a 1.

A.3 Frequency Shift Keying (FSK)

The design could be improved by using not one carrier frequency and varying its amplitude, but by picking two carrier frequencies and keeping the amplitude constant instead.

Since noise (more or less) affects all frequencies equally, this technique, called Frequency Shift Keying (FSK), is more robust in the presence of additive noise.

A typical design for the voiceband would pick one frequency, for example, 1200 Hz to represent "0" and another frequency, 2200 Hz to represent a "1" (or the other way around, if you prefer). These two frequencies are the signals that are sent over the line. The transmitter shifts back and forth between these two signals to represent 1s and 0s.

The rate at which it is possible to shift back and forth between the frequencies and reliably detect the result is limited by the width of the pass-band and the shifting technique.

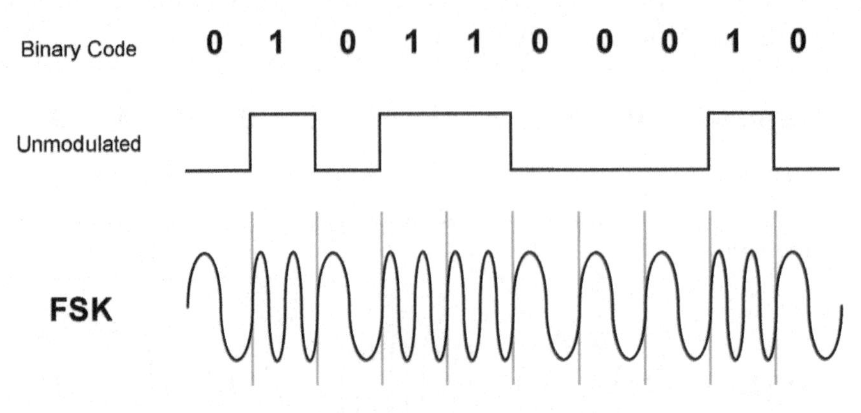

FIGURE 229 FREQUENCY SHIFT KEYING

Since the maximum frequency allowed on the voiceband is 3300 Hz, the maximum rate for signaling, that is, shifting back and forth between the two signals, would be 3,300 times per second.

However, by back and forth between two frequencies at a third frequency, unwanted frequencies called *harmonics* are created, which has the effect of reducing the maximum practical signaling rate to about 2,400 times per second on a voiceband circuit.

Since there is one bit communicated every time a signal is sent, this means a data rate of 1 bit per signal x 2,400 signals per second = 2,400 bits/second = 2.4 kb/s.

A.4 Phase Shift Keying (PSK)

Phase shifting works better than amplitude shifting or frequency shifting.

The *phase* of a signal is its position with respect to a time reference. To perform phase shifting, one carrier frequency is used, with a constant amplitude, but the position of the signal is shifted back and forth to convey information. In effect, there is one single pure tone at one amplitude, and a *jitter* added to carry the information. It is easy to detect *jitters*.

Figure 230 illustrates *Differential Phase Shift Keying (DPSK)*, which is the simplest technique: the phase of the signal is changed 180 degrees (it is shifted by half a period) to indicate a "1", and nothing to indicate a "0".

Since the frequency is constant, the design of the receiver is simpler, and there is less harmonic noise. Since the amplitude is constant, the technique is less susceptible to added noise.

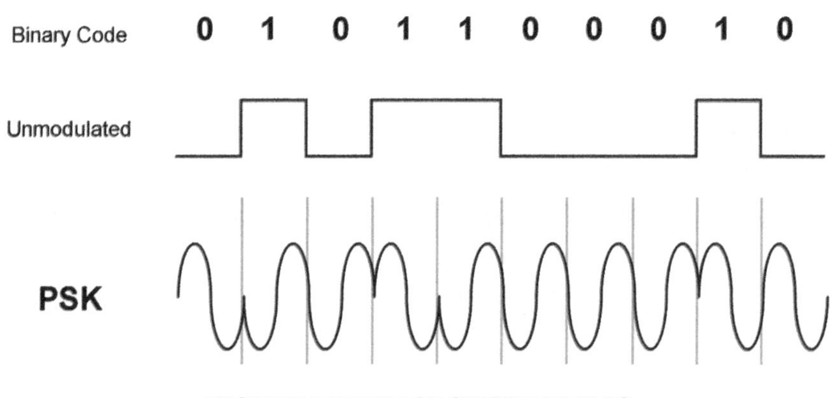

FIGURE 230 PHASE SHIFT KEYING

Overall, the error performance of PSK is about twice as good as FSK.

A.4.1 Baud Rate vs. Bit Rate

Voiceband modems that can establish a data rate of more than 2.4 kb/s exist. V.91 standard modems can achieve 53 kb/s in the voiceband.

Can this be achieved with FSK and switching back and forth between the two frequencies 53,000 times per second?

No. The maximum *signaling* rate on a voiceband circuit is around 2,400 signals per second. Any more often is trying to use more bandwidth than 300-3300 Hz, which is not passed by the filters, and the received signal will be distorted to the point where the receiver cannot reliably detect it.

The rate at which signals, also called *symbols* or *bauds*, can be transmitted is limited by the width of the pass-band, and is called the *baud rate*.

Once the baud rate is implemented, the key to greater data rate is not signaling more often, but **more signals.**

A.5 Quadrature PSK (QPSK)

If four phase shifts are defined, each of 90 degrees (1/4 of a period), this yields four possible signals that might be conveyed to the receiver.

Since there are four signals, each can be used to represent 2 bits. Writing out the numbers between 0 and 3 in binary will demonstrate this, as illustrated in Figure 231.

By making a choice of one of the four signals in particular and transmitting that signal, two bits are communicated with one signal.

Since there are four signals, this modulation scheme is called Quadrature Phase Shift Keying (QPSK).

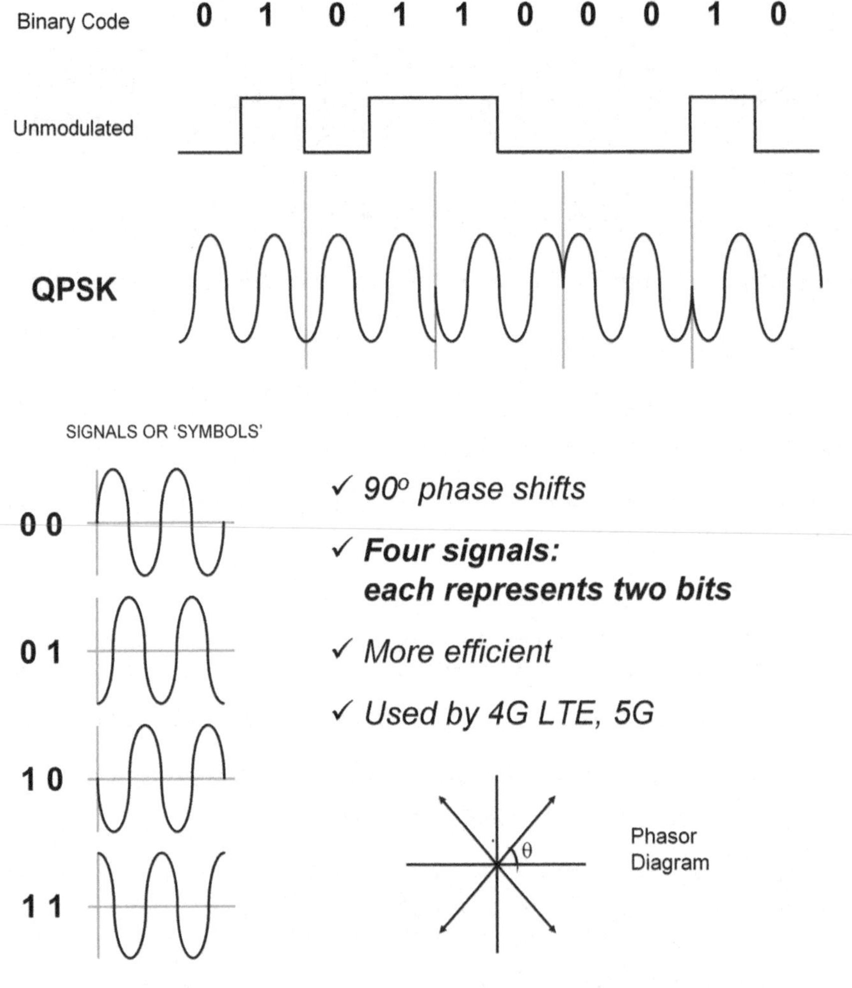

FIGURE 231 QUADRATURE PHASE SHIFT KEYING (QPSK)

In the example of the voiceband channel, we could continue signaling at the baud rate of 2,400 signals per second, but now with 2 bits conveyed by each signal, the *data rate* or bit rate is 4,800 bits/second.

On a 4G cellular (LTE) system, the signaling rate is 15,000 symbols per second per subcarrier. QPSK, one of the allowed modulation schemes for LTE, would yield 30 kb/s per subcarrier.

As can be seen on the lower left side of Figure 231, drawing pictures of signals with different phases using Cartesian (x,y) coordinates becomes tedious and uninformative, particularly as the number of signals increases.

More often, *polar* coordinates are used to represent the signals. As illustrated on the lower right of Figure 231, each signal is represented by an arrow, where the length of the arrow is the amplitude and the rotational angle represents the phase shift.

The diagram is called a phasor diagram; the arrows are the phasors. This is the source of the Star Trek expression "phasors on stun".

A.6 Quadrature Amplitude Modulation (QAM)

Why stop at a repertoire of four signals? Why not define a million signals, using combinations of amplitude, frequency and phase shifting, so each signal conveys $\log_2(1 \text{ million}) = 20$ bits?

The answer is errors. The more signals defined, the higher the probability of making an error at the receiver deciding which signal was transmitted, and so the effective error-free data rate decreases.

Quadrature Amplitude Modulation (QAM) is a technique that combines phase shifting and amplitude shifting to generate a repertoire of many signals that could be transmitted, to increase the number of bits indicated by each signal.

QAM-16 uses combinations of phase and amplitude shifting to define 16 signals, evenly spaced in a square as illustrated in Figure 232.

Since $16 = 2^4$, each signal conveys 4 bits.

In the voiceband, QAM-16 at 2400 signals per second would yield a data rate of 9,600 bits per second, a popular modem standard implemented in a fax machine.QAM-16 and QAM-64 are modulation techniques specified in the 4G and 5G cellular standards.

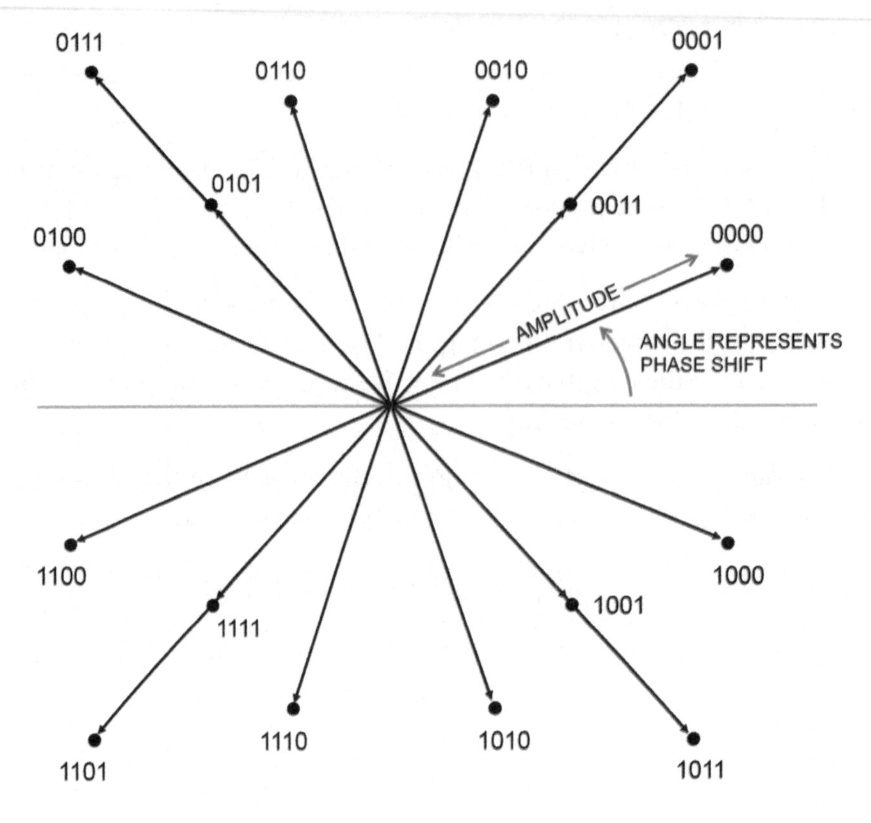

FIGURE 232 QUADRATURE AMPLITUDE MODULATION (QAM)

A.7 Constraints on Achievable Bit Rate

The rate at which bits can be transmitted is proportional to the frequency bandwidth times the signal to noise ratio.

Noise sources are in many cases external and cannot be controlled.

Signal strength is limited due to *crosstalk* - interfering with other signals on other pairs in the same cable.

The remaining variable is the width of the frequency band.

To allow the representation of more bits per second, a wider frequency band is required.

Appendix B Legacy Channelized Transmission Systems

From about 1960 to 2000, the telecom network was constructed of high-capacity transmission systems shared amongst users by employing *Channelized Time-Division Multiplexing* (TDM), also known as *Synchronous Time-Division Multiplexing.*

These systems are now referred to as *legacy systems,* meaning left over from a previous era... but that does not mean they have disappeared. Telephone companies tend to keep existing systems running for as long as possible.

B.1 The Digital Hierarchy: Legacy Channelized Transmission Speeds

A *legacy* is something inherited from a previous generation. A *legacy transmission system* is one that moves traffic in channels, which are bytes transmitted in fixed time slots on a high-bit-rate system.

There is a huge installed base of legacy channelized transmission systems. In fact, a small – but appreciated – fraction of a telephone company's revenue is monthly recurring billing for expensive legacy circuits that are no longer being used... but the customer has never canceled them.

These systems were designed to be voice trunk carrier systems, and so operate at multiples of 64 kb/s, the standard bit rate for digitized voice, referred to as Digital Service Level Zero or DS0 for short.

To allow interoperability of systems, standardized multiples of DS0 channels were defined. These standard multiples and the resulting line speeds are known as the *digital hierarchy*.

Equipment manufacturers made products operating at these standard bit rates. Telephone companies purchased this equipment and integrated it to form networks operating at these bit rates. Their marketing departments created products and services at these standardized bit rates.

B.1.1 Kilo, Mega, Giga, Tera

Abbreviations are used to refer to data rates:

10^3 = thousand = **kilo**
kilobits per second (**kb/s**)

10^6 = million = **Mega**
Megabits per second (**Mb/s**)

10^9 = billion (US), thousand million (UK) = **Giga**
Gigabits per second (**Gb/s**)

10^{12} = trillion (US), billion (UK) = **Tera**
Terabits per second (**Tb/s**)

DS0	**64 kb/s**	**digitized voice**
DS1	**1.5 Mb/s**	**24 DS0**
E1	2.0 Mb/s	32 DS0
~~DS2~~	~~6.3 Mb/s~~	~~4 DS1 = 3 E1~~
DS3	**45 Mb/s**	**28 DS1**
STM	155 Mb/s	3 DS3

FIGURE 233 THE CHANNELIZED DIGITAL HIERARCHY

B.1.2 DS0

Channelized digital transmission systems move the 64 kb/s DS0 rate for historical (voice) reasons. Multiple DS0 channels are combined or aggregated into higher bit-rate streams for transmission.

Anything below 64 kb/s is referred to as a *subrate*.

B.1.3 DS1 and E1

The first step above a DS0 is the *DS1* rate. This rate is equal to 24 DS0s, or if you prefer, 24 times as fast as the DS0 rate.

Note that the mathematics does not quite work out. Multiplying 64 kb/s by 24 does not equal 1.544 Mb/s. This is due in fact to some overhead added in by the T1 carrier system (the framing bits).

There is also an E1 rate, used in Europe, which is 32 DS0s.

B.1.4 DS2

The next rate up is the DS2 rate. This rate is not interesting, and is hardly ever offered commercially. It is the least common denominator between DS1 and E1, and was used as a stepping stone to the DS3 rate on old multiplexing systems.

B.1.5 DS3

The next rate of real interest is the *DS3* rate. North American carriers' legacy backbone transmission systems operate at *multiples of DS3* rates.

Both SONET fiber optic systems and point-to-point microwave radio systems were used to implement *n x DS3 circuits.*

B.1.6 STM and SDH

In the rest of the world, transmission systems conformed to the European *Synchronous Digital Hierarchy* (SDH), which moves *Synchronous Transport Modules* (STM).

STM is a frame size, which transmitted 8,000 times per second results in a data rate of about 155 Mb/s. This is also called an STS-3C in North America.

B.2 Digital Carrier Systems: Legacy Transmission Technologies

This section provides an overview of widely-deployed technologies used to implement the channelized digital hierarchy.

B.2.1 Technologies

The chart of Figure 233 does not include any mention of copper wires, fiber optics or radio: it is not showing *how* these bit rates are implemented; it only lists the standard channelized system *line speeds* in the industry.

An actual way of implementing these bit rates is called a *technology*.

Technology is two Greek words, meaning in English "knowledge of methods". In this case, actual methods of implementing the line speeds and channelized transmission systems from the previous section.

B.2.2 Carrier Systems

Digital Carrier System is the name given to the technologies for implementing the Digital Hierarchy of Figure 233, since they end up carrying multiple DS0-rate circuits on higher-rate DS1 and DS3 aggregate circuits.

B.2.3 T1

T1 was in the past a popular carrier system technology. T1 carries 24 DS0 channels, which is a DS1-rate signal (1.5 Mb/s), over four copper wires.

B.2.4 T3 and Bit-Interleaved Multiplexing

There were two methods of multiplexing up to the DS3 rate (45 Mb/s): *T3 multiplexing,* often also called *asynchronous DS3 multiplexing* is the old method.

This involves multiplexing in three stages, from DS0 to DS1, then DS1 to DS2, then finally DS2 to DS3. At the middle stage, the data is scrambled, resulting in a DS3 data stream with bits scattered all over the place.

This is called *bit-interleaved multiplexing,* and means that to drop out a particular DS0 channel, the entire DS3 has to be demultiplexed. This is not a good idea.

B.2.5 SONET and Byte-Interleaved Multiplexing

SONET (Synchronous Optical Network), also called *synchronous DS3 multiplexing,* was the newer method. This involves *byte-interleaved multiplexing* right from DS0 or DS1 to DS3 rates and beyond.

This means that it is easy to *drop and insert* individual channels out of the DS3. In addition, extra signaling and control for end-to-end error checking is included in overhead bits.

B.2.6 SDH

The *Synchronous Digital Hierarchy* (SDH) is a European standard for multiplexing that moves multiples of 155 Mb/s, called *Synchronous Transport Modules* (STMs).

Where a SONET system moving multiple DS3s would be used in North America, an SDH system moving multiple STMs is used in Europe.

B.2.7 Line Speed vs. Technology

It is important to distinguish between the line speed or data rate of a circuit and the particular technology employed to provide a circuit with that data rate.

A common mistake is to always refer to 1.5 Mb/s as "a T1". It ain't necessarily so. T1 is a particular technology for providing a DS1-rate service, which is 1.5 Mb/s, using four copper wires and a particular scheme for pulses to represent 1s and 0s.

There are other ways of moving 1.5 Mb/s, including HDSL on copper, on fiber and wireless. It would be most accurate to refer to 1.5 Mb/s as "a DS1".

That said, keep in mind that most people erroneously interchange "T1" and "DS1", making statements like "we've got a T1 coming into the building".

To avoid this mistake, say "we have a 1.5 Mb/s circuit" or "we have a DS1-rate service" or "we have a DS1 coming into the building".

It is the rate – the *line speed* – not the technology that is usually of most interest.

B.3 Framing

In this section, we understand *framing*. Framing is extra information transmitted with the data, allowing the demultiplexer at the far end to direct bits in the incoming aggregate to the correct output port.

B.3.1 Synchronous Time-Division Multiplexing

To recap: in a channelized transmission system, the traffic for many users is aggregated onto a high-bit-rate transmission system using *synchronous Time Division Multiplexing* (TDM).

Multiplexing means sharing. Time Division means that the sharing is done in time. Synchronous means that the time-sharing is performed in a strict order in time, resulting in each user being assigned a fixed time slot on the transmission system, called a channel.

The TDM is implemented by network equipment called multiplexers, which send a byte from each user in a strict order, one after another, across the transmission system. This happens 8,000 times per second, and so moves 64 kb/s per channel: DS0 channels.

The stream of bytes from a particular user is interspersed with bytes from other users – other channels – on the transmission system.

B.3.2 Framing and Transmission Frames

To allow the demultiplexer at the far end to direct the correct bits to the correct low-speed output, it is necessary to also send control information.

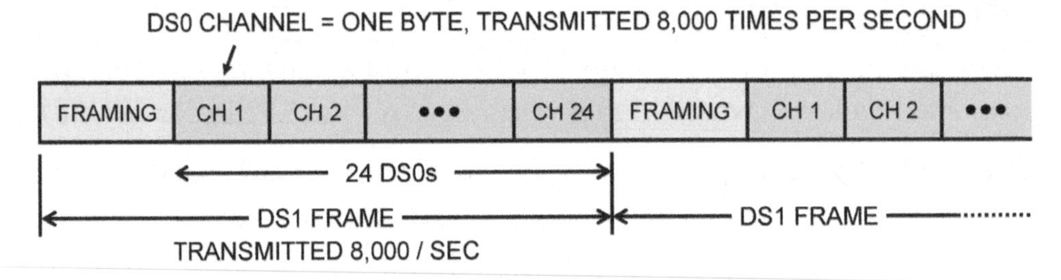

FIGURE 234 DS1 FRAME

Since the users are sending bytes in a strict order, like a batting order at a baseball game, the control information is minimal: it is only necessary to mark the beginning of the batting order with *framing bits*.

When the far-end detects this information marking the beginning of the batting order, it then knows that the next byte goes out to user 1, the byte following goes to user 2 and so forth… in a strict order.

The information marking the beginning of the batting order, i.e., marking the beginning of the frame is called the *framing*.

The framing, plus a byte from each user, is called a *frame* in the transmission business. Frames are transmitted synchronously, 8,000 times per second.

B.3.3 DS1 Frame

The DS1 frame is the lowest level, the smallest frame, containing bytes from 24 channels plus one bit for framing, as illustrated in Figure 234 and at the top of Figure 235.

A discussion of the actual DS1 framing bits and the framing patterns, Superframe and Extended Superframe formats is included in Section C.4.

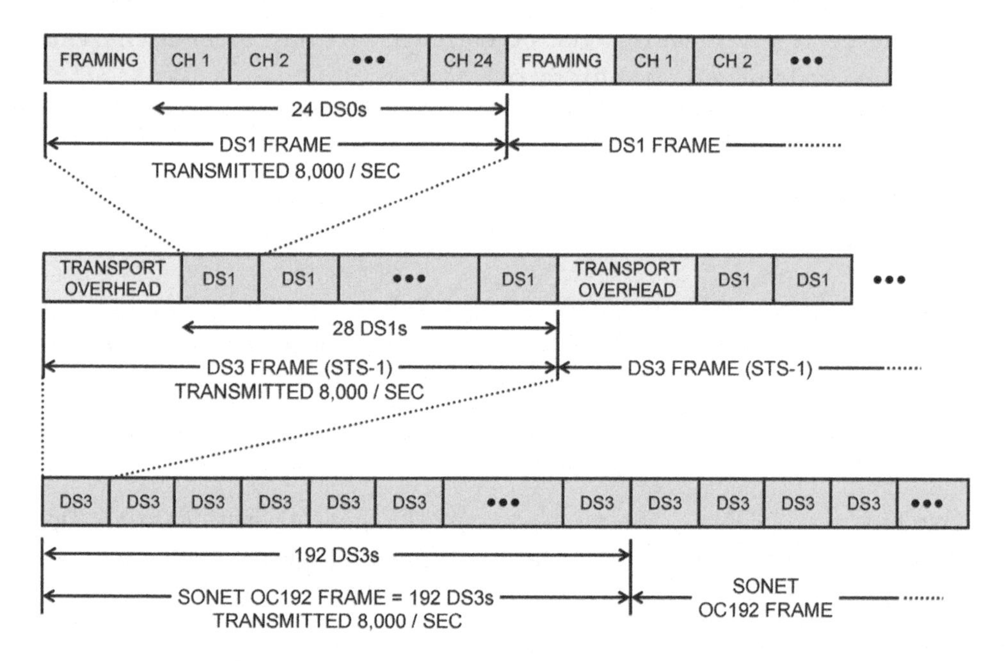

FIGURE 235 STS-1 DS3 FRAMES AND SONET OC192 FRAMES

B.3.4 STS-1 (DS3) Frames

Frames are packaged together into larger frames for high-bit-rate transmission systems. As illustrated in the middle of Figure 235, the next frame size up from DS1 is the Synchronous Transport Signal 1 (STS-1), which carries 28 DS1 frames.

The STS-1 carries a DS3, plus more framing called the transport overhead, and is commonly called a DS3 frame. These larger DS3 frames are also transmitted 8,000 times per second.

B.3.5 SONET Optical Carrier Frames

The SONET Optical Carrier (OC) system moves multiple DS3 frames. For example, as illustrated at the bottom of Figure 235, a SONET OC192 system transmits 192 DS3 frames 8,000 times per second.

B.3.6 Advantages and Disadvantages of Channels

Implementation of channelized TDM results in "pipes", that is, the capability to move a fixed number of bits per second between A and B.

The main advantage of channelizing is that each user knows exactly what capacity they are going to get.

The downside is that if a user has nothing to transmit, their channel is nonetheless reserved and cannot be employed by any other users. This makes the system inefficient for carrying bursts of data.

OC3 (3 DS3 frames), OC12 (12 DS3s), OC48 (48 DS3s) and OC192 products were commonly deployed.

B.4 ISDN

Integrated Services Digital Network (ISDN) is another technology for carrying DS0 channels. Unlike T1 and SONET, which are essentially point-to-point transmission technologies, ISDN also includes network addressing and circuit-switching: being able to specify where the DS0s are to be terminated on a call-by-call basis.

Two flavors of ISDN are *Basic Rate Interface* (*BRI*) and *Primary Rate Interface (PRI)*. These are two very different technologies and must be distinguished.

B.4.1 Basic Rate Interface (BRI)

BRI was designed for residences, running over the same twisted pair currently used for analog POTS. This was to become the foundation for a new "basic" digital telephone service that everyone would have. This did not happen.

ISDN BRI provides two 64 kb/s DS0 channels plus a 16 kb/s signaling channel.

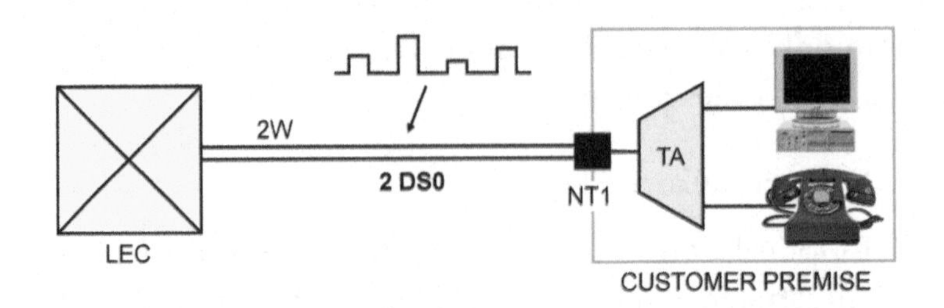

FIGURE 236 ISDN BASIC RATE INTERFACE (BRI)

The DS0 channels are called bearer or "B" channels in ISDN lingo, and can be used to communicate voice or data.

The 16 kb/s channel is called a delta or "D" channel, and is used for signaling functions such as call setup and release.

Since the user gets two bearer channels and one D channel, ISDN BRI is sometimes referred to as *2B+D* service.

The two DS0s can be used for two voice calls, or one voice and one data connection at 64 kb/s, or two data connections at 64 kb/s each, or two channels bonded together to form one data connection at 128 kb/s.

Combining B channels to form a 128 kb/s data connection for telecommuters was one of the applications for ISDN BRI.

The equipment needed to connect devices to an ISDN line must include the circuit terminating function, a Network Termination Type 1 (NT-1), as well as a Terminal Adapter (TA).

These two functions usually come together in a single device, which has a jack for the phone line on the phone company side, and a POTS jack and data equipment connector on the other side.

B.4.2 Obsolescence of BRI

People have been talking about ISDN BRI for about 50 years. It was slow getting off the ground and never gained much momentum. It now appears that ISDN BRI will join telegraphs in the dustbin of history. Technologies like DSL, IP and Optical Ethernet have made it obsolete.

B.4.3 Primary Rate Interface (PRI)

ISDN PRI is not yet obsolete. PRI is a service that turns a DS1-rate access into 23 DS0 channels plus a signaling channel. *PRI trunk* means a DS0 that has an associated PRI signaling channel.

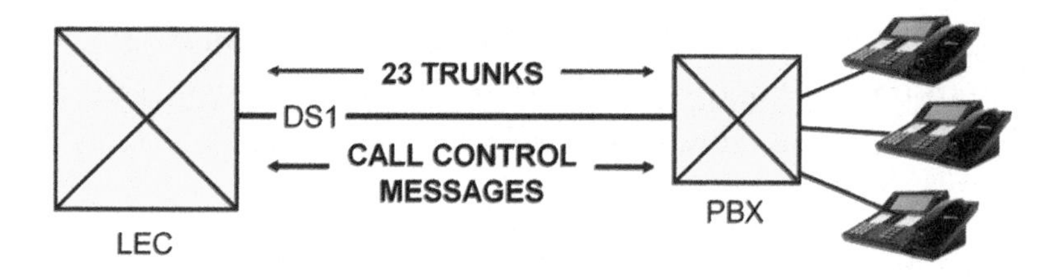

FIGURE 237 ISDN PRIMARY RATE INTERFACE (PRI)

The signaling channel allows the communication of call control messages between switches. This is the main value-added feature of PRI.

The messages that can be exchanged are called the ISDN User Part (ISUP).

An outbound phone call connection request is an example of a message from the PBX to the CO switch.

Messages from the CO switch to the PBX include

- Automatic Number Identification (ANI), which is Caller ID for PBXs and 911 systems,
- Direct Inward Dial (DID) **called** number identification messages, telling the PBX where to terminate the incoming call,

and many others.

As described in Section 13.7, call centers use this signaling capability to dynamically change the routing of 800 numbers to different locations depending on load.

B.4.4 PRI Physical Connection

Unlike BRI, ISDN PRI does not specify the physical connection. The standard implementation was to run PRI on a T1 access.

PRI service can also be provisioned on a fiber-based access that carries multiple DS1s. One signaling channel can support up to 5 DS1s.

B.4.5 T1 vs. PRI

Note that T1 can be used to carry PBX trunks from a customer's building to the Local Exchange Carrier's Central Office, or WATS lines from the Customer Premise through the LEC to an Inter-Exchange Carrier.

This is not a "dedicated T1"; it is carrying multiple PBX trunks on a single physical access circuit; a convenience for the carrier. The T1 in itself does not change or add any value to the story.

If ISDN PRI is ordered as a service using that T1, then PRI adds the capability to transmit and receive control messages between the CO and the PBX via the PRI signaling channel.

Appendix C All About T1

This appendix provides detailed information on the carrier system technology called T1. These discussions used to be a principal part of telecommunications courses, but are now relegated to the back of the book, as T1 is a copper-wire technology running at 1.5 Mb/s.

That said, there are thousands of T1 circuits installed and in use, and some readers of this reference book have picked it up precisely to learn about T1 because they have been tasked with supporting it, or auditing an existing installation.

Power companies, the military and government still have T1s in place, along with T1s at big organizations that are no longer being used, but have been forgotten and are still being paid for each month... a small but appreciated part of any phone company's revenue.

This chapter on T1 also provides a detailed explanation of synchronous Time-Division Multiplexing, framing and channels. Since the principles of operation of legacy SONET fiber-optic transmission systems are the same as T1, learning about T1 is also learning about SONET.

Appendix B should be read before this one.

> If you do not need to know about T1 or channelized time-division multiplexing, feel free to skip this Appendix.

C.1 T1 History and Applications

The T1 Carrier System was designed in 1958 by Bell Labs. The main requirement for this system was to increase the *circuit density* on existing copper wiring between Central Offices… to increase the density of phone calls and thus revenue on existing copper long-distance circuits.

In the old days, this might have been called a *pair gain system,* because it increases the number of circuits actually carried on each pair of wires.

T1 carries 24 digitized voice signals on a single set of copper wires. It was originally designed and deployed by AT&T for use by AT&T for long-distance voice calls.

T1 was popular in the previous millennium as an access circuit installed from the customer premise to the service provider. It implemented "high speed" data access at 1.5 Mb/s or 24 DS0 channels to carriers' voice and data services for business, government and other organizations

T1 was used for:

- "Dedicated T1s", a DS1-rate (1.5 Mb/s) point-to-point connection across a carrier's network all the time,
- PBX trunks and ISDN PRI service, carrying circuit-switched connections to a LEC and/or IXC for the duration of a communication session, and
- Frame Relay, VPN and Internet data service connections at 1.5 Mb/s.

Today, T1 is not used much at all. Optical Ethernet has taken its place.

C.2 T1 Circuit Components

As illustrated in Figure 238, a basic T1 system consists of multiplexers, Channel Service Units (CSUs) and the T1 circuit, which is four copper wires with *repeaters* every mile or so.

The multiplexer (one at each end) is variously referred to as a T1 multiplexer, a *mux*, or a *channel bank*. Since T1 was designed to carry 24 trunks over 4 copper wires, the mux has 24 hardware ports, each one running at 64 kb/s.

The multiplexer's *aggregate* or high-speed output connects to a CSU.

The CSU is the interface device connecting the T1 multiplexer and the actual T1 circuit. This device is the one that represents binary digits on the physical T1 circuit. It performs the same functions as a modem – but since it is a digital device, it is not called a modem (Section 3.1).

The T1 circuit is four copper wires, two for each direction.

Binary digits are represented on these copper wires using pulses of voltage following a line code called AMI, covered in Section C.5.

Repeaters are spaced every 6 kft (1 mile / 1.6 km) along the T1 circuit.

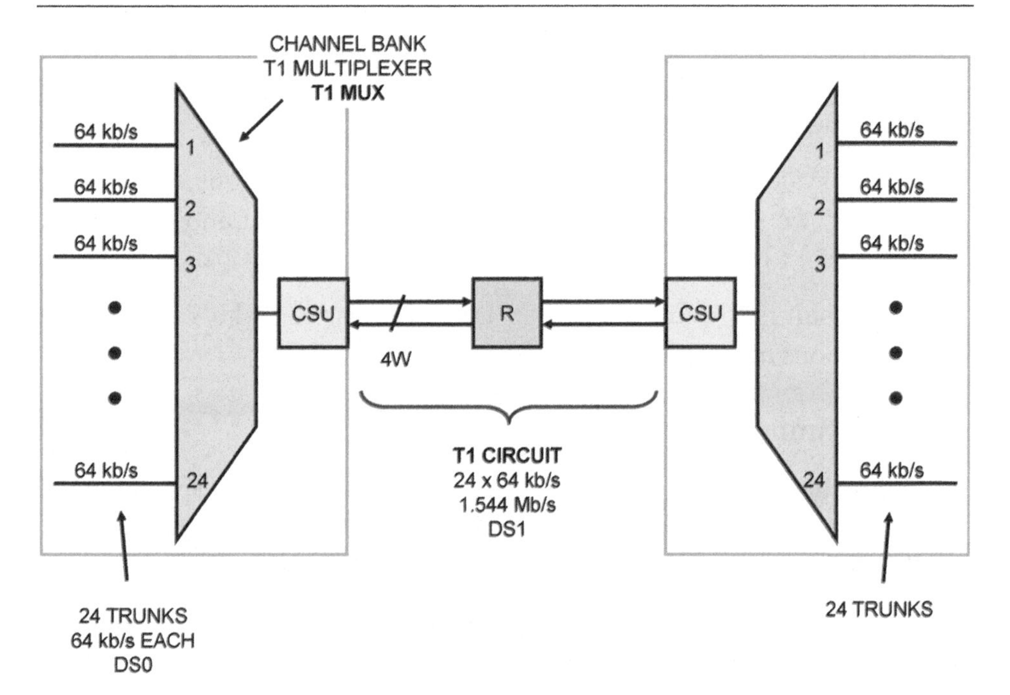

FIGURE 238 T1 CIRCUIT COMPONENTS

The data rate on the T1 circuit is 1.544 Mb/s, which is a DS1-rate signal. 1.544 Mb/s includes extra bits added in by the multiplexers for framing.

C.3 Operation

A T1 system works in a strict, ordered rotation, where each user transmits one byte at a time.

First, the user attached to hardware port 1 on the multiplexer gets to use the outgoing high-speed aggregate circuit to transmit a byte to the corresponding port 1 output at the other end.

Then, port 2 sends a byte, then port 3, and so on down the line in strict order, until port 24 sends a byte.

Then, **one** extra bit called the *framing bit* is transmitted, marking the end of the batting order, and the process repeats: port 1 sends a byte, port 2 sends a byte and so on. The process repeats 8,000 times per second.

The bytes from each port are interspersed or interleaved on the T1, and so it is called a *byte-interleaved* system.

At the far end, the high-speed aggregate circuit is plugged into a demultiplexer, which directs each byte to the correct output hardware port one at a time.

The entire system is two-way simultaneous: both directions at the same time. When we say "input", we should really say "input and output" ... but it is easier to discuss it one direction at a time.

The end result is to communicate 24 DS0s, that is, 24 64 kb/s channels in both directions at the same time over four copper wires.

C.4 T1 Framing

To ensure that each user ends up with a fixed fraction of the capacity of the high-speed circuit, a *channel*, the users transmit one byte at a time, one after another in a strict order as illustrated at the top of Figure 239.

After each cycle, a framing bit is transmitted. The framing bit is used at the far end to locate each user's bytes in the incoming bit stream, to direct each byte to the correct output.

A byte from each port plus a framing bit makes up a *T1 frame*.

A T1 frame is 24 channels x 1 byte/channel x 8 bits/byte + 1 framing bit = 193 bits long. Frames are transmitted 8,000 times per second.

C.4.1 Superframe Format

Every 193rd bit coming down the line is a framing bit, sent to mark the beginning of the frame... but what good is sending **one** framing bit per frame?

After all, a single bit could be either a zero or a one; it looks just like all the other bits... is it not necessary to have a unique pattern of bits that can be recognized at the far end instead of one bit?

Yes, a pattern of framing bits is required; but instead of putting the pattern as a clump at the beginning of the frame, which would increase the overhead, the designers were clever, and considered a number of frames in a row to be a larger unit called a *superframe*.

This way, there is not just one framing bit, there is a group of bits. But they are not in a clump: they are spread out, one framing bit at the beginning of each of the frames in the larger group of frames.

The original design took groups of 12 T1 frames together to make a *superframe*. Since there are 12 frames, there are 12 framing bits (spread out, one at the beginning of each of the 12 frames).

Now that there are 12 framing bits, we use a *framing pattern* defined by Bell Labs, 12 bits long, and insert it where the framing bits occur, one bit from the pattern every 193rd bit in the data stream.

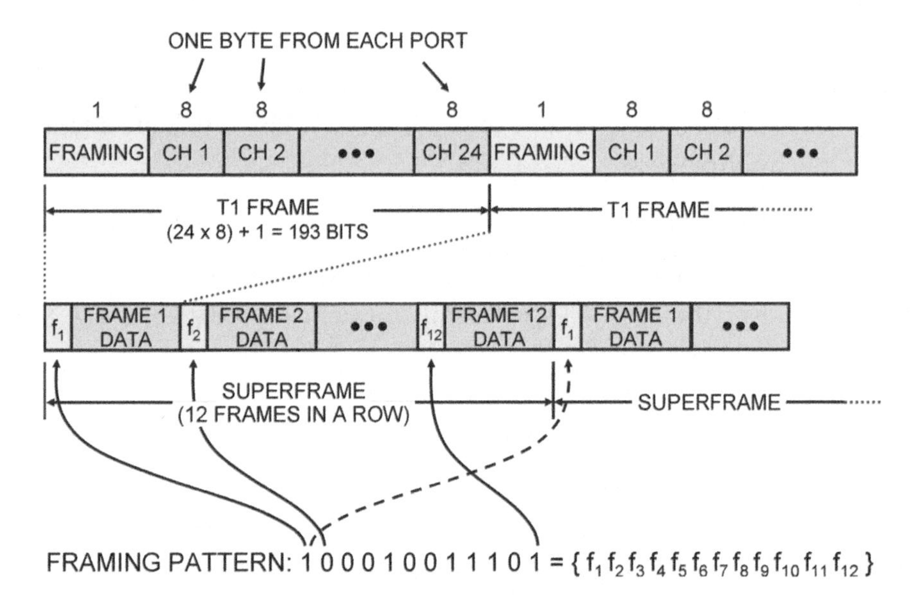

FIGURE 239 FRAMING AND SUPERFRAME FORMAT

At the far end, when first powered up, the receiving multiplexer brings in 12 frames' worth of data and looks every 193rd bit to see if it sees the framing pattern.

Chances are 1:193 that it is not looking in the right place, so it shifts over one bit and looks again, and keeps shifting over until it sees that specific pattern sitting in the incoming data stream every 193rd bit.

The receiving multiplexer has now found the framing bits, and *frame synchronization* is achieved.

The receiving multiplexer now knows which byte goes to which output, and at this moment, all of the data on the outputs becomes valid.

This process usually happens once, when the circuit is turned up.

This idea is referred to in general as *framing*, and this technique is called *Superframe Format* or *D4 format* in particular after a type of AT&T equipment.

C.4.2 ESF

An improvement called *Extended Superframe Format* (ESF) or *D5 format* was made to the original design, for more efficient use of the framing bits.

The rationale behind Extended Superframe Format is that it was not really necessary to have one bit per frame = 8 kb/s for framing, so only some of those framing bit positions would actually be used for framing, and the rest could be used for other functions, like error checking and reporting.

ESF groups T1 frames in groups of 24, and uses 2 kb/s for framing, 2 kb/s to perform a CRC-6 on each frame and provide a 4 kb/s free data channel between the multiplexers.

This data channel is often referred to as the Facility Data Link (FDL).

It is used to report the results of the CRC check and other performance parameters down the line. AT&T Technical Publication 54016 and ANSI T1.403 are standards for use of the FDL.

C.5 Pulses and Line Code: AMI

Here, we examine how the binary digits actually represented on the copper wires. Even though this is a "digital" system, we still need a method of representing the binary digits (1s and 0s) on the physical medium, which in the case of T1 are pairs of copper wires.

Digital transmission means applying energy to the communication circuit, or not, for a pre-determined length of time to represent 1s or 0s.

The burst of energy is called a *pulse*, and the strategy for representing 1s and 0s using pulses employed by a particular technology is called its *digital line code*.

The line code used for T1 circuits is called *Alternate Mark Inversion* (AMI).

1s are represented as *pulses*, that is, charging the line to 3 volts for a short period. 0s are represented by doing nothing for a short period of time.

Also, the pulses must alternate in polarity: +3 volts, -3 volts, +3 volts, ...

The CSUs are the devices that perform this function.

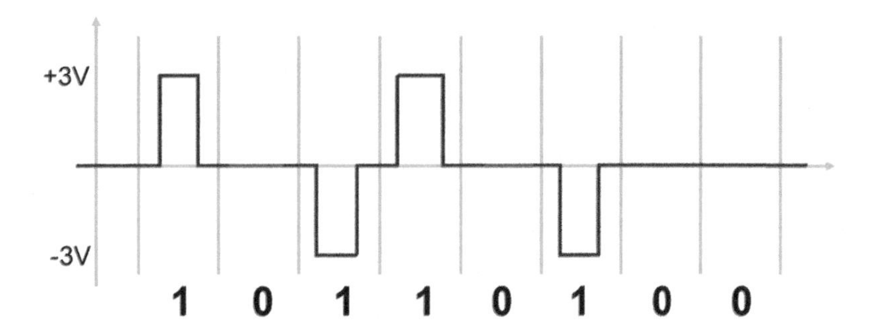

FIGURE 240 ALTERNATE MARK INVERSION LINE CODE

Mark is an obsolete term from telegraphy meaning an active condition on the line. Thus, we have 1s represented as pulses (marks), zeros as nothings (spaces), and the marks alternate in polarity +, -, +, - ,... Alternate Mark Inversion.

The pulses alternate in polarity so that the average value on the line is zero, and more importantly, giving a simple error detection mechanism.

Interested observers will note that the pulses occupy only half of each bit time slot. This is called a *50% duty cycle return-to-zero line code*. This was the choice made for the T1 system in 1958.

C.5.1 Repeaters

A pulse is energy applied to a circuit by the transmitter for a pre-determined length of time. At the receiver, we wish to make a simple decision: whether a pulse is happening or not.

However, as the energy which is the pulse travels over the physical medium between transmitter and receiver, it will be degraded due to the imperfect nature of the physical medium.

On a T1 system, a pulse is voltage carried on copper wires... and the voltage is *attenuated* by the resistance of the copper wires. This is exactly the same problem encountered when discussing analog techniques and maximum loop lengths in Section 11.4.

The shape or *envelope* of the pulse will also be distorted, with the corners rounded by the capacitance of the wires.

If the distance between the transmitter and receiver is such that the pulse will become so badly degraded that it is not possible to make a reliable decision whether a pulse is happening or not, it is necessary to *regenerate* the pulse at intermediate points using a *repeater*.

Repeaters are binary devices that make a decision. If they decide they detect a pulse on the input, they *regenerate* a new noiseless square pulse on the output to send down the next cable segment.

Repeaters are required every 6,000 feet on a T1 circuit.

As discussed in Section 3.3, it is possible to boost analog signals using an amplifier, to be able to transmit information more than the maximum loop length; however, the amplifiers boost both the signal and the noise, making analog transmission noisy and distance-limited.

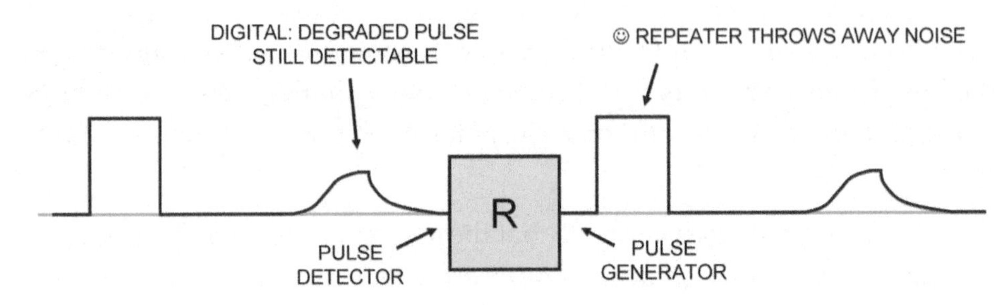

FIGURE 241 REPEATERS REGENERATE PULSES

As illustrated in Figure 241, the advantage of using pulses is that the repeaters do not boost the incoming signal, they **regenerate** it, essentially discarding the noise at each intermediate point and only transmitting the signal.

Using this idea, information can be coded into 1s and 0s, which are in turn represented as pulses on the line and transmitted long distances via regeneration of the pulses at regular intervals.

The result is communicating the information without adding in any noise; so quiet, you could hear a pin drop.

C.6 Synchronization: Bit-Robbing

Framing is for synchronization at the byte level, so the demultiplexer can determine what goes where.

Synchronization of the start and end of pulses, at the bit level, is also required.

When the sending CSU charges up the line to 3 volts to indicate a "1", it is necessary that the receiving repeater is looking at the line at that exact same time to decide if there is a voltage there or not.

We do not want the far end looking at the line after the sender has finished doing a pulse on the line. Both ends must be in synchronization.

In the 1980s and 1990s, transmission systems were synchronized by clocks derived from a master clock operated by the Federal Government. We were all marching to the Federal beat of the drum.

Later, carriers started using the Global Positioning System (GPS) satellites to derive their own master clocks.

In 1958, there were not any central network clocks nor GPS satellites. Timing was kept by re-synchronizing on the rising edges of the pulses that are the AMI line code.

Every time a pulse happened, its rising edge was used to pull wayward devices back into alignment if they had drifted a bit.

Since the system uses pulses for synchronization, there is a requirement to send a certain number of pulses down the line; and since pulses are caused by sending 1s, this boils down to a requirement to send a certain number of 1s down the line.

This is referred to as the *Ones Density Rule*. A simplified version of this rule is that there must be at least one 1 per byte to keep synchronization.

The designers of the T1 system in 1958 came up with an inelegant solution to this requirement: they made the multiplexers always set the least significant bit of most of the channels to a 1.

This was called *bit robbing*... the network appropriated one of the eight bits in every byte for network clocking purposes.

Except in frames 6 and 12 of the Superframe; these positions are reserved for supervision signaling for voice trunks.

The interested voice communications reader may want to note that these bit positions usually hold digitized versions of the E&M signaling leads from the analog trunks that the T1 carrier system replaced. These bits are referred to as the A and B signaling bits respectively.

If the system is carrying digitized voices, which was its original intent, the effect is to add in a bit more quantization noise to the voice signal, because half the time the received signal is in error by one level.

A human being cannot hear this happening on a voice call. Since T1 was designed for voice only, it was deemed at the time that this robbing of one of the bits for network synchronization purposes was acceptable.

C.7 56 kb/s for Data

The problem with this bit-robbing scheme to synchronize all of the equipment at the bit level is that when we try to use this system for communicating data, we find that the least significant bit of every byte is always set to 1... and so cannot be used.

We have only 7 bits per byte useful for data communications. Bytes are always communicated at the rate of 8,000 per second.

7 bits per byte 8,000 times per second yields 56,000 bits per second useful data bandwidth per channel, or 56 kb/s.

This is why 56 kb/s was historically a popular data circuit speed from service providers.

> This has nothing to do with "56K" modems. "56K" modems and 56 kb/s digital data circuits are different things.

C.8 B8ZS and 64 kb/s Clear Channels

Bit robbing was fine as long as the requirement for T1 was overwhelmingly voice. Then applications for T1 became as much data as anything else, and so a design modification was needed to eliminate the robbing of the least significant bit.

The only byte that really causes trouble with the Ones Density Rule is the byte that has 8 zeros in it (00000000). All of the other bytes have at least one 1 in them.

The design modification was to stop setting the LSB of every byte to a 1 and let the bytes pass through the T1 system unmolested, except that when a byte with 8 zeros was to be transmitted, instead of sending nothing for 8 bit periods, as the AMI code would require, a special line code normally never seen would be substituted instead.

For example, the "special" line code to represent eight zeros in a row for the case when the previous pulse was negative is illustrated in Figure 242. If the previous pulse was positive, this special line code is inverted.

This code is special, because it causes pulses of the same polarity to occur one after another: in effect, we have created an exception to the alternating rule to mean eight zeros in a row.

When this special code is received at the far end, it is interpreted to mean a byte with 8 zeros in it. This technique is known as *Bipolar Eight Zero Substitution (B8ZS)*.

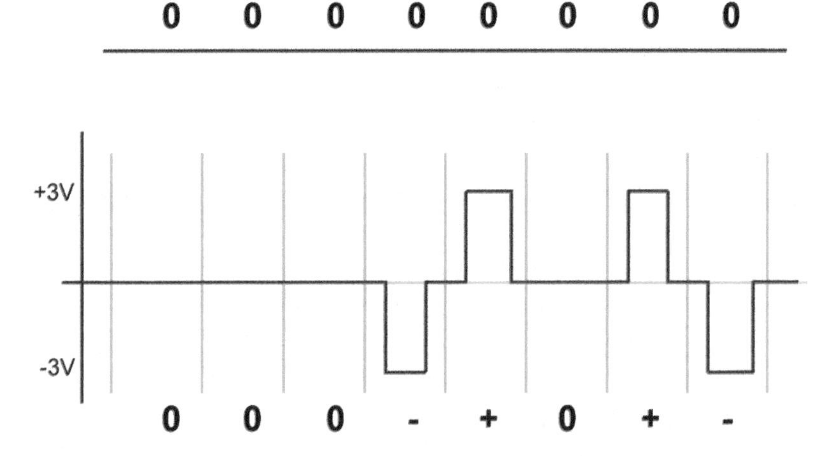

FIGURE 242 BIPOLAR EIGHT-ZERO SUBSTITUTION

The result is that the LSB in each byte is not molested by the transmission system, and so the user can employ this bit to transmit data, resulting in being able to employ all eight bits of each byte for data transmission.

This capability was referred to as *clear channels* in the business. At eight thousand per second, the result is 64,000 bits per second or 64 kb/s per channel for data communication.

This is 12.5% better performance than the original bit-robbed scheme described on the previous page, where one could only use 7 bits per byte for data communications.

The special code causes pulses of the same polarity in a row, which is normally considered a *Bipolar Violation* (BPV) by the system, since this violates the Alternate Mark Inversion rule.

This use of a normally illegal code means that all of the repeaters and CSUs have to be upgraded in software to know about this special code and not think it a BPV error.

The installed base of T1 circuits did not historically support B8ZS and clear channels, so it was imperative to specify "64 kb/s *clear channels*" when ordering T1 circuits... otherwise you could get 56 kb/s bit-robbed circuits, since this was the original plan and only "new" (1980+) facilities supported the new plan.

C.9 How T1 Is Provided

In practice, T1 was an *access* technology, used only for the last mile or two: copper wires used to get a DS1-rate service from the customer premise to the fiber systems. The information is transported on fiber long-distance.

As illustrated in Figure 243, T1 is a 4-wire copper circuit running from the local phone company's building (usually a Central Office) to the customer premise.

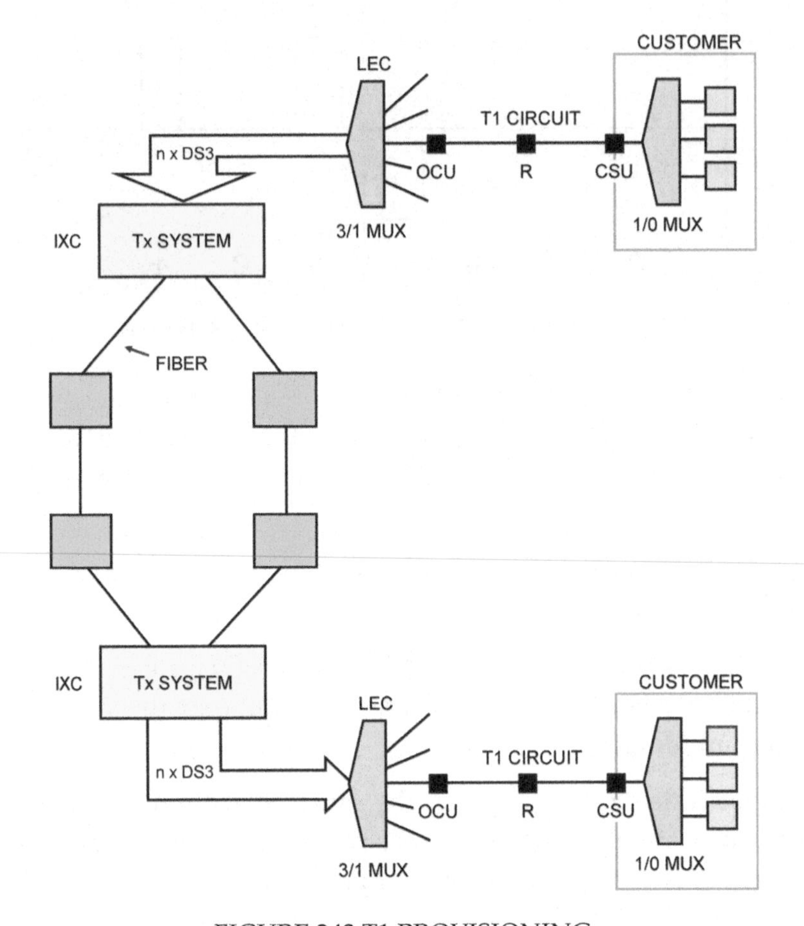

FIGURE 243 T1 PROVISIONING

At the customer side, the wires are terminated on a Channel Service Unit (CSU), which provides an interface for the customer's multiplexer.

At the Central Office, the wires are terminated on an Office Channel Unit, which performs the same functions. If obtaining service from an Inter-Exchange Carrier for a long-distance circuit, the service will be carried through the local phone company's CO to the IXC's Point of Presence.

A *1/0* multiplexer, with DS1 on one side and DS0s on the other, can be located at the customer premise. The T1 carries the DS1 to the CO. Fiber backbone transmission systems carry multiples of DS3-rate signals, not DS1s, and so at the CO, the information on the T1 will be combined with many other DS1-rate streams to form a DS3-rate stream by a 3/1 multiplexer.

These DS3-rate streams are then moved long distance over fiber.

At the far end, the reverse process takes place with similar equipment and cabling. If the other end is in Europe, the signal may be delivered as an E1 over the CEPT-1 carrier system.

C.9.1 HDSL

For advanced readers: T1 as such was not actually used for many T1 services (!) An issue with T1 is that it requires repeaters: the first one at 3,000 feet, and every 6,000 feet thereafter.

Repeaters are expensive to install and maintain. Variations on T1 that do not require repeaters up to 12,000 feet were developed.

These technologies are called High-Speed Digital Subscriber Line (HDSL). [not related to residential DSL].

When someone said, "we have a T1" from here to there, this might have been wholly inaccurate. They had HDSL access and SONET transport, and no T1 technology at all.

It would be more accurate to say, "we have a full-period DS1-rate service" from here to there. But "T1" is short and catchy-sounding…

C.10 Fractional T1, DACS and Cross-Connects

If the customer does not require the capacity of a full T1, but rather only a few of the DS0 channels, a service generically called *fractional T1* was available.

From the customer premise to the CO, the equipment and cabling is identical to full T1 service. The only difference is that the customer's T1 multiplexer can only put data on the channels that the customer purchased.

The rest are not connected, and the carrier will ignore any data received on the extra channels.

The carrier requires additional equipment to provide fractional T1. Since these customers are only using some of the channels in their DS1-rate signal, it is necessary to drop out the DS0 channels that are being used, and insert them into a DS1 with other customers, to make up a DS1 with all channels used.

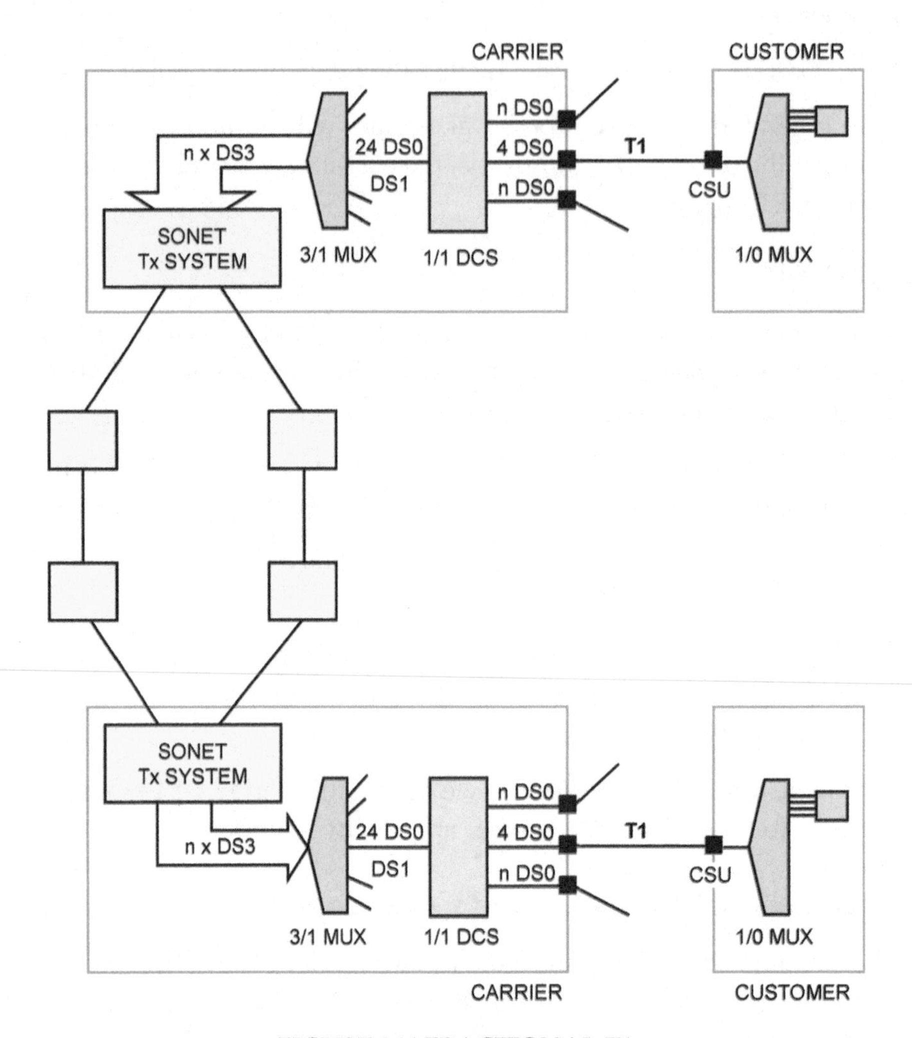

FIGURE 244 FRACTIONAL T1

This piece of equipment is called a Cross-Connect, a Digital Cross-Connect System (DCS) or Digital Access and Cross-Connect System (DACS).

It takes fractional DS1s from a number of customers to make up a full DS1. This DS1 then is multiplexed into a DS3 and carried over the backbone just as for full T1 service in the previous section.

At the far end, the same equipment is required, and the reverse process happens, picking out the right DS0s and sending them over a T1 to the far end customer premise.

This is a good illustration of difference between "T1" and "DS1". T1 is the physical layer protocol for physically cabling together the OCU and CSU. DS1 is the rate of signal that it carries. The DS1 can be split into individual DS0s. The T1 is four wires.

C.11 Subrate Data Circuits 1.2 kb/s to 56 kb/s

If the customer did not require even fractional T1 service, but needed only a *subrate* data circuit at 9.6 kb/s or 56 kb/s, services were available for that.

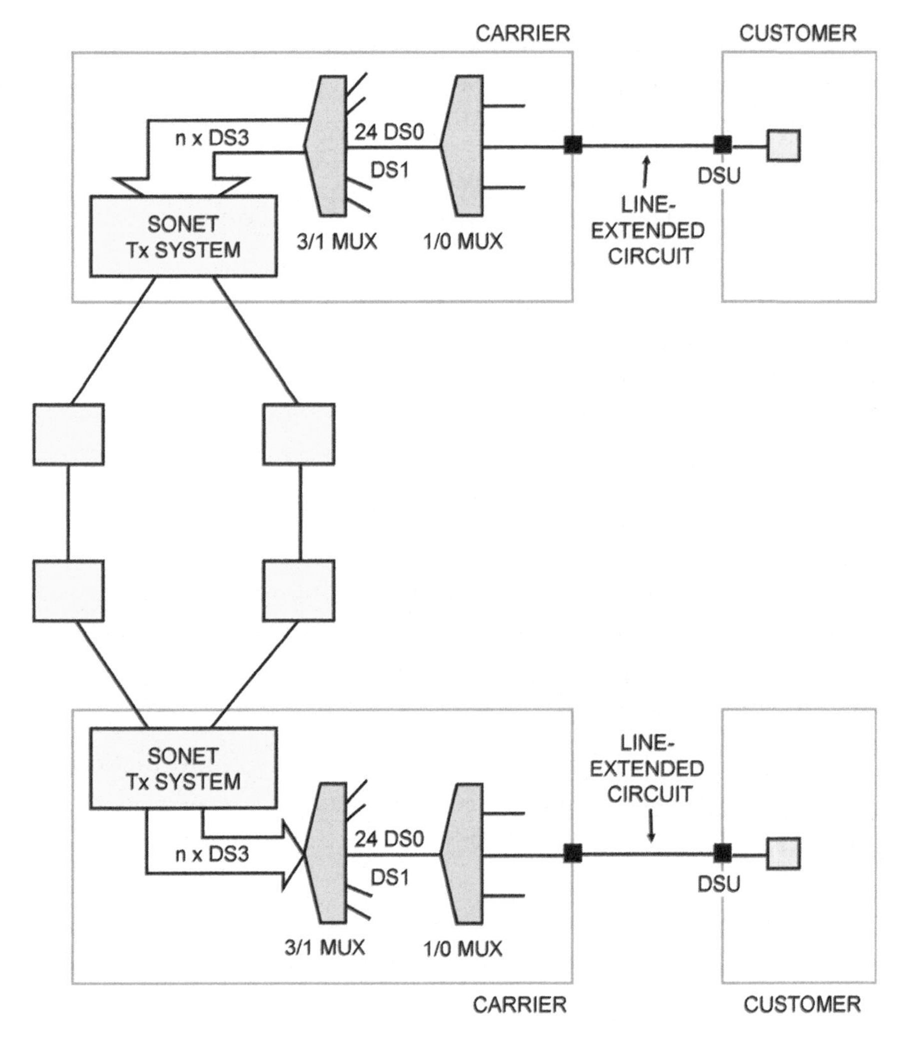

FIGURE 245 DSU

From a conceptual point of view, the equipment is much the same. The fundamental difference is that the service is provided from a T1 multiplexer in the carrier's building instead of a T1 multiplexer on the customer premise.

Since it is usually a far distance from the Central Office to the Customer Premise, a *line extender* system is required to extend this single channel from the CO to the customer.

Carriers typically installed a 4-wire circuit from the CO to the CP, and put line extender devices at each end. At each end, a standard interface like EIA-232, EIA-422 or V.35 is presented to the network equipment.

C.11.1 CSUs, DSUs and CSU/DSUs

There are numerous terms in use for T1-related circuit-terminating devices. The most common names are Data Service Unit (DSU), Digital Service Unit (DSU), and Channel Service Unit (CSU).

Other names for this device include Digital Terminating Unit (DTU), Terminal Interface Unit (TIU), Terminal Interface Equipment (TIE), and Network Channel Terminating Equipment (NCTE).

They all perform the same function.

Switched services require a combined CSU/DSU, which can perform signaling functions as well as line extension.

The terms CSU, DSU and CSU/DSU are often interchanged. Most people in the business called all three devices a "CSU/DSU" without knowing the official definitions:

- A CSU was the circuit-terminating equipment for a T1.
- A DSU was a line extender for extending single channels from the CO to the customer. This is called a *tail circuit,* since the DSUs are on the slow side of the multiplexer.
- A CSU/DSU is a DSU that can also signal control information to the network, used for switched 56 kb/s services, which were called Dataphone Digital Services (DDS).

Appendix D Legacy Data Communications Technologies

This chapter describes data communication technologies that were formerly mainstream but no longer in wide use.

D.1 "Asynchronous": Start/Stop/Parity

Formatting and packaging characters to be transmitted one at a time used to be called *asynchronous*.

As this method was used mostly for connecting modems through serial ports and dumb terminals, it has diminished greatly in importance and the details of start, stop and parity bits are no longer part of the required knowledge in the telecom business.

D.1.1 Asynchronous Communications

It would have been more precise to refer to this as "a type of asynchronous communications", because it is possible to transmit many things asynchronously.

From the Greek, asynchronous means "not timed". The word asynchronous was associated with transmitting characters one at a time, because pressing a key is not a timed event. The time between keystrokes is random.

More generally, another way of thinking of the meaning of asynchronous is from the communication circuit's point of view: statistically speaking, if someone is doing asynchronous communications, most of the time, they are doing nothing. Every once in a while, they send some information, then go back to doing nothing.

In this older method of communications, characters were coded into seven-bit patterns using the US-ASCII codes of Section 7.12.

D.1.2 Framing: Start and Stop Bits

Communications is for the most part serial, that is, there is one data circuit established, and when there is a group of bits to communicate, the bits are transmitted one after another in a sequence in time on the circuit.

When communicating asynchronously, normally nothing is happening, and data can be randomly transmitted. Because of this, there must be a mechanism for the receiver to detect the beginning of a serial group of bits, capture the group and re-package it into a parallel form at the receiver. The general term for this is *framing*.

For transmitting keystrokes, the framing is performed on each byte. First, a *start bit* is transmitted to warn the receiver that a byte is coming, then the character code, followed by a *stop bit*.

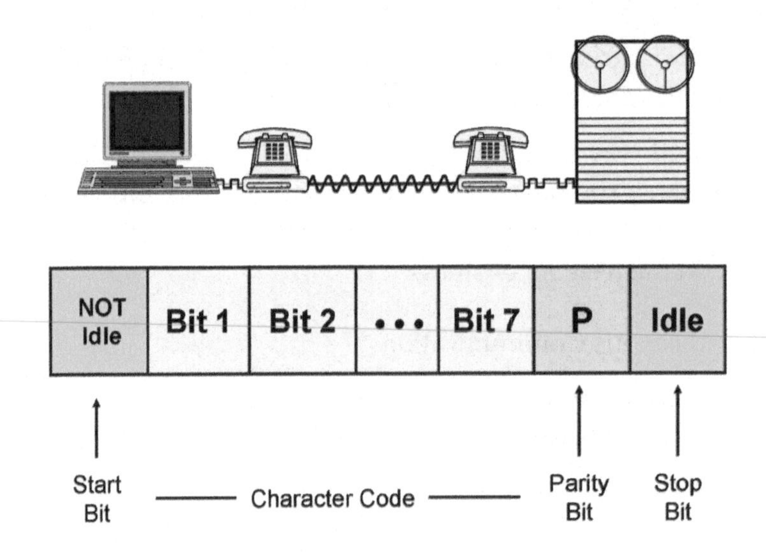

FIGURE 246 START/STOP/PARITY FORMAT

To understand what a start bit is, first, consider that most serial communication systems are also binary communication systems: they have only two levels. This is not a tri-state system that has three levels – high, low and disconnected. This is a binary system that has only two levels – high and low.

Since the circuit is never disconnected, it is always communicating one or the other. One of these levels must be chosen as the idle condition, i.e., the condition that is presented to the receiver when nothing is happening.

Understanding this, deciding what a "start bit" should be is simple. It is NOT the state chosen to be the idle condition. This idle to not-idle transition warns the receiver that a byte is coming.

A stop bit is the idle condition to guarantee an idle to not-idle transition when the next start bit happens.

D.1.3 Parity Checking

To perform error detection at the receiver, a single extra bit called the parity bit is appended to the 7-bit ASCII code and sent over the communication circuit. Just one bit is used, to implement a simple technique while minimizing the overhead in extra bits.

This is one reason why ASCII was designed as a 7-bit code: 7 bits of data and one bit of error detection to make up an 8-bit byte.

There are two parity rules: *even parity* and *odd parity*. The transmitter and receiver must decide in advance and agree upon which rule they will use, then stay with that rule.

Under the even parity rule, the extra bit is set at the transmitter so that the total of all of the bits including the parity bit is even. The receiver checks to see if the total is even. If not, it knows an error happened in transmission, and can flag a parity error. If the total is even then there were no errors.

The odd parity rule works in the same manner.

Unfortunately, this even/odd scheme does not work if there are two bits in error, or four, or six... and errors happen in bursts.

The probability that if there is one bit in error, that the bit beside it is also in error is between 20 and 50% depending on the physical medium being used. For this reason, parity checking is almost useless.

Many systems ended up using *no parity*, and instead of using one bit for parity, use it for data. This is often represented as a code "8N1" that has to be typed into a setup screen on software.

Using what was the parity bit for data instead extends the ASCII code to 8 bits as discussed in Section 7.12.

D.2 X.25: Packet-Switching using Virtual Circuits

The next few sections cover legacy technologies for virtual circuits, carrier packet networks and network services.

We begin with the first technology, X.25, then its improvement, Frame Relay, then what was supposed to be the "final answer" ATM. These technologies have been replaced with MPLS. These four technologies are all essentially the same thing: virtual circuits – with different jargon and buzzwords.

We'll use X.25 to establish a graphical method that illustrates the protocol stacks on each device, and how packets travel in frames over physical circuits from one system to another, within the framework of the OSI 7-layer reference model.

X.25 was a widely-accepted standard protocol for packet networks, standardized by the CCITT (now ITU) in 1976 and deployed by all telephone companies for business and government data communications. Some precursors of what we know today as the Internet, for example, CompuServe, ran on X.25 networks.

X.25 typically offered at most 56 kb/s access speed, and less throughput. It was not scalable to higher line speeds, and did not have any QoS mechanisms to provide service level guarantees. It is now obsolete.

D.2.1 X.25 Network Structure and Operation

At the left of Figure 247 is the customer equipment, a terminal or computer (called Data Terminal Equipment (DTE) by the ITU), plus another function called a Packet Assembler-Disassembler (PAD). The far-end customer equipment is at the right of the diagram. In the middle is an X.25 network operated by the phone company, the military or another private company.

This X.25 network is composed of routers (called *packet switches*) in cities connected long-distance. This is called *the core* and *regional rings* in other chapters.

The Customer Edge is the PAD. On one side, terminals are connected by data circuits. On the other side, it runs the X.25 protocol stack to communicate with the packet switch. The PAD assembles keystrokes or file records received from the terminal into packets, carried in frames, signaled over an access circuit to the packet switch.

First, this packet communication capability is used for a control function. The terminal communicates via the PAD to network control equipment, requesting that a connection to a particular far-end terminal be set up.

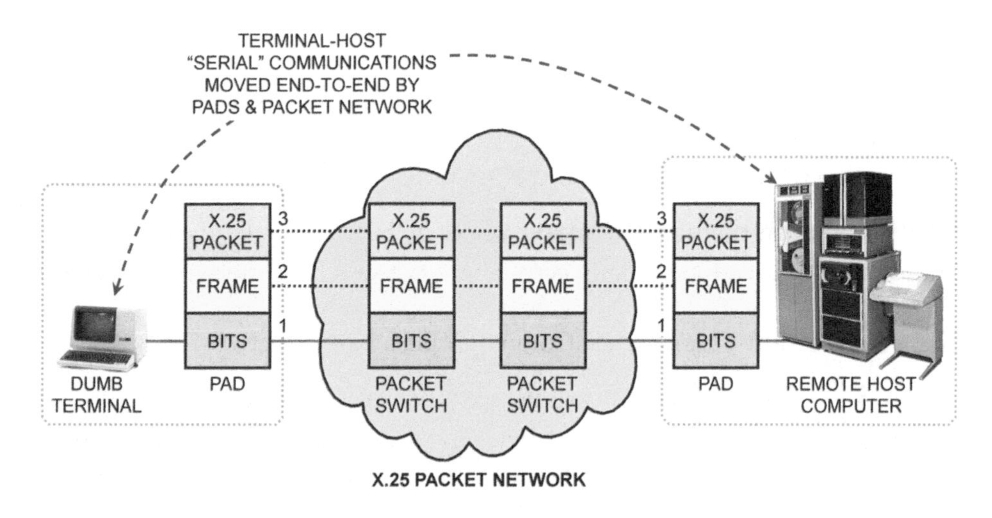

FIGURE 247 X.25

The network sets up a virtual circuit to that terminal, populating routing tables all along the line, and returns the virtual circuit ID. Then the packet communication function is used by the near end to communicate to a far-end PAD by populating the far end's virtual circuit ID in the packet header and transmitting it to the packet switch.

The packet switch receives the frame one bit at a time, and once it has the entire frame, layer 2 software on the packet switch performs an error check, verifies its link address is on the frame, does any resequencing or error recovery necessary at the frame level, and when all of this is complete, extracts the packet from the frame and gives it to a second piece of software, the layer 3 network software running on the X.25 packet switch.

The layer 3 network software on the packet switch performs error recovery or resequencing at the packet level, then uses the virtual circuit ID to look up in its routing table where to send the packet.

This software passes the packet back to the layer 2 software on the packet switch, along with the link address of the next hop. The layer 2 software on the packet switch revises the frame address, recalculates the frame error check value then transmits the frame on the appropriate physical output to the next packet switch.

This repeats until the packet is delivered in a frame to the far-end PAD, which extracts the data from the packet and passes it to the far-end application.

D.2.2 Reliable Network Service: Guaranteed Delivery

X.25 implements an error-recovery mechanism, retransmitting missing or errored data on individual links, to guarantee delivery of user data (supporting dumb terminals). There are no timing or delay guarantees.

X.25 in effect implements a single Class of Service that might be called guaranteed data delivery. This is referred to as a *reliable* network service.

D.2.3 Connection-Oriented vs. Connectionless Network Service

Connection-oriented communications means in general that there is communication with the far end before a file transfer begins; the sender gets an acknowledgment that the receiver is online and ready to accept data. Since a virtual circuit is set up before communications begins, X.25 implements connection-oriented communications at the network level. X.25 was reliable, connection-oriented packet communications from the Phone Company.

Contrast that with unreliable, connectionless network service. Unreliable means that there is no guarantee from the network that a packet will be delivered, and no acknowledgment of transmission of a packet is provided by the network.

Connectionless means that there is no communication with the far end before a file transfer begins. To use an unreliable, connectionless network service, the user must perform the reliability and connection functions. The Postal Service is an example of an unreliable, connectionless network service. The Internet is another.

Business, government and the military liked X.25, because it was cheaper and much more flexible than dedicated lines, allowing communication to many locations over one access circuit.

However, X.25 was not scaled to higher throughputs to support LAN-LAN communications, and did not support any kind of class of service other than "data". It was necessary to deploy other packet network and virtual circuit technologies that supported higher line speeds and more sophisticated classes of service.

D.3 Frame Relay

Frame Relay was the standard mainstream solution for business and government data wide-area networking popular in the 1990s and 2000s.

Carriers have put a cap on new deployment of Frame Relay, but there are organizations still using it, pending a migration to IP and Optical Ethernet. In this section, we review Frame Relay, its jargon and buzzwords.

D.3.1 Elimination of a Layer of Software

Frame Relay is a bandwidth on demand service that was **faster** than X.25. It could provide communications at rates of up to 1.5 Mb/s (and from some carriers, 45 Mb/s) and so was suitable for client-server wide area networking.

Frame Relay is faster than X.25 mainly because of the elimination of a layer of software.

In a packet-switched network, information on the packet header is used for routing decisions. At every packet switch, layer 2 software has to receive the frame, perform the error detection, examine the frame address, do any error recovery necessary, then extract the packet from the frame and pass it to a second software program, the layer 3 routing software.

This software receives the packet, does any network-level error recovery, then looks at the network address on the packet, uses that to make a route decision, then passes the packet back to the layer 2 software along with an indication of the outbound link. The layer 2 software changes the link address in the frame and recalculates the frame check sequence before sending it off.

This would be like sending a parcel via UPS and putting the address **inside** the box. Any time a UPS employee had to make a routing decision, they would have to open the box, find the address, look at it, make a decision, tape the box shut and then throw it in the appropriate pile.

Wouldn't it be faster to put the address on the **outside** of the box?

This passing of the packet between two sets of software and the duplication of functions takes a relatively long time, reducing the end-to-end *throughput*.

The main idea behind Frame Relay is to have the network equipment use information in the layer 2 header (the *frame* header) to make routing decisions. Then, it is not necessary to have a second (layer 3) software program on the network equipment, and not necessary to pass a packet from one piece of software to another then get it back on every piece of network equipment.

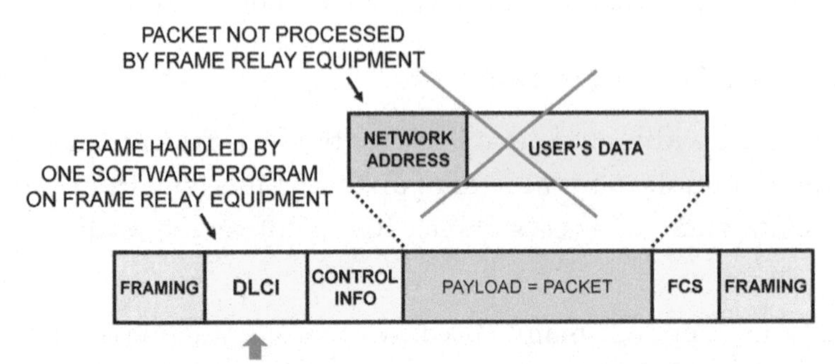

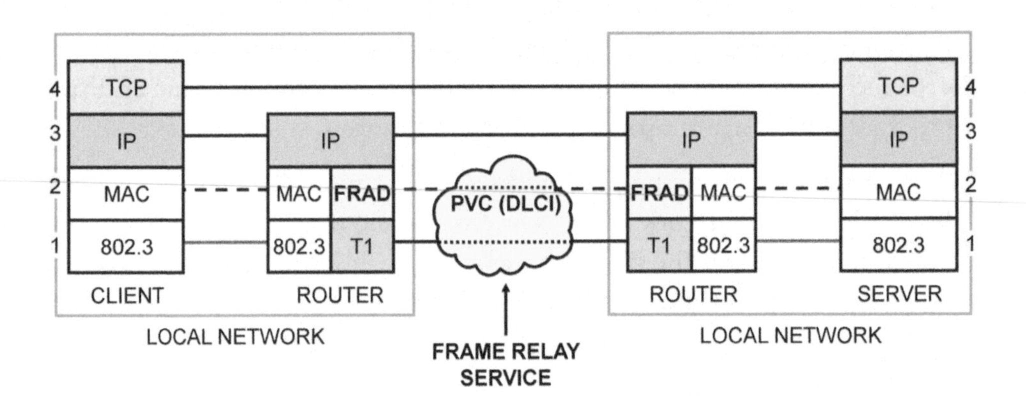

FIGURE 248 FRAME RELAY

In this way, a layer of software is eliminated, speeding up routing, and changing the service from being a packet-switched service (layer 3 + layer 2) to a frame-relay service (layer 2 only).

This is implemented by defining virtual circuits and associating them with a virtual circuit ID called a *Data Link Connection Identifier* (DLCI), and populating this in the **frame** address field.

A control system configures routing tables in the network elements with the routing for each DLCI. When a frame arrives at the network element, the DLCI on the frame is used to look up the next hop.

On a private network, the DLCI can be the same across the network. On a public network, the DLCI will change from link to link, and so the end-to-end connection is a virtual circuit made up of a sequence of DLCIs.

D.3.2 Unreliable Service

Another reason why Frame Relay is faster is because it is does not provide reliable network service... the delivery guarantee and error recovery protocols of X.25 are replaced with mere error detection.

If a frame is corrupted, or if the network gets busy, the network discards the frame. The network does not retransmit the frame (like X.25 does). This also eliminates overhead and redundancy, improving throughput.

D.3.3 Network Structure and Operation

Frame Relay was designed for LAN-to-LAN client-server data communications between locations of a business or government. Permanent Virtual Circuits identified with a sequence of DLCIs are established between every two of the locations.

To communicate from one customer location to another, the customer must have edge equipment that relates DLCIs to destinations, and packages the customer data into the same frame format used by the service provider.

The customer-premise equipment that performs this function is called a *Frame Relay Access Device* (FRAD). Typically, IP subnets, that is, blocks of IP addresses, will be assigned to each customer location. The routing table in the FRAD is populated with the DLCI corresponding to a subnet.

Since there is a possibility that frames will be discarded during transit, users must run an end-to-end error-checking and retransmission protocol to implement reliability. TCP is normally used.

D.3.4 No Guarantees for Voice

Frame Relay provides no guarantee of end-to-end delay in delivery of frames, and no guarantee of the maximum variability of delay, called *jitter*. This means that while it is possible to communicate digitized speech in packets in frames over a Frame Relay network, it is not possible to **guarantee** the quality of the reconstructed speech in the case of a live telephone call.

A 20-ms-long segment of digitized speech arriving 300 ms after the previous segment usually has the same effect as not arriving at all:

reconstructed speech that has parts of syllables missing and noticeable clicking noises.

A technology that can guarantee transmission characteristics such as delay and jitter is required to be able to guarantee the quality of delay- and loss-sensitive applications like telephone calls and live television.

D.4 ATM

For a long time, *Asynchronous Transfer Mode* (ATM) was thought to be the answer to all of the requirements for guaranteeing different transmission characteristics for voice, video and data interspersed on the same circuit.

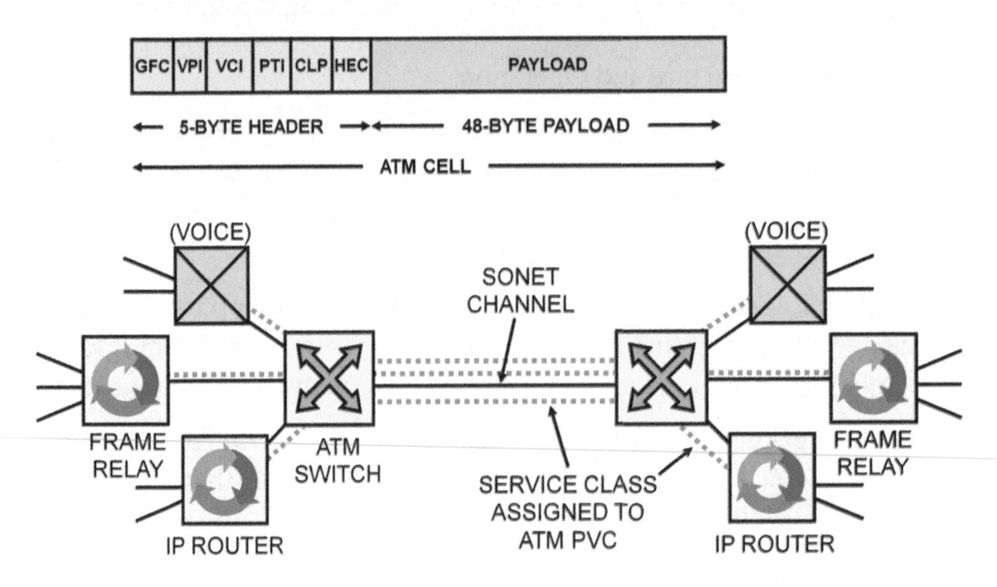

FIGURE 249 ATM

Unfortunately, ATM became very complicated and very expensive and is no longer used for new deployments. This section provides an overview of ATM and its jargon. It has been replaced by MPLS.

ATM is similar to X.25 and Frame Relay, transferring *cells* (instead of packets or frames) of information over virtual circuits. The difference is that ATM was supposed to be able to guarantee suitable transmission characteristics for any type of traffic: telephone calls, television, business data, web pages, e-mail or anything else, to achieve the goal of *integration or convergence*, everything on one network.

D.4.1 Future-Proof Technology (Not)

Some even were claiming that ATM was *future-proof*, supporting switched virtual circuits for flexibility, scalable to arbitrarily high line speeds and supporting any type of traffic. There would be no technology after ATM.

Unfortunately, this did not happen. If it had, people would be discussing "Voice over ATM" instead of "Voice over IP". ATM was used on carrier networks to achieve integration of all data services, but it was never deployed on the PSTN to carry delay-sensitive telephone calls.

The establishment and management of Switched Virtual Circuits in ATM is so complicated that it was rarely implemented – most often, manually-configured Permanent Virtual Circuits were used. ATM became so cumbersome and expensive that it is headed for the dustbin of history, replaced by MPLS and Differentiated Services.

D.4.2 ATM Cells

ATM packages data into 53-byte packets (called *cells* to confuse the innocent), consisting of 48 bytes of data and a 5-byte header. Three of the bytes in the header are a virtual circuit ID used to route the cell.

D.4.3 Service Classes

ATM implemented *Quality of Service* (QoS), allowing the specification of service classes for virtual circuits. This allowed the integration of many services over the physical circuits that make up a network (e.g., 10 Gb/s OC192) by establishing ATM virtual circuits between numerous types of network and edge equipment across the physical circuits and assigning a service class to each.

On an ATM virtual circuit, the user agrees to a traffic profile: the number of bits per second steady-state, the maximum short-term burst rate, how long it can last and how often it can happen.

In exchange, the network guarantees transmission characteristics like delay, variability in delay, number of errored bits and so forth.

ATM Service Classes were a set of standardized choices for traffic profile and transmission characteristics:

- *Continuous Bit Rate* for constant bit rate traffic with fixed timing, typically for full-period service emulation.
- *Variable Bit Rate - Real Time* for variable bit rate traffic with fixed timing; for example, phone calls and television. This was never deployed in practice on the PSTN, and here ATM failed to meet the objective of service integration and convergence.
- *Variable Bit Rate - Non-Real Time* for variable bit rate traffic with no timing relationship between data samples, but requiring guaranteed average bandwidth. This was used to integrate traffic for all data services, including Frame Relay and IP, on the core.
- *Available Bit Rate* (ABR): "best efforts" service, where flow control is used to increase and decrease the capacity allowed to the user based on available network capacity. Designed to transport LAN-LAN communications, which opportunistically use as much bandwidth as is available from the network.
- *Unspecified Bit Rate* (UBR): no guarantees. The user is free to send any amount of data up to a specified maximum while the network makes no guarantees at all on the cell loss rate, delay, or delay variation that might be experienced.

When setting up a virtual circuit, ATM switches implement an algorithm called *Connection Admission Control* (CAC) to determine if it is possible to deliver a requested service class. Using link parameters and end-to-end connection metrics, the switch determines whether accepting "just one more" connection would impact its ability to meet Service Class guarantees for existing virtual circuits.

The network can enforce a traffic profile by traffic policing. The ATM switch will meter an incoming stream to confirm it is respecting the agreed traffic profile. A switch can either discard out-of-profile cells or tag them by setting a *Cell Loss Priority* bit in the cell header, marking the cell to be first to be discarded should there be congestion in the network.

ATM switches from companies like Nortel, Lucent and Cisco that implemented CAC and traffic policing cost $800,000 or more in 1995 dollars each, plus yearly license fees and upgrades.

Acronyms and Abbreviations

100BASE-T 100 Mb/s Ethernet LAN using Twisted Pair
10BASE-2 Thinwire Ethernet LAN
10BASE-5 Ethernet LAN
10BASE-T 10 Mb/s Baseband Ethernet LAN using Twisted Pair
1xEV-DO 1X Evolution, Data-Optimized
3GPP Third Generation Partnership Project
ACD Automated Call Distributor
ACK Acknowledgment
ADSL Asymmetric Digital Subscriber Line
AES Advanced Encryption Standard
AIN Advanced Intelligent Network
AM Amplitude Modulation
AMI Alternate Mark Inversion
ANSI American National Standards Institute
ARP Address Resolution Protocol
ARPA Advanced Research Projects Agency
ASK Amplitude Shift Keying
ATM Asynchronous Transfer Mode
BEL Bell character in ASCII
BGP Border Gateway Protocol
BH Begin Header character in ASCII
bis Second version
BRA ISDN Basic Rate Access
BRI ISDN Basic Rate Interface
BSC Base Station Controller
BST Base Station Transceiver
BTS Base Transceiver Station Subsystem
CAT Category
CAT5e Enhanced Category 5 Twisted Pair
CATV Community Antenna Television
CC Competitive Carrier
CCITT Comité Consultatif International de Téléphone et de Télégraphe
CDMA Code Division Multiple Access
CIF Common Interface Format
CLEC Competitive Local Exchange Carrier
CO Central Office

COO Chief Operating Officer
CP Customer Premise
CPE Customer Premise Equipment
CR Carriage Return
CRC Cyclic Redundancy Check
CRLF Carriage Return Line Feed pair
CSMA-CD Carrier Sensing Multiple Access with Collision Detection
CSU Channel Service Unit
CTS Clear To Send
DACS Digital Access and Cross-Connect System
DARPA Defense Advanced Research Projects Agency
DB Data Bus
DCE Data Circuit-terminating Equipment
DCS Digital Cross-connect System
DDS Dataphone Digital Service
DEC Digital Equipment Corporation
DES Data Encryption Standard
DF Don't Fragment
DHCP Dynamic Host Configuration Protocol
DISA Direct Inward System Access
DLE Data Link Escape
DMS Digital Multiplex Switch
DMT Discrete Multi-Tone
DNS Domain Name System
DOD Department of Defense
DPSK Differential Phase Shift Keying
DS0 Digital Service Level 0: 64 kb/s
DS0A Subrate multiplexing scheme "A"
DS0B Subrate multiplexing scheme "B"
DS1 Digital Service Level 1: 1.544 Mb/s
DS2 Digital Service Level 2: 6.3 Mb/s
DS3 Digital Service Level 3: 44.7 Mb/s
DSL Digital Subscriber Line
DSLAM Digital Subscriber Line Access Multiplexer
DSP Digital Signal Processor
DSU Data Service Unit
DTE Data Terminal Equipment
DTMF Dual Tone Multiple Frequency
DVD Digital Versatile Disk
DWDM Dense Wave Division Multiplexing

EBCDIC Extended Binary Coded Decimal Interchange Code
EGP Exterior Gateway Protocol
EH End of Header character in ASCII
EIA Electronic Industries Association
eNB Enhanced Network Base Station, Enhanced NodeB
ENQ Inquiry
EOT End of Transmission
EPC Evolved Packet Core
ESN Electronic Serial Number
ETSI European Telecommunications Standards Institute
ETX End of Transmission
FCS Frame Check Sequence
FEC Forward Error Correction
FDM Frequency Division Multiplexing
FDMA Frequency Division Multiple Access
FEC Forwarding Equivalence Class
FM Frequency Modulation
FPPL Full Period Private Line
FRAD Frame Relay Access Device
FSA Fiber Serving Area
FSK Frequency Shift Keying
FTP File Transfer Protocol
FPPL Full Period Private Line
FRAD Frame Relay Access Device
FSA Fiber Serving Area
FSK Frequency Shift Keying
FTP File Transfer Protocol
Gb Gigabit
GHz Gigahertz
Giga (G) 10^9 = Billion (US), Thousand Million (UK)
GPON Gigabit Passive Optical Network
GPS Global Positioning System
GSM Global System for Mobile Communications
HD High Definition
HDLC High-level Data Link Control protocol
HDR High Data Rate
HDSL High-Speed Digital Subscriber Line
HFC Hybrid Fiber-Coax
HLR Home Location Register
HMI Human Machine Interface

HSDPA High Speed Downlink Packet Access
HSPA High Speed Packet Access
HSUPA High Speed Uplink Packet Access
HTTP Hypertext Transport Protocol
HTTPS Secure Hypertext Transport Protocol
Hz Hertz = cycles per second
IAB Internet Advisory Board
IANA Internet Assigned Numbers Authority
ICMP Internet Control Message Protocol
IEEE Institute of Electrical and Electronic Engineers
IETF Internet Engineering Task Force
ILEC Incumbent Local Exchange Carrier
IMT International Mobile Telecommunications
IMT-2000 International Mobile Telecommunications 2000 (3G)
IMT-DS IMT-Direct Spread (UMTS, W-CDMA)
IMT-MC IMT-Multicarrier (CDMA2000, 1X)
InterNIC Internet Network Information Center Agency
IP Internet Protocol
IP-PSTN IP Packet-Switched Telecommunications Network
IPsec IP Security
IPv4 IP version 4
IPv6 IP version 6
ISA Industry Standard Architecture computer bus
ISDN Integrated Services Digital Network
ISO International Organization for Standardization
ISP Internet Service Provider
ISUP ISDN User Part
ITU International Telecommunications Union
IVR Interactive Voice Response System
IX Internet Exchange
IXC Inter Exchange Carrier
JPEG Joint Photographic Experts Group
k $2^{10} = 1024$
kB, KiB Kilobyte = 1,024 bytes
kilo (k) 10^3 = thousand
kb kilobit = 1,000 bits
L2 Layer 2
L3 Layer 3
LAN Local Area Network
LATA Local Access and Transport Area

LEC Local Exchange Carrier
LEO Low Earth Orbit
LF Line Feed
LLC Logical Link Control
LSB Least Significant Bit
LSR Label-Switching Router
LSP Label-Switched Path
LTE Universal Terrestrial Radio Access Network Long Term Evolution
LZW Lempel – Ziv –Welch
M13 DS1-DS3 Multiplexer
MAC Media Access Control
MAN Metropolitan Area Network
Mega (M) 10^6 = Million
Mb Megabit = 1,000,000 bits
MB,MiB Megabyte = 2^{20} bytes = 1,048,576 bytes
MF Mainframe
micro (m) 10^{-6}
milli (m) 10^{-3}
MIME Multipart Internet Mail Extensions
MIMO Multiple-Input, Multiple-Output
MIPT Managed-IP Telephony
MPEG Moving Picture Experts Group
MPLS Multiprotocol Label-Switching
MSB Most Significant Bit
MTP Message Transfer Part
MTSO Mobile Telephone Switching Office
MUX Multiplexer
NAK Negative Acknowledgment
nano (n) 10^{-9}
NAT Network Address Translator
NEXT Near-End Crosstalk
NFF No Fault Found
NHLFE Next Hop Label Forwarding Entry
NIC Network Interface Card
NIST National Institute of Standards and Technology
NMT Nordic Mobile Telephone System
NNI Network – Network Interface
NRT Non-Real Time
NSF National Science Foundation
NSFNET National Science Foundation Network

NT1 Network Termination type 1
NTF No Trouble Found
NTSC National Television Standards Committee
NUL Null = 0
OC Optical Carrier (SONET)
OC3 OC level 3 = 3 DS3s
OC48 OC level 48 = 48 DS3s
OCU Office Channel Unit
OFDM Orthogonal Frequency-Division Multiplexing
OFDMA Orthogonal Frequency-Division Multiple Access
ONU Optical Network Unit
OPI Outside Plant Interface
OPX Off-Premise Extension
OSI Open Systems Interconnect
OSPF Open Shortest Path First
PBX Private Branch Exchange
PC Personal Computer
PCM Pulse Code Modulation
PCS Personal Communication Services
PEM Privacy Enhanced Mail
PGP Pretty Good Privacy
PIC Preferred Interexchange Carrier
pico (p) 10^{-12}
POP Point of Presence
POP Post Office Protocol
POS Point of Sale
POTS Plain Ordinary Telephone Service
PPP Point to Point Protocol
PPTP Point-To-Point Tunneling Protocol
PPV Pay-Per-View
PRI ISDN Primary Rate Interface
PSK Phase Shift Keying
PSTN Public Switched Telephone Network
PSTN Packet-Switched Telecommunications Network
PTI Payload Type Identifier
PTT Post Telephone and Telegraph
PVC Permanent Virtual Circuit
PVG Packet Voice Gateway
QAM Quadrature Amplitude Modulation
QoS Quality of Service

QPSK Quadrature Phase Shift Keying
RF Radio Frequency
RFC Request for Comments
RFQ Request for Quote
RIP Routing Information Protocol
RMS Root Mean Square
RPR Resilient Packet Ring
RSA Rivest Shamir Adelman
RT Real Time
RTC Remote Terminal Controller
RTP Real-Time Control Protocol
RTP Rusty Twisted Pair
RTS Request to Send
RXD Receive Data
SAC Subscriber Area Concept
SAP Service Advertising Protocol
SC-FDMA Single-Carrier Frequency-Division Multiple Access
SCP Service Control Point
SD Standard Definition
SDH Synchronous Digital Hierarchy
SDN Software-Defined Network
SDSL Symmetric Digital Subscriber Line
SIP Session Initiation Protocol
SLA Service Level Agreement
SMTP Simple Mail Transfer Protocol
SNMP Simple Network Management Protocol
SNR Signal to Noise Ratio
SOA Start of Authority
SOH Start of Header
SONET Synchronous Optical Network
SRC Source
SSP Service Switching Point
STM Synchronous Transport Module
STP Signal Transfer Point
STS Synchronous Transport Signal
STX Start of Transmission
SVC Switched Virtual Circuit
SYN Synchronization
T1 Trunk Carrier System 1
TACS Total Access Communication System

TCP Transmission Control Protocol
TDM Time Division Multiplexing
TDMA Time Division Multiple Access
tera (T) 10^{12} = Trillion (US), Billion (UK)
TIA Telecommunications Industries Association
TIU Terminal Interface Unit
TSB Technical Service Bulletin
Tx Transmit
TXD Transmit Data
UART Universal Asynchronous Receiver/Transmitter
UBR Unspecified Bit Rate
UDP User Datagram Protocol
UMTS Universal Mobile Telecommunications Service
U-NII Unlicensed National Information Infrastructure
UNI User-Network Interface
UNIX Harem Guards
URI Uniform Resource Indicator
URL Uniform Resource Locator
UTF Unicode Transformation Format
UTRAN Universal Terrestrial Radio Access Network
UUCP Unix-Unix Copy Protocol
VAX Virtual Address Extension
VBR Variable Bit Rate
VC Virtual Circuit
VCI Virtual Channel Identifier
VDSL Very High Bit Rate Digital Subscriber Line
VLAN Virtual Local Area Network
VoIP Voice Over IP
VPI Virtual Path Identifier
VPN Virtual Private Network
VSAT Very Small Aperture Terminal
VSP VoIP Service Provider
W-CDMA Wideband CDMA (= IMT-DS, UMTS)
WAN Wide Area Network
WPA Wireless Protected Access
WATS Wide Area Telephone Service
WDM Wave Division Multiplexer
XDSL Any DSL Technology
XML Extensible Markup Language

About the Author

Eric Coll is an international expert in telecommunications, broadband and networking, and has been actively involved in the telecom industry since 1983. He holds Bachelor of Engineering and Master of Engineering (Electrical) degrees.

Mr. Coll has broad experience, and broad knowledge of telecom developed working as an engineer in the telecommunications industry.

He has used his knowledge of telecom to develop and teach telecommunications technology training seminars to wide acclaim across North America since 1992... and answering questions at seminars for companies and organizations ranging from Bell Labs to the Department of Justice keeps things up to date. In his spare time, Mr. Coll authors textbooks and online courses based on the latest updates to the seminar courses.

Mr. Coll also provides consulting services as a Subject Matter Expert in telecommunications to government, carriers, and their customers.

About Teracom

Public Seminars

Instructor-led training is the best you can get, allowing you to ask questions and interact with classmates. Teracom's public seminars are in-person and live online instructor-led courses geared for the non-engineering professional needing a comprehensive overview and update, and those new to the business needing to get up to speed.

Private Onsite and Online Seminars

Since 1992, we have provided high-quality on-site and live online private instructor-led training in telecommunications, data communications, IP, networking, VoIP and wireless at hundreds of organizations ranging from Bell Labs to the Defense Systems Information Agency.

We have built a solid reputation for delivering high-quality training programs that are a resounding success. We would like to do the same for you! Please contact us via teracomtraining.com for more information.

Online Courses and TCO Certifications

Upgrade your knowledge - and your résumé - with high-quality telecom training courses by Teracom coupled with certification from the Telecommunications Certification Organization:

- Certified Telecommunications Network Specialist (CTNS)
- Certified Telecommunications Subject Matter Expert (CTSME)
- Certified Telecommunications Analyst (CTA)
- Certified Wireless Analyst (CWA)
- Certified VoIP Analyst (CVA)

TCO Certification is proof of your knowledge of telecom, datacom and networking fundamentals, jargon, buzzwords, technologies and solutions.

Guaranteed to Pass and repeat courses anytime with the Unlimited Plan!

Join our thousands of satisfied customers including:

the FBI Training Academy, US Marine Corps Communications School, US Army, Navy, Air Force and Coast Guard, the NSA and CIA, DoJ NSD, IRS, FAA, DND, CRA, CRTC, RCMP, banks, power companies, police forces, manufacturers, government, dozens of local and regional phone companies, broadband carriers, individuals, telecom planners and administrators, finance, tax and accounting personnel and many more from hundreds of companies.

Teracom's GSA Contract GS-02F-0053X for supplying this training to the United States is your assurance of approved quality and value.

Visit us at teracomtraining.com to get started today!

Made in the USA
Las Vegas, NV
16 March 2023